K. Hostettmann · A. Marston · M. Hostettmann
Preparative Chromatography Techniques

Springer

Berlin
Heidelberg
New York
Barcelona
Budapest
Hong Kong
London
Milan
Paris
Santa Clara
Singapore
Tokyo

K. Hostettmann · A. Marston ·
M. Hostettmann

Preparative Chromatography Techniques

Applications in Natural Product Isolation

Second, Completely Revised and Enlarged Edition

With 59 Figures

Springer

Professor Dr. K. Hostettmann
Dr. A. Marston
Dr. M. Hostettmann
University of Lausanne
Institute of Pharmacognosy and Phytochemistry
CH-1015 Lausanne – Dorigny
Switzerland

ISBN 3-540-62459-7 2nd ed. Springer-Verlag Berlin Heidelberg New York
ISBN 3-540-16165-1 1st ed. Springer-Verlag Berlin Heidelberg New York

Die Deutsche Bibliothek – CIP-Einheitsaufnahme

Hostettmann, Kurt:
Preparative chromatography techniques : applications in natural product isolation / K. Hostettmann ; A. Marston ; M. Hostettmann. – 2., completely rev. and enl. ed. – Berlin ; Heidelberg ; New York ; London ; Paris ; Tokyo : Springer 1997
ISBN 3-540-62459-7

Production Editor: Christiane Messerschmidt, Rheinau
Typesetting: Fotosatz-Service Köhler OHG, Würzburg
Cover design: Design & Production GmbH, Heidelberg

SPIN: 10470948 52/3020 – 5 4 3 2 1 0 – Printed on acid-free paper

Preface

Over the past few years, increasing attention has been paid to the search for bioactive compounds from natural sources. The success of plant-derived products such as paclitaxel (Taxol) in tumor therapy or artemisinin in the treatment of malaria has provided the impetus for the introduction of numerous research programmes, especially in Industry. A great deal of effort is being expended in the generation of novel lead molecules of vegetable, marine and microbial origin by the use of high throughput screening protocols. When interesting hits are found, it is essential to have methods available for the rapid isolation of target compounds. For this reason, both industry and academia need efficient preparative chromatographic separation techniques and experience in their application.

Purified natural products are required for complete spectroscopic identification and full characterization of new compounds, for biological testing and for the supply of pharmaceuticals, standards, and starting materials for synthetic work. Obtaining pure products from an extract can be a very long, tedious and expensive undertaking, involving many steps. Sometimes only minute amounts of the desired compounds are at hand and these entities may be labile. Thus it is an advantage to have access to as many different methods as possible in order to aid the isolation process. Although a certain amount of trial and error may be involved, nowadays there is the possibility of devising suitable rapid separation schemes by a judicious choice of the different techniques available.

While a vast amount of literature is available about analytical procedures, very few monographs specifically treat preparative chromatography. The aim here is to fill this gap and provide a guide to isolation strategy.

Since the first edition of this book in 1986, there have been important advances in the study of compounds from natural sources. The rapid development of spectroscopic techniques, including 2D-NMR methods, automated instrumentation and routine availability of X-ray crystallography, has greatly simplified the structure elucidation of natural products. Consequently, the main challenge is now the efficient isolation and purification of bioactive components from different organisms (plants, animals, microorganisms etc.). These range from highly polar to very lipophilic compounds. Bioactivity can often be lost during chromatography and it is essential to have techniques available which provide gentle separation conditions. New and improved chromatographic processes are constantly being introduced and these have to be employed judiciously in an effort to solve the separation problems at hand. The aim of this volume is to build on the material covered in the first edition and to bring together the rather disperse applications of the new chromatographic techniques so that the reader has the means of selecting an approach for his own separation problem. For this reason, a critical look has been taken at the different methods available. This is supplemented by numerous examples of each chromatographic procedure, which can be used for reference purposes. No attempt has been made to provide an exhaustive literature survey but consultation of the bibliography given with each section will cover most of the ground necessary for an appreciation of the possibilities available.

The first part of the book describes the important area of sample preparation. Much time can be saved by pre-treatment of a sample before chromatography. Liquid-solid chromatographic procedures are then covered, with emphasis on techniques which involve the application of pressure. All-liquid chromatography is introduced and some idea of given of its advantages in the separation of problematic labile compounds. Chapters on the separation of macromolecules and chiral substances are also provided. These two themes are of vital importance for the isolation of new pharmaceuticals, gene products and even primary metabolites. Finally, there is an insight into strategy when designing separation protocols.

Lausanne, July 1997

K. Hostettmann
A. Marston
M. Hostettmann

Contents

List of Symbols and Abbreviations

CCC	countercurrent chromatography
CCCC	centrifugal countercurrent chromatography
CPC	centrifugal partition chromatography
CSP	chiral stationary phase
CTLC	centrifugal thin-layer chromatography
DCCC	droplet countercurrent chromatography
DMSO	dimethylsulphoxide
DNP	dinitrophenyl
GC	gas chromatography
HIC	hydrophobic interaction chromatography
HPLC	high-pressure liquid chromatography
HSCCC	high-speed countercurrent chromatography
IR	infra-red
LPLC	low-pressure liquid chromatography
MPLC	medium-pressure liquid chromatography
PTLC	preparative thin-layer chromatography
RLCC	rotation locular countercurrent chromatography
RPC	reversed-phase chromatography
SEC	size exclusion chromatography
SFC	supercritical fluid chromatography
SFE	supercritical fluid extraction
TBME	tertiary butyl methyl ether
TFA	trifluoroacetic acid
THF	tetrahydrofuran
TLC	thin-layer chromatography
VLC	vacuum liquid chromatography

CHAPTER 1

Introduction

The origins of chromatography are closely linked with the study of natural products. Following the pioneering work of Tswett in 1906, the first true preparative liquid chromatographic separations in the 1930s were of plant metabolites – pigments such as chlorophylls and carotenes. This association has continued and it would be inconceivable to imagine the advances in the purification of biomolecules without considering chromatography. In the area of HPLC alone, applications are to be found in the pharmaceutical industry, biotechnological, biomedical and biochemical research, energy, food, cosmetics, environmental fields, drugs and vitamins. With the increasing interest in the discovery of new lead compounds from microorganisms and both marine and terrestrial higher organisms, there is a constant need to separate both large and small quantities of mixtures efficiently, rapidly and inexpensively. Although it is seldom that all three of these requirements are satisfied in commonly used chromatographic techniques, it is important to know how to choose the right methods. The first edition of this book was produced in order to group together the different available preparative separation methods and provide a general reference work. Since 1986, there have been tremendous advances in the different chromatographic instruments and numerous new applications. For this reason, it was deemed important to provide an update and include techniques relevant to the separation of biomolecules, a field which is seeing an explosion in size at the present time.

Other details of the extraction and isolation of bioactive (and other) compounds can be found in numerous review articles in journals such as "Journal of Chromatography", "Journal of Liquid Chromatography and Related Technologies", "Analytical Chemistry", "Natural Product Reports", "Phytochemical Analysis", "LC-GC Magazine" and the specialist journals of the different areas of research. Useful chapters on the isolation of natural products are provided by E.L. Ghisalberti in "Bioactive Natural Products: Detection, Isolation and Structural Determination" (eds SM Colegate, RJ Molyneux, CRC Press, Boca Raton, 1993), T.A. Van Beek in "Chemicals from Plants" (ed NJ Walton, World Scientific Publishing, London, 1997) and in "Advances in Natural Products Chemistry: Extraction and Isolation of Biologically Active Compounds" (eds S Natori, N Ikekawa, M Suzuki, John Wiley, New York, 1981).

The objective of this book is not to provide a comprehensive theoretical background to the techniques, more fittingly covered by authors elsewhere, but to emphasise the *applications*. Most of the examples come from the areas of secondary metabolites and general natural product chemistry because the rôle of preparative chromatography is so important for the isolation of the substances concerned. In the purification of macromolecules, chromatographic operations are also a requisite for the production of pharmaceutically important bioactive proteins (including those obtained by recombinant technology) and for obtaining biotherapeutics and diagnostics in the required purity. For this reason, some of the specific techniques for biomolecules are included, e.g. affinity chromatography and size exclusion chromatography. The optical purity of therapeuticals is another area of increasing importance and thus separations of chiral molecules need to be considered. This aspect is treated in Chap. 9.

CHAPTER 2

Sample Preparation and Purification

Preparation of the sample is of vital importance before undertaking complex separation and purification procedures. A suitable sample pretreatment can save much time and effort in subsequent steps and make isolation considerably easier. Whether the probe is of biological origin with contaminating proteins, or from industrial processes and contains residual catalyst, or from a plant source with interfering matrix elements, a simple preliminary step is often useful to remove most of the undesired material.

Sample preparation is of special regard when performing preparative HPLC on costly columns and, although the aims are not necessarily the same, many of the procedures applied are identical to those encountered in analytical HPLC. When handling biological samples, the problems are particularly acute because of the complex nature of the material under investigation – the sample may contain not only complex macromolecules but also buffers, salts and detergents from previous purification steps. To avoid rapid contamination of HPLC columns, these contaminants should preferably be removed.

In devising initial purification protocols, it is usual to have early steps which combine high capacity and low resolving power. Classical techniques of great utility include selective extraction, filtration, precipitation, dialysis, centrifugation and simple open column chromatography.

Newer approaches which give rapid results and good recoveries include supercritical fluid extraction and solid-phase extraction.

Biomolecules (and, more particularly, biopolymers) have properties distinct from the most frequently encountered classes of secondary metabolites and require special handling. Their pre-purification can be tackled by specialized methods such as dialysis and ultrafiltration, together with more routine techniques like centrifugation, precipitation (e.g. salting-out with ammonium sulphate for proteins), ion exchange and gel filtration (Wehr 1990).

2.1 Extraction

A judicious choice of extraction solvent or solvents provides the first and most obvious means of correct sample preparation. Initial extraction with low-polarity solvents yields the more lipophilic components, while alcoholic solvents give a larger spectrum of apolar and polar material. If a more polar solvent is used for the first extraction step, subsequent solvent partition allows a finer division into different polarity fractions.

There are also different approaches to the actual extraction procedure. While stirring or mechanical agitation are the most common methods, percolation or even pressurized solid-liquid extraction are possible. In the latter approach, a column is filled with dried plant material and the extraction solvent is pumped through the bed (Härmälä et al. 1992). An instrument is now available from Dionex for pressurized solid-liquid extraction. This is known as the ASE 200 (Accelerated Solvent Extraction); it is claimed to complete an extraction within 20 min. The temperature can be regulated up to 200 °C and three extraction cell sizes are possible: 11-, 22- and 33-ml. Although not cheap, the ASE 200 represents an important advance in extraction technology.

Extract preparation from the raw material is a crucial step in the process of protein purification. During the extraction of proteins, a number of precautions have to be taken and it may be necessary to add buffer, detergent, co-factors and other agents.

2.2 Supercritical Fluid Extraction (SFE)

This is especially suitable for thermally or chemically unstable compounds and is mainly employed for the extraction of lower polarity components from a matrix (McHugh and Krukonis 1986; Westwood 1993; Modey et al. 1996). SFE is a valuable alternative to steam distillation or Soxhlet extraction, for example. The method is fast and the solvent strength can be modulated by varying the pressure. Greater extraction *selectivity* can be achieved and cleaner products obtained by careful choice of temperature and pressure conditions. Recovery rates are higher than conventional liquid-liquid and liquid-solid extractions. The safety aspect is important since extraction with supercritical fluids eliminates the need for large amounts of organic solvents. The environmental implications are also worth stressing, especially concerning the reduction in use of chlorinated solvents. Many supercritical fluids have suitable mass transfer characteristics (their low viscosities and high diffusion rates, compared with those of liquids, make them ideal for the extraction of plant tissues) and are gases at room temperature, relatively inert, non-toxic and inexpensive. Examples are carbon dioxide, nitrous oxide and ammonia. They can be modified by the addition of co-solvents such as methanol, ethanol, propanol, acetonitrile, dichloromethane and water.

The list of natural products extracted by SFE is rapidly getting longer (Bevan and Marshall 1994; Castioni et al. 1995) and both plant samples and microbial products (Cocks et al. 1995) have been investigated. Extraction of hops, of nicotine from tobacco and of caffeine from coffee are well-known processes (King and Bott 1993). SFE of flavours, spices and essential oils is of great commercial importance (King and Bott 1993).

Anethole (>90% purity) was obtained from star anise with supercritical carbon dioxide – a procedure considerably more efficient than the classical solvent extraction method, which is tedious, time-consuming and less economic (Liu 1996).

In the extraction of alkaloids from the Amaryllidaceae, several parameters for the most efficient yields were studied. The supercritical fluid used was nitrous oxide, with SFE following a preliminary percolation step (Queckenberg and Frahm 1994).

It has been found that extraction of phenolic anacardic acid, cardol and cardanol derivatives from cashew nut shells by supercritical carbon dioxide gave a better quality product than the corresponding extraction with pentane (Shobha and Ravindranath 1991).

Extraction of bioactive sesquiterpenes from the leaves of *Magnolia grandiflora* (Magnoliaceae) with supercritical carbon dioxide has been compared to near-critical extraction with propane and standard extraction with dichloromethane. The carbon dioxide method was much more selective for the three sesquiterpenes and gave an extract virtually clear of chlorophyll and fats. Thus, a charge of 20 g of leaves in the extractor furnished 36.85 mg extract containing 194 mg/kg parthenolide (**1**), 37.5 mg/kg costunolide (**2**) and 604.5 mg/kg cyclocolorenone (**3**) (Castaneda-Acosta et al. 1995).

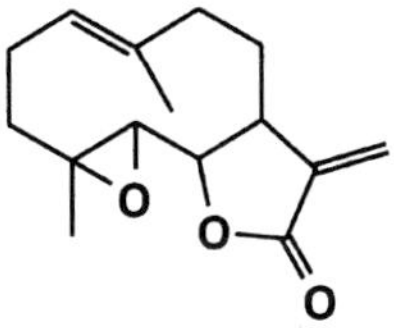

1

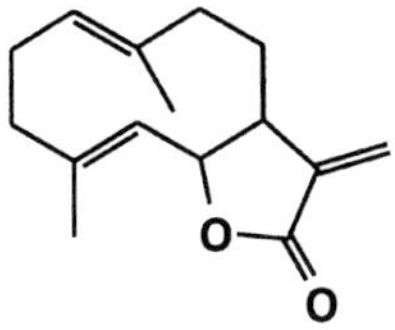

2

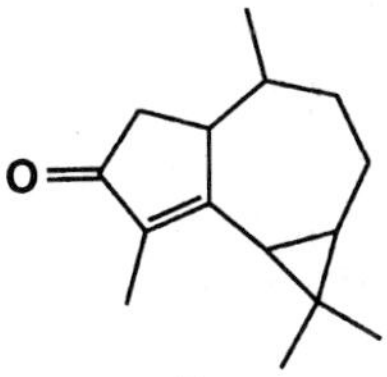

3

In the bioassay-guided isolation of anti-inflammatory triterpenoids from *Calendula officinalis* (Asteraceae), initial extraction of the ground flowers (360 g) was performed with supercritical carbon dioxide, giving a total of 15 g waxy product. This was chromatographed by a combination of vacuum liquid chromatography, LPLC and semi-preparative HPLC (Della Loggia et al. 1994).

Other extractions include antimigraine sesquiterpene lactones from feverfew (*Tanacetum parthenium*, Asteraceae) (Smith and Burford 1992), cucurminoids from the rhizomes of *Curcuma longa* (Zingiberaceae) (Sanagi et al. 1993), furanocoumarins from the fruits of *Cnidium formosanum* (Apiaceae) (Miyachi et al. 1987) and paclitaxel (Taxol) from *Taxus brevifolia* (Taxaceae) stem bark (Jennings et al. 1992). In the latter study, carbon dioxide-ethanol mixtures were more efficient at extracting paclitaxel than supercritical carbon dioxide alone.

The scale of SFE considered here falls in between the commercial industrial processes (removal of caffeine from coffee etc.) and analytical SFE. Several manufacturers produce laboratory-size equipment (Isco, Supelco, Dionex, Suprex etc.) – see Fig. 2.1 for a typical configuration. Although expensive large-scale SF extractors are commercially available, typical volumes of the extraction vessel range from 1 to 50 ml. For the majority of instruments, the size of the extraction chamber is approximately 10 ml, which does introduce some restriction on sample quantity, even when several chambers are interconnected.

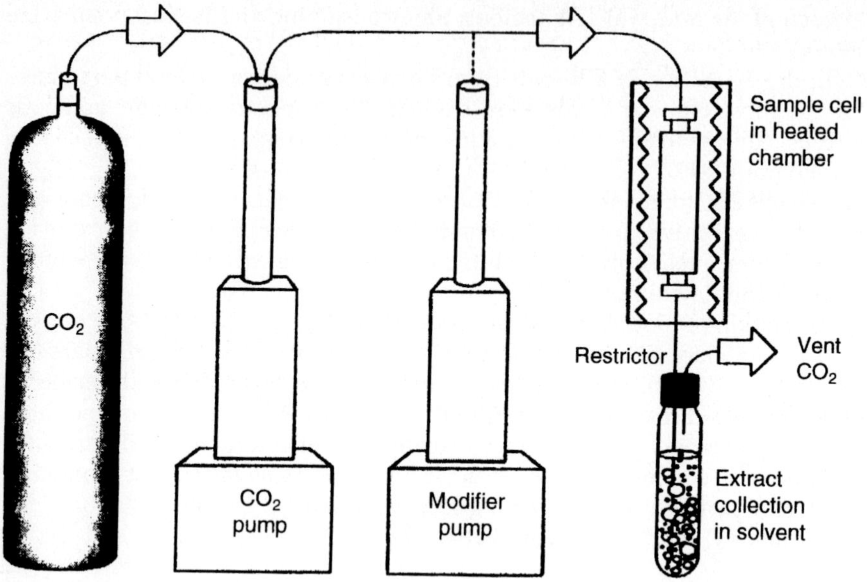

Fig. 2.1. Typical setup for SFE (Isco)

While on-line SFE is mainly used for analytical purposes, off-line SFE allows direct collection of extracted analytes. This is achieved by several methods:

a) passing the supercritical fluid through a column packed with chromatographic material;
b) depressurizing the fluid into a small amount of solvent;
c) allowing the supercritical fluid-containing sample to expand directly into an empty container with or without cryogenic cooling. Extracts thus obtained contain little or no solvent and evaporation processes are minimal.

2.3 Solvent Partition

Once an extract has been obtained, straightforward solvent partition methods remove a large proportion of extraneous material and, especially when used in conjunction with a bioassay, fractions enriched in the sought-for constituents are rapidly obtained. For example, such solvent partition schemes have been used while searching for antitumour (Wall et al. 1976; Pettit et al. 1995) and anti-HIV (Gustafson et al. 1992) agents from plant sources and for the detection and isolation of acetogenins from the Annonaceae (Rupprecht et al. 1990). A procedure for the preliminary purification of paclitaxel by a similar solvent partition has been reported (Cardellina 1991).

While searching for new HIV-inhibitory compounds from plants, extraction has been performed with dichloromethane-methanol 1:1, followed by methanol. A solvent partitioning protocol was then applied (Grode et al. 1983). This was done as follows: the crude extract was distributed between hexane and 10% aqueous methanol; the polar phase was increased to 25% water and extracted with carbon tetrachloride; the upper phase was increased to 35% water and extracted with chloroform; after evaporation of methanol from the aqueous phase, this was extracted with ethyl acetate. In the case of the calanolides, coumarins from the Malaysian tree *Calophyllum lanigerum* (Guttiferae), anti-HIV activity was found in the hexane and carbon tetrachloride soluble fractions. Isolation of the HIV inhibitors proceeded via fractionation of these two fractions (Kashman et al. 1992).

Prostratin (**4**), a relatively polar 12-deoxyphorbol ester, which inhibits cell killing by HIV-1 was isolated from the Samoan plant *Homalanthus nutans* (Euphorbiaceae) by a procedure which involved a similar preliminary partitioning scheme (Gustafson et al 1992).

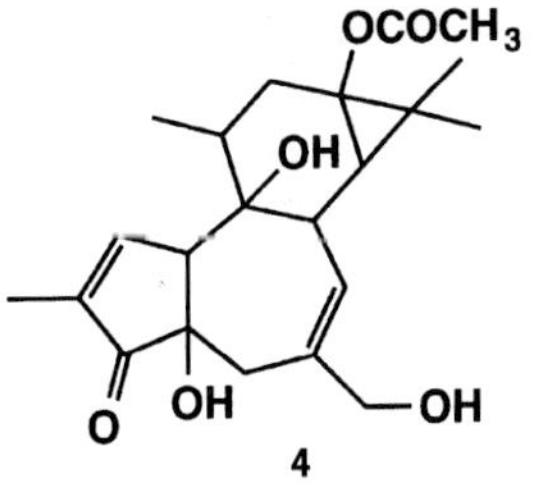

4

In the separation of saponins from plant material, a single butanol-water partition step often suffices to concentrate the saponins in the butanol fraction and provide a preliminary clean-up step (Hostettmann and Marston 1995). Normally the saponin-containing plant is defatted with petrol ether or dichloromethane and then extracted with a polar solvent such as methanol. The resulting extract is partitioned between n-butanol and water to remove sugars and other polar components in the aqueous layer. The organic layer is then chromatographed. This was the procedure used, for example, in the isolation of a triterpene glycoside from *Tetrapleura tetraptera* (Leguminosae), a plant from Nigeria with molluscicidal activity. Silica gel open-column chromatography of the butanol extract, followed by MPLC and gel filtration, gave the desired saponin (Maillard et al. 1992).

Multiple partition steps provide another possibility for the preliminary purification of samples which are to be separated by liquid chromatography. This is generally achieved by countercurrent methods. A Craig countercurrent distribution, generally with a restricted number of transfers (e.g. the isolation of cocarcinogenic phorbol esters from *Euphorbia cooperi* (Euphorbiaceae); Gschwendt and Hecker 1973) or a droplet countercurrent separation (e.g. in the isolation of steroidal glycoside sulphates from starfish; Riccio et al. 1985) is useful in this respect, especially when separating lipophilic from more polar constituents. For the isolation of the ester iridoid glucoside ebuloside (**5**) from the roots of *Sambucus ebulus* (Caprifoliaceae), the butanol extract was subjected to countercurrent distribution before final chromatographic purifica-

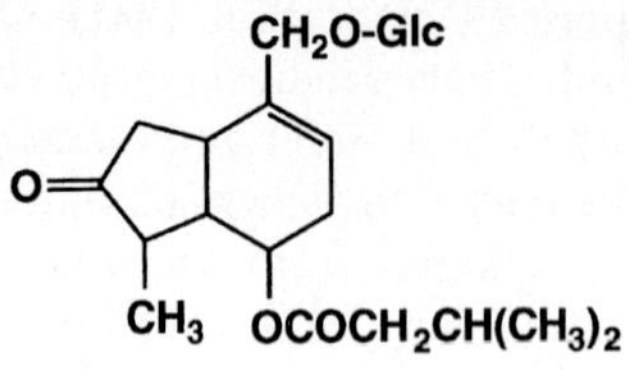

5

tion. A total of 200 transfers with the solvent system $CHCl_3$-MeOH-H_2O 43:37:20 were performed to obtain crude ebuloside (Gross et al. 1986).

2.4 Filtration

This provides the easiest and most obvious method of sample preparation necessary for countercurrent chromatographic, low-, medium- and high-pressure liquid chromatographic separations. Filtration can take the form of passage of a sample solution through a filter paper or sintered glass funnel in order to remove particulate and insoluble material.

A further degree of purity can be achieved by filtering the solution through a short column of silica gel or other suitable packing material. This has the effect of removing strongly adsorbing contaminants which may prove awkward during column chromatography (Hostettmann et al. 1977).

2.5 Gel Filtration

Initial chromatography on size exclusion gels such as Sephadex LH-20 is frequently used as a clean-up step for further purification (see also Chap. 10).

2.6 Precipitation

This preliminary purification method is very often employed in work on saponins: a concentrated methanol solution of an extract containing saponins (after butanol-water partition, for example) is poured into a large volume of diethyl ether. The precipitated saponins are collected by filtration or centrifugation. For better results, the precipitation can be repeated several times (see, for example, Hostettmann and Marston 1995).

2.7 Solid Introduction in Chromatography

When the sample that is to be introduced onto a chromatographic column (flash, dry-column, vacuum liquid chromatography etc.) is not very soluble in the eluent, a solid introduction may be performed. The material is dissoved in a suitable solvent and mixed with about five times its weight of deactivated adsorbent (or Celite). The mixture is evaporated in a rotary evaporator at 30–40 °C and the resulting powder is distributed on the top of the column (Loev and Goodman 1967). This may then be covered with a shallow layer of sand or glass beads before elution.

2.8 Removal of Chlorophyll

As long as there are no solubility problems, a convenient method of freeing extracts of chlorophyll is to include a clean-up step over octadecylsilica. Prior to the separation of flavonoid glycosides from *Dryas octopetala* (Rosaceae), for example, a silica gel cartridge was used to eliminate tannins from an ethanol extract and then chlorophyll was removed by elution on a C-18 cartridge (De Bernardi et al. 1984). Similarly, before medium-pressure and high-pressure liquid chromatographic purification of antiinflammatory polyacetylenes from *Bidens campylotheca* (Asteraceae), chlorophyll was eliminated by chromatography of a methanol suspension of the hexane extract of the aerial parts on Sep-Pak C-18 cartridges with methanol as eluent (Redl et al. 1994).

2.9 Removal of Waxes

Treatment with acetonitrile provides a suitable means of removing waxes. In one case, a chloroform extract (with antifeedant activity) of *Petunia integrifolia* (Solanaceae) was suspended in boiling acetonitrile with stirring for 1 h. A waxy, solid material formed after cooling to 5 °C. This could be separated by decantation, leaving about 50% of the original extract in solution. Chlorophyll and some less polar lipid material were removed by chromatography of this solution on C-18 material, eluting with acetonitrile. The active petuniolides (ergostane derivatives, e.g. **6**, petuniolide C) were obtained after semi-preparative HPLC (Elliger et al. 1990). Cytotoxic flavonoids and terpenoids from *Artemisia annua* (Asteraceae) were isolated by a procedure involving an acetonitrile precipita-

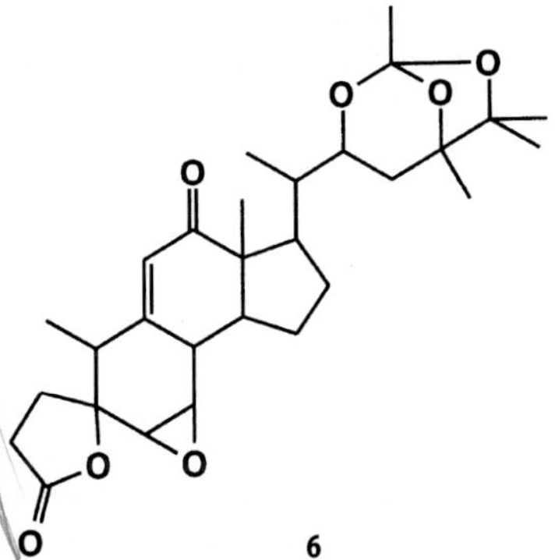

6

tion step. Aerial parts of the plant were extracted with hexane. This extract was dissolved in chloroform (20 ml) and waxes were precipitated with acetonitrile (180 ml), before subsequent chromatographic steps (Zheng 1994).

2.10 Removal of Tannins

It is sometimes required that tannins be removed from plant extracts or fractions before submission to biological testing. For this purpose, there are several methods described: precipitation with gelatin/NaCl solution, treatment with soluble polyvinylpyrrolidone (PVP), caffeine or hide powder, polyamide column chromatography. Of these possibilities, the last method is the most effective but suffers from the disadvantage that it is not very selective and can remove other polyphenolics besides tannins. The polyamide procedure was employed during the evaluation of plant extracts for inhibition of human immunodeficiency virus type 1 reverse transcriptase (HIV-1 RT) (Tan et al. 1991). These same authors have also compared the various methods for tannin removal.

The mechanism by which tannins are eliminated in all these examples involves a precipitation phenomenon, due to the formation of hydrogen bonds between the tannin phenolic hydroxyl groups and the amide function of the precipitating agent. This causes the production of insoluble complexes. The problem is that some of the above-mentioned procedures can also remove non-tannin compounds with phenolic hydroxy groups, e.g. certain flavonoids. Furthermore, quinones can be removed by covalent interaction with the reagents concerned.

If it is necessary to remove all phenolics, precipitation with lead (II) acetate is the most effective way. A solution of 10 g lead acetate in water (100 ml) is suitable for this purpose.

Polyamide. This tannin removal procedure was established by Wall et al. (1969). For small quantities of plant extracts suitable for biological testing, 3 mg is dissolved in a minimum volume of water and applied to a glass column (10×0.6 cm) packed with polyamide powder (400 mg). Elution is performed with water (2 ml) followed by 50% methanol (2 ml) and finally absolute methanol (5 ml). The total eluate is collected (Tan et al. 1991). Washing with methanol elutes non-tannin compounds with two or three phenolic hydroxyl groups, i.e. most flavonoids can be recovered. In another procedure, solid phase extraction columns were prepared by packing 1 g of Macherey Nagel polyamide SC6 (preswollen in water), into a 12-ml syringe fitted with a glass wool plug. The substrate (3–6 mg) was dissolved in a minimal amount of water (< 500 µl), applied to the column and eluted with water (2 ml), followed by methanol-water 1:1 (2 ml) and methanol (2×5 ml) (Cardellina et al. 1993).

Polyvinylpyrrolidone (PVP). This method has been described by Loomis and Battaile (1966); 500 ml of a 10% w/v PVP solution is sufficient for the complete removal of tannins from 2 mg of plant extract (dissolved in 500 µl water). This corresponds to an effective concentration of 50 mg/ml (5% w/v) PVP to 2 mg/ml plant extract (Tan et al. 1991).

Hide powder. The European Pharmacopoeia specifies boiling 0.75 g powdered plant material with 150 ml water for 30 min. The solution is filtered and 100 ml of the filtrate is shaken for 60 min with 1 g of hide powder, in order to rid the aqueous phase of tannins (see also Stahl and Jahn 1984). In our experience, if an extract (plant or otherwise) is concerned, tannins can be removed by mixing the extract (100 mg) with 25% ethanol or water (10 ml) and stirring for 60 min with 200 mg hide powder. The solution is filtered and solvent can be evaporated.

2.11 Solid-Phase Extraction

Pre-packed cartridges for solid-phase extraction operate on the principle of liquid-solid extraction and may be used in one of two modes: a) the interfering matrix elements of a sample are retained on the cartridge while the components of interest are eluted; b) the components of interest are retained while interfering matrix elements are eluted. In the latter case, a concentration effect can be achieved. The required compounds are then eluted from the cartridge by changing the solvent. Cartridges with a variety of packings, both normal- and reversed-phase, are obtainable (e.g. Sep-Pak, Millipore Waters, Milford, Massachusetts; Bond-Elut, Varian Analytichem, Harbor City, California; Alltech Extract-Clean tubes, Deerfield, Illinois; Hamilton Chrom-Prep cartridges, Hamilton Bonaduz, Bonaduz, Switzerland; Bakerbond spe cartridges, J. T. Baker Inc., Phillipsburg, New Jersey; Supelclean extraction tubes, Supelco).

Solid-phase extraction lends itself particularly well to automation and is especially helpful when large numbers of samples have to be routinely purified.

Preliminary purification on a C-18 Bond Elut cartridge was carried out on an extract of the mollusc *Philinopsis speciosa* (Cephalaspidea) before semi-preparative HPLC separation of the C_{16}-alkadienone-substituted 2-pyridine **7** (Coval and Scheuer 1985).

A petroleum ether extract of tomato purée was freed of carotenoids other than lycopene (**8**) (red) by passage through a silica cartridge (Bond Elut) and elution with petroleum ether. Lycopene was eluted with chloroform and under-

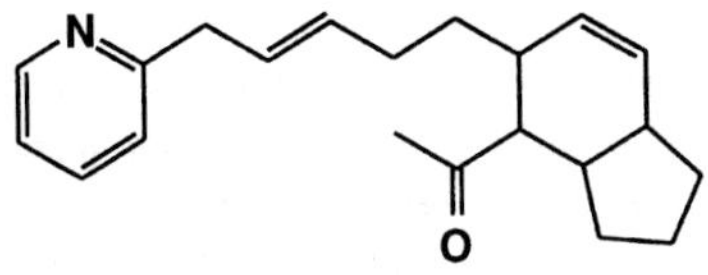

7

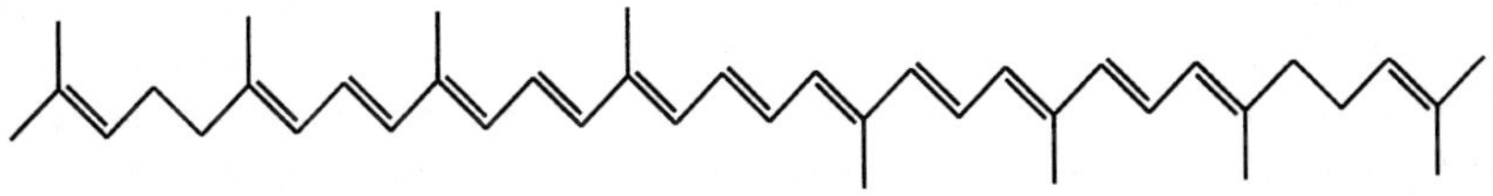

8

went a final purification by semi-preparative HPLC. The maximum capacity of the silica cartridge was estimated to be 31 mg of lycopene in 105 ml petroleum ether (Hakala and Heinonen 1994). Sep-Pak C-18 cartridges were used to remove chlorophyll from a lipophilic extract of *Bidens campylotheca* (Asteraceae), as mentioned above (Redl et al. 1994).

2.12
Preliminary Purification for Preparative High-Performance Liquid Chromatography

Filtration is a necessity before HPLC injection. Syringe-mounted filters provide a convenient and inexpensive way to remove sample particulates which could damage HPLC valves, block transfer lines or clog inlet frits on the column. Filters can be used with a stainless steel or plastic filter holder but most frequently, disposable filters in polypropylene housings are purchased. These generally have a female Luer fitting on the inlet side and a male Luer taper on the outlet side so that they can be attached to a syringe. The filters themselves come in a variety of materials such as PTFE, cellulose acetate, nylon, paper or inorganic membranes but to ensure solvent compatibility, some care must be exercised in selecting the appropriate medium. This is especially important for tetrahydrofuran-containing solvents. Filter porosity can range from 0.1 to 2 μm – a porosity of 0.45 μm is suitable for most applications, using Millex HV filter units or similar membranes of carefully controlled pore size.

Sample preparation with techniques such as medium-pressure LC is not too important a problem when employing a silica gel stationary phase, since the packing material is generally rejected after the separation and impurities left on the stationary phase are consequently also disposed of. However, in the case of preparative HPLC or in work using reversed-phase packing material, the columns are very expensive and careful sample preparation is necessary to avoid contamination with slow-running impurities. This initial work may take the form of an off-line clean-up or an on-line clean-up, the former method involving preliminary purification by, e.g. open-column chromatography, simple filtration through coarse silica gel (Burton et al. 1982) or cartridges (Sep-Pak, Bond Elut etc.).

Guard columns inserted between the injector and chromatography column are recommended for the removal of particles and/or strongly retained sample components. They are usually packed with a small volume of the same support used in the main column and do not greatly decrease system efficiency if correctly packed.

Silica pre-columns deal with problems produced by mobile phases containing buffer salts or bases. These dissolve the silica backbones of bonded-phase packing materials, often causing voids to form at the top of the column. The pre-column is placed between the pump and injector so that, although the mobile phase becomes saturated with silica gel, no dead volume is introduced in the path of the sample.

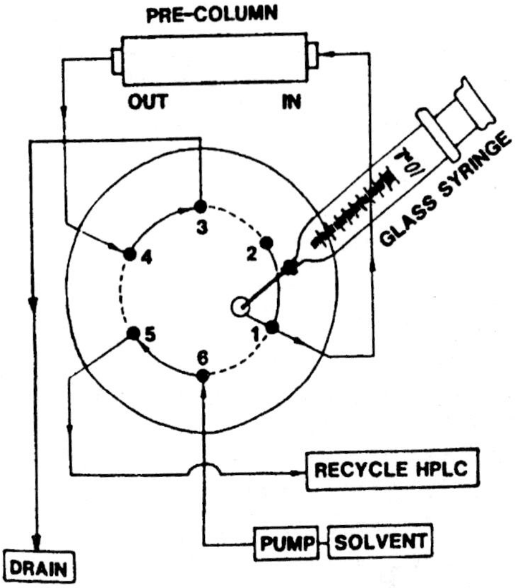

Fig. 2.2. On-line sample preparation by use of a pre-column. (Reprinted with permission from Okano et al. 1984)

On-line clean-up can also be accomplished by column switching, illustrated by the separation of 25-hydroxyvitamin D_2 (Okano et al. 1984). In this method, the pre-column attached to the sample injector (Fig. 2.2) was used to retain 25-hydroxyvitamin D_2 while less polar components were eluted. The six-port valve was then turned to the dotted position, the polarity of the solvent increased and the purified sample eluted into the recycle HPLC system. This method fulfils the same clean-up function as a Sep-Pak cartridge but is especially useful for loading a large volume of sample solution directly onto the HPLC column, i.e. there is a *concentration* effect.

Other examples of column switching are given by Henschen et al. (1985) but it should be noted that with these on-line systems there are quite severe restrictions on the solvent systems that can be employed.

2.13 References

Bevan CD, Marshall PS (1994) Nat Prod Rep 11:451
Burton G, Veleiro AS, Gros EG (1982) J Chromatogr 248:472
Cardellina JH (1991) J Liq Chromatogr 14:659
Cardellina JH, Munro MHG, Fuller RW, Manfredi KP, McKee TC, Tischler M, Bokesch HR, Gustafson KR, Beutler JA, Boyd MR (1993) J Nat Prod 56:1123
Castaneda-Acosta J, Cain AW, Fischer NH, Knopf FC (1995) J Agric Food Chem 43:63
Castioni P, Christen P, Veuthey JL (1995) Analysis 23:95
Cocks S, Wrigley SK, Chicarelli-Robinson MI, Smith RM (1995) J Chromatogr A 697:115
Coval SJ, Scheuer PJ (1985) J Org Chem 50:3025
De Bernardi M, Uberti E, Vidari G (1984) J Chromatogr 284:269
Della Loggia R, Tubaro A, Sosa S, Becker H, Saar S, Isaac O (1994) Planta Med 60:516
Elliger CA, Wong RY, Waiss AC, Benson M (1990) J Chem Soc Perkin Trans I 525
Grode SH, James TR, Cardellina JH, Onan KD (1983) J Org Chem 48:5203

Gross GA, Sticher O, Anklin C (1986) Helv Chim Acta 69:156
Gschwendt M, Hecker E (1973) Z Krebsforsch 80:335
Gustafson KR, Cardellina JH, McMahon JB, Gulakowski RJ, Ishitoya J, Szallasi Z, Lewin NE, Blumberg PM, Weislow OS, Beutler JA, Buckheit RW, Cragg GM, Cox PA, Bader JP, Boyd MR (1992) J Med Chem 35:1978
Hakala SH, Heinonen IM (1994) J Agric Food Chem 42:1314
Härmälä P, Vuorela H, Hiltunen R, Nyiredy Sz, Sticher O, Törnquist K, Kaltia S (1992) Phytochem Anal 3:42
Henschen A, Hupe KP, Lottspeich F, Voelter W (1985) High performance liquid chromatography in biochemistry. VCH, Weinheim
Hostettmann K, Marston A (1995) Saponins. Cambridge University Press, Cambridge
Hostettmann K, Pettei MJ, Kubo I, Nakanishi K (1977) Helv Chim Acta 60:670
Jennings DW, Deutsch HM, Zalkow LH, Teja AS (1992) J Supercrit Fluids 5:1
Kashman Y, Gustafson KR, Fuller RW, Cardellina JH, McMahon JB, Currens MJ, Buckheit RW, Hughes SH, Cragg GM, Boyd MR (1992) J Med Chem 35:2735
King MB, Bott TR (1993) Extraction of natural products using near-critical solvents. Chapman and Hall, London
Liu LK (1996) Anal Commun 33:175
Loev B, Goodman MM (1967) Chem Ind (London) 2026
Loomis WD, Battaile J (1966) Phytochemistry 5:423
Maillard M, Adewunmi CO, Hostettmann K (1992) Phytochemistry 31:1321
McHugh M, Krukonis V (1986) Supercritical extraction: principles and practice. Butterworth, Boston, MA
Miyachi H, Manabe A, Tokumori T (1987) Yakugaku Zasshi 107:367
Modey WK, Mulholland DA, Raynor MW (1996) Phytochem Anal 7:1
Okano T, Masuda S, Kusunose S, Komatsu M, Kobayashi T (1984) J Chromatogr 294:460
Pettit GR, Singh SB, Boyd MR, Hamel E, Pettit RK, Schmidt JM, Hogan F (1995) J Med Chem 38:1666
Queckenberg OR, Frahm AW (1994) Pharmazie 49:159
Redl K, Breu W, Davis B, Bauer R (1994) Planta Med 60:58
Riccio R, Greco OS, Minale L (1985) J Nat Prod 48:97
Rupprecht JK, Hui YH, McLaughlin JL (1990) J Nat Prod 53:237
Sanagi MM, Ahmad UK, Smith RM (1993) J Chromatogr Sci 31:20
Shobha SV, Ravindranath B (1991) J Agric Food Chem 39:2214
Smith RM, Burford MD (1992) J Chromatogr 627:255
Stahl E, Jahn H (1984) Arch. Pharm. 317:573
Tan GT, Pezzuto JM, Kinghorn AD, Hughes SH (1991) J Nat Prod 54:143
Wall ME, Taylor H, Ambrosio L, Davis K (1969) J Pharm Sci 58:839
Wall ME, Wani MC, Taylor H (1976) Cancer Treat Rep 60:1011
Wehr CT (1990) In: Gooding KM, Regnier FE (eds) HPLC of biological macromolecules: methods and applications. Marcel Dekker, New York, p. 215
Westwood SA (1993) Supercritical fluid extraction and its use in chromatographic sample preparation. Chapman and Hall, New York
Zheng GQ (1994) Planta Med 60: 54

CHAPTER 3

Planar Chromatography

Several preparative planar chromatographic techniques are available to the natural products chemist. Some of these involve mobile phase migration through a stationary phase by capillary forces (classical preparative TLC) and some are forced-flow methods, such as centrifugal TLC and overpressured layer chromatography (Nyiredy 1990; Erdelmeier and König 1991).

3.1 Preparative Thin-Layer Chromatography

One of the separation methods requiring the least financial outlay and using the most basic of equipment is classical preparative thin-layer chromatography (PTLC) (Sherma and Fried 1987). Although gram quantities of material can be separated by PTLC, most applications (and these are very numerous) involve milligram quantities. PTLC, in conjunction with open-column chromatography, is still to be found in many publications covering the isolation of natural products, especially in work from those laboratories without access to modern separation techniques. However, as will be explained below, there are numerous inconveniencies connected with PTLC.

3.1.1 Adsorbents

Various studies have been performed to investigate the effect of the thickness of adsorbent on the quality of separation (Stahl 1967) but the most frequently employed thicknesses are 0.5–2 mm. The format of the chromatography plate is generally 20×20 cm or 20×40 cm. Limitations to the thickness of the layer and on the size of the plates naturally reduce the amount of material that can be separated by PTLC. The maximum sample load for a silica layer 1.0 mm thick is about 5 mg/cm^2. Silica gel is the most common adsorbent and is employed for the separation of both lipophilic and hydrophilic substance mixtures. A number of commercially-available sorbents are recommended for the preparation of crack-free layers. Their particle and pore sizes approximate to those of the equivalent TLC grades.

PTLC plates can either be self-made or purchased with the adsorbent already applied (the so-called "Fertigplatten"). The advantage of preparing plates oneself is that any thickness (up to 5 mm) or composition of plates can be accommodated. Thus, silver nitrate, buffers etc. can be incorporated in the adsorbent. Application of the required sorbent can be performed with one of a number of commercially-available spreaders, e.g. from Camag, Desaga. Instructions for the preparation of these plates are supplied with the relevant adsorbent.

3.1.2 Sample Application

This is one of the most critical aspects of PTLC. Plates should preferably be prewashed to minimize impurities that may be recovered when the compound of interest is removed from the sorbent. The sample is dissolved in a small quantity of solvent before application to the PTLC plate. A volatile solvent (hexane, dichloromethane, ethyl acetate) is preferred since problems of band-broadening occur with less volatile solvents. The concentration of the sample should be about 5–10%. The sample is applied as a band, which must be as narrow as possible because the resolution depends on the width of the band. Application can be by hand (pipette) but an automatic applicator (Camag, Desaga, etc.) is preferred. For a band which is too broad, concentration can be achieved by allowing the migration of a polar solvent to about 2 cm above the applied band. The plate is then dried and eluted with the desired solvent (Stahl 1967). Special pre-coated plates with concentration zones (Whatman) are also available. These zones can take the form of an inert preadsorbent strip or a chemically bonded C-18 layer. Analtech produces tapered plates with a silica gel layer varying in thickness from 300 μm at the bottom to 1700 μm at the top. These wedge-shaped plates aid resolution of sample zones.

3.1.3 Choice of Mobile Phase and Development of the PTLC Plate

There are many variables in PTLC but as a general guideline, 10–100 mg of sample can be separated on a 1 mm thick 20×20 cm silica gel or aluminium oxide layer (Székely 1983). Doubling the thickness allows the application of 50% more sample.

Choice of eluent is determined from a preliminary investigation by analytical TLC. Since the particle sizes of the adsorbents are approximately the same, the analytical TLC solvent is directly transferable to PTLC. The standard reference work on thin-layer chromatography by Stahl (1967) gives a large selection of solvent systems for many different classes of compounds.

A method (the "PRISMA" model) based on Snyder's solvent selectivity triangle (Snyder 1978) has been described to aid mobile phase optimisation (Nyiredy et al. 1988).

The following binary mobile phases (in varying proportions) are very often applied to PTLC separations: n-hexane-ethyl acetate, n-hexane-acetone, and chloroform-methanol. Addition of acetic acid or diethylamine in small amounts is useful for the separation of acidic and basic compounds, respectively.

Elution of PTLC plates generally takes place in glass tanks, which may hold several plates at one time. The tank is kept saturated with mobile phase by the presence of a sheet of filter paper dipping into the solvent.

The separation efficiency can be increased by *multiple development*. When a PTLC separation has been completed, the plate is dried and then re-introduced into the development tank. Depending on the Rf of the band, this process can be repeated several times, albeit with a corresponding time penalty.

3.1.4 Isolation of Separated Substances

Most PTLC adsorbents contain a fluorescent indicator which aids localisation of the separated bands, as long as the separated compounds absorb UV light. A problem with some indicators, however, is that they may react with acids – occasionally even with acetic acid.

For non-UV-absorbing compounds, there are several alternatives:

- spraying the plate with water (e.g. saponins)
- covering the plate with a sheet of glass and spraying one edge with a spray reagent
- addition of a reference substance

The bands, having been localised, are scraped off the plate with a spatula or with a tubular scraper connected to a vacuum collector. This latter method is not very practicable for sensitive substances because the adsorbent containing the purified product is in constant contact with a stream of air and there is a risk of autoxidation. Whatever the collection method, the substance has to be extracted from the adsorbent – with the least polar solvent possible (ca. 5 ml solvent for 1 g sorbent). It should be noted that the longer the substance is in contact with the adsorbent, the more likely is decomposition to occur. The extract is filtered through a porosity 4 glass frit and then through a 0.2–0.45-μm membrane.

Methanol is not recommended for elution of separated compounds from the TLC plate because it can solubilise silica gel and some of its impurities. Good alternatives are acetone, ethanol or chloroform.

3.1.5 Impurities in Substances Separated by PTLC

PTLC adsorbents contain binders and fluorescent indicators, the chemical compositions of which are not generally known. During extraction of substances separated as bands on the PTLC plates, these and other impurities are quite possibly extracted as well. In fact, the higher the polarity of the extraction solvents, the greater the quantity of unwanted material. A further problem is that these extraneous substances often do not absorb UV light and are missed when carrying out a final TLC anaysis of the purified product. Székely (1983) has performed the gravimetric, IR and ^{1}H-NMR analysis of impurities extracted from blank silica gel plates and the presence of phthalates and polyesters was quite clearly shown. A final purification step by gel filtration on Sephadex LH-20 is therefore highly recommended.

3.1.6
Overpressured Layer Chromatography (OPLC)

In overpressured layer chromatography a horizontal thin-layer chromatography plate is covered by an elastic cushion. Pressure is applied to the cushion so that separations are performed on the TLC plate in the *absence* of a vapour phase (Fig. 3.1) (Tyihak et al. 1979). Under forced-flow conditions of eluent, sample development is rapid. High efficiencies can be achieved due to the use of fine-particle sorbents and longer plates than those found in capillary-controlled systems. As separation times are short, diffusion effects are reduced. It is also possible to employ mobile phases with poor solvent wetting characteristics.

The following instruments are available for OPLC separations: Chrompres 10 and Chrompres 25 (with a higher cushion pressure) from Labor MIM Works, Budapest-Esztergom, Hungary; KB 5121, KB 5125, KB 5129 from Cobrabid, Warsaw, Poland.

Both off-line (Nyiredy et al. 1986a) and on-line (Nyiredy et al. 1986b) preparative separations have been reported with OPLC apparatus, but on-line applications require special preparation of the plates by scraping inlet and outlet channels for the eluent and coating edges with a polymer to prevent solvent leakage. The technique is best suited to the isolation of small (50–100 mg) amounts of partially purified samples but separations of up to 1 g of sample have been performed within 45 min.

OPLC separations are usually initiated on dry plates in a non-equilibrated system. This procedure often leads to solvent demixing and the consequent formation of ill-shaped bands. However, by on-plate injection, the chromatographic layer can be pre-equilibrated with the solvent system, thus avoiding the solvent demixing effect (Erdelmeier et al. 1987).

Selection of solvent systems by the “PRISMA” optimization procedure has been described, with special reference to the OPLC of polar natural products (ginsenosides and flavonoid glycosides) (Dallenbach-Tölke et al. 1986).

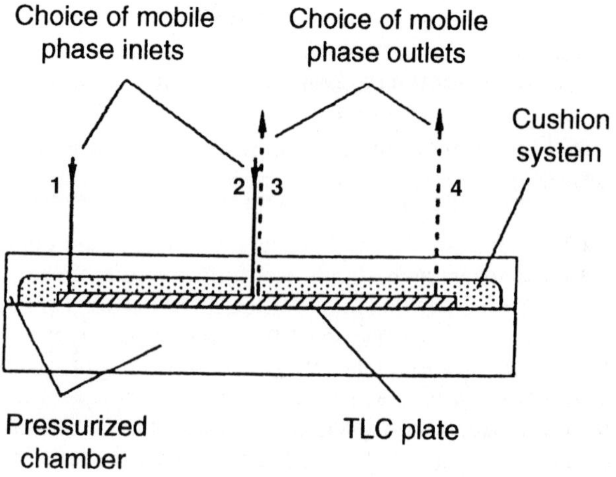

Fig. 3.1. Principle of preparative OPLC. (Reprinted with permission from Tyihak and Mincsovics 1988)

Applications to the separation of natural products include the following:

Anthraquinones: frangula-emodin from *Rhamnus frangula* (Rhamnaceae) (Nyiredy et al. 1986b)
Furocoumarins: methoxylated furocoumarins from *Heracleum sphondylium* (Apiaceae) (Nyiredy et al. 1986b); furocoumarins from *Peucedanum palustre* (Apiaceae) (Vuorela et al. 1988)
Secoiridoid glycosides: amaropanin, amarogentin, amaroswerin and gentiopicrin from *Gentiana purpurea* (Gentianaceae) (Nyiredy et al. 1986b)
Sesquiterpenes: synthetic hernandulcin (an intensely sweet sesquiterpene) was purified in 100 mg quantities on normal TLC plates with the mobile phase hexane-ethyl acetate 10:3. Minor reaction products were removed in this manner (Fullas et al. 1989)
Diterpenes: phorbol diesters from croton oil (Erdelmeier et al. 1988)
Cucurbitacins: from *Hemsleya gigantha* (Cucurbitaceae) (Yang et al. 1988)
Steroid saponins: polypodoside A from *Polypodium glycyrrhiza* (Polypodiaceae) (Fullas et al. 1989)
Essential oils: in the characterization of fungicidal components of the essential oil from the fresh bark of *Ocotea usambarensis* (Lauraceae), OPLC was used (solvent: *t*-butyl methyl ether-ethyl acetate-dichloromethane-n-hexane 5:3:20:72; plate: Si 60 200×200×2 mm; flow rate: 1.8 ml/min; sample: 130 mg) (Terreaux et al. 1994)

3.2 Centrifugal Thin-Layer Chromatography

Classical preparative thin-layer chromatography suffers from several drawbacks, the main disadvantage being the removal of purified substance from the plate and its subsequent extraction from the sorbent. When toxic products are being scraped from the plates, serious problems can arise (e.g. Adolf et al. 1982). Other drawbacks include the length of time required for a separation and the presence of impurities and residues from the plate itself, found after solvent extraction of the zones containing the product (Székely 1983).

In order to overcome some of these problems, a number of approaches involving centrifugal chromatography have been attempted. The technique of centrifugal chromatography is, in principle, classical PTLC with an accelerated flow of mobile phase produced by the action of a centrifugal force – a forced-flow method.

3.2.1 Historical Development

Caronna (1955) introduced centrifugal *paper* chromatography, in which a rotor consisting of two plexiglass plates sandwiched a circular sheet of chromatography paper. This was used for the separation of inorganic and organic substances. There have been several modifications, notably by McDonald et al. (1957), Herndon et al. (1962) and Dauphin et al. (1960), all summarized in a review by Deyl et al. (1964). All these methods employed horizontal rotors and did not meet with general acceptance because of numerous problems associated with the delivery of mobile phase, the flow of mobile phase through the paper, collection of eluent and restrictions on the amount of sample.

Meanwhile, Hopf (1947) had developed a centrifugally-accelerated apparatus for the separation of 100 mg sample quantities – the *chromatofuge*. This consisted of a perforated cylinder filled with adsorbent (aluminium oxide, barium carbonate etc.) and a central tube, down which sample and eluent were introduced. Radial migration of solvent in the basket was achieved by rotation about its axis. The method was modified by Heftmann et al. (1972), such that preparative separations could be more easily performed. Using this modification, the purines caffeine, theobromine, theophylline and xanthine were quickly separated in 5-mg amounts on silica gel. However, the shape of the collection vessel caused some re-mixing of the fractions. A scaled-up chromatofuge enabled the preparative separation of amino-acids on a gram scale (Finley et al. 1978). The cylindrical basket was filled with ion-exchange resin and a mixture containing 5 g each of alanine, histidine and tryptophan was separated at a rotor speed of 1000 rpm. Despite the satisfactory results, there was no improvement in resolution over conventional column chromatography.

Although prototypes of other centrifugal chromatographic instruments have been reported (Deyl et al. 1964; Lepoivre 1972; Pfander et al. 1976), it was not until the introduction of two commercially available devices, the Chromatotron (Harrison Research, Palo Alto, California, USA) and the Hitachi CLC-5 Centrifugal Chromatograph (Hitachi Koki Ltd., Takeda Katsuta City, Japan), that centrifugal TLC (CTLC) really became of more general interest. The latter two chromatographs overcome most of the difficulties encountered with the earlier methods and, in addition, their very simplicity of operation explains their widespread acceptance.

3.2.2 Apparatus

3.2.2.1 *Hitachi CLC-5 Centrifugal Chromatograph*

This apparatus is so constructed that a slurry of adsorbent in the eluting solvent is introduced into the space between two *horizontal* 30 cm diameter plates, rather like a sandwich. Excess solvent is removed by centrifugal force and the sample mixture is introduced onto the still damp layer. Elution of bands is carried out in the normal fashion.

The CLC-5 has approximately the same load capacity as the Chromatotron for an equal layer thickness (Hunter and Heftmann 1983) but whereas the Chromatotron cannot accommodate sorbent layers thicker than 4 mm, the CLC-5 can be packed to a much greater thickness.

Applications of this instrument are less numerous than those for the Chromatotron. One example involves the determination of solasodine in fruits of *Solanum khasianum* (Solanaceae) (Nes et al. 1980). Before HPLC determination, an extract of the fruits was fractionated on the CLC-5 with a 3-mm spacer. Silica gel (Woelm, 50 g) was used as adsorbent and both the rotational speed and flow rate were varied so as to keep the right amount of solvent between the plates.

Tanaka et al. (1982) reported the isolation of confusoside, a dihydrochalcone glycoside, from the leaves of *Symplocos confusa* (Symplocaceae) with the aid of the CLC-5 instrument. The thickness of the silica gel layer was 3 mm, rotation speed 600 rpm and solvents EtOAc-EtOH-H_2O 100:8:1 and then EtOAc saturated with water.

The Hitachi Chromatograph has also been used in the isolation of neutral lipids from rice-bran oil (Shimasaki and Ueta 1983). A 3-mm layer of silica gel was eluted with hexane-benzene 85:5, hexane-diethyl ether 95:5, hexane-benzene-acetic acid 94:5:1, hexane-diethyl ether 85:15, 70:30 and diethyl ether at 1000 rpm, to obtain the different lipid fractions. The flow rate was maintained at 15 ml/min.

3.2.2.2
Chromatotron

The big difference between the Chromatotron and its predecessors lies in the fact that the rotor is *inclined* and not horizontal (Fig. 3.2).

The heart of the apparatus is a 24 cm diameter circular glass plate which is covered with a suitable sorbent, to provide the thin layer for preparative separations. In order to prevent breaking up of the thin layer, the sorbent is mixed with a binder. This is generally TLC-grade calcium sulphate hemihydrate (plaster of Paris, dried gypsum), such as supplied by J. T. Baker (catalogue no. 1463).

For the majority of separations on silica gel, the following recipe for a single 2-mm plate can be used. Silica gel 60 F_{254} for TLC (60 g; Merck cat. no. 7730) is

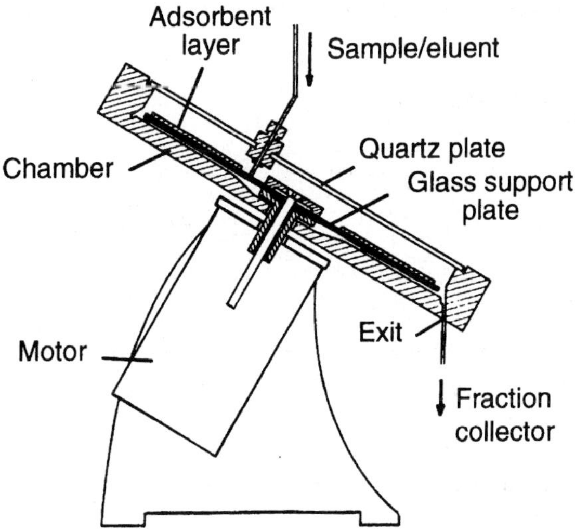

Fig. 3.2. Schematic view of the Chromatotron (Model 7924)

mixed with calcium sulphate hemihydrate (8 g) in a 250-ml Erlenmeyer flask. Water (110 ml) is added and vigorously shaken for 30 s. The slurry is poured onto the plate and allowed to spread out into a relatively uniform layer. The sorbent is prevented from running over the edges by a collar of masking tape around the circumference of the plate. Overnight drying at room temperature is followed by oven heating at 60–70 °C for ca. 1 h. A procedure has been described for drying plates by slow rotation at 70 °C (Vieth and Sloan 1986). After cooling, the uneven surface of the sorbent is smoothed by a scraping tool fixed to the centre of the plate. A small area is left free in the middle to allow introduction of eluent.

Other recipes for the preparation of plates are provided in the Chromatotron Handbook (Harrison Research Inc., 840 Mona Court, Palo Alto, CA 94306, USA). These include silica gel plates with silver nitrate, and aluminium oxide (procedure based on the method of Desai et al. (1985, 1986)). Binding reversed-phase material to the glass plates is a much more difficult task.

The prepared plate (1, 2 or 4 mm sorbent thickness) is screwed onto the hub of an electric motor and rotated at 800 rpm. Eluent is introduced onto the sorbent-free centre of the plate via a piston pump capable of delivering 1–10 ml/min, and passes across the thin layer under the influence of the centrifugal force. The rotor is first washed for several minutes to remove impurities present in the adsorbent. Following this step, two options are available: direct introduction of the sample or preliminary drying of the plate before introduction of sample. Elution is then continued, either with a solvent of constant composition or with a step gradient, at ca. 3–6 ml/min.

The rotor unit is housed in a chamber covered with a quartz glass plate. This cover enables the observation of colourless but UV-active substance zones with the aid of a UV lamp. A steady flow of nitrogen is passed through the chamber to prevent condensation of the eluent and to avoid oxidation of the sample.

Introduction of sample, followed by solvent elution gives concentric bands of the components (Fig. 3.3). At the periphery, the bands are spun off and collected through an exit tube in the chamber. Fractions of eluate thus obtained are analysed by TLC.

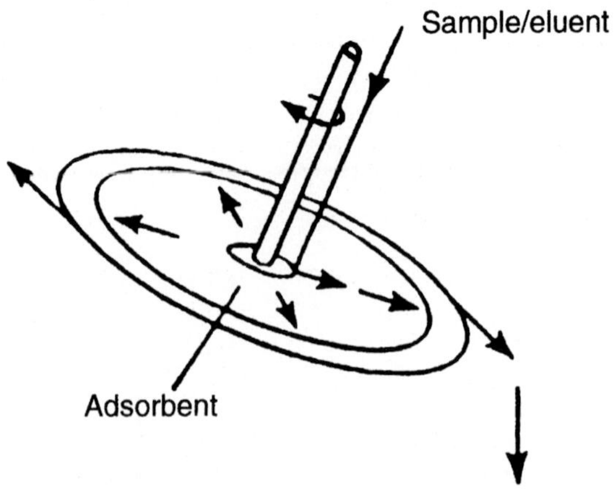

Fig. 3.3. The principle of the Chromatotron

Table 3.1. Advantages and disadvantages of the Chromatotron

Advantages
- simple operation
- rapid (separations generally completed within 30 min)
- no scraping of bands necessary
- low consumption of solvent
- coated rotors can be regenerated and re-used
- less impurities extracted from sorbent than preparative TLC
- straightforward sample application
- step gradients possible
- oxidation of sensitive substances less likely than PTLC
- higher recoveries of product than PTLC
Disadvantages
- restricted choice of stationary phases
- coated rotors not commercially available
- resolution limited
- limited range of detection methods
- contamination possible during collection

In general, 50–500 mg of a mixture can be separated on a 2-mm layer, as long as the R_f on an analytical TLC plate lies somewhere between 0.2 and 0.5 (Hostettmann et al. 1980). Weakly polar solvent systems should be used at the beginning of the separation, with a gradual increase in polarity during the run. A more detailed investigation of the parameters important for CTLC separations is given by Stahl and Müller (1982).

A summary of the advantages and disadvantages of the Chromatotron instrument is given in Table 3.1.

3.2.2.3 *Rotachrom*

The Rotachrom or "Rotationsplanarchromatograph" is another commercially available instrument for CTLC (Petazon Ltd., Zug, Switzerland), with a rotor speed which can be regulated between 100 and 1500 rpm. It is a modular instrument which allows CTLC, sequential centrifugal layer chromatography (SCLC) or centrifugal planar column chromatography (CPCC) to be performed on the same machine (Erdelmeier and König 1991).

SCLC has been introduced in an attempt to separate mixtures, e.g. isomers, which are not successfully resolved by CTLC. In this method there are basically two refinements over the conventional Chromatotron:

a) the solvent delivery system can be moved to any position between the centre of the plate and the periphery;
b) a circular solvent application system of variable radius can be introduced onto the plate at any time when it is *horizontal* and *motionless*.

This annular device introduces solvent to the plate in such a fashion that developed bands are pushed back towards the centre of the rotor by capillary action.

By a judicious choice of a) and b), difficult separations can be performed on a single plate without having to recycle mixtures onto a second CTLC plate (Nyiredy et al. 1986a). The separation of furocoumarin isomers provides an example of this technique (Nyiredy et al. 1985). However, it should be noted that when changes in solvent polarity are made, care must be taken to dry the plate thoroughly before eluting with the new solvent. This is especially important when methanol is used. The drying process is itself a possible hazard, as some substances may undergo chemical reactions at the same time (Stahl 1968). Application of sample must also be performed with precision to avoid irregular zones. Finally, the different operations – elution, drying, non-centrifugally-accelerated elution – imply that development of the zones is a lengthy process.

CPCC operates with a closed circular chamber and employs a non-uniform thickness of thin layer adsorbent to increase resolution of the separated bands. Since the layers are wedge-shaped, a sharpening effect of the bands is produced during their radial development. The preparative separation of furocoumarins has also been reported by this method (Botz et al. 1990).

3.2.3 Applications of the Chromatotron

Advantages of CTLC over preparative TLC are numerous and, if the price of the Chromatotron was not so high, this instrument would surely be even more extensively exploited. As far as loading capacities are concerned, one Chromatotron plate can separate the same amount of sample as three 20×20 cm preparative TLC plates of the same thickness (Székely 1983). In one study, CTLC was shown to have advantages over preparative LC for the separation of diastereoisomeric 2-arylpropionic acid derivatives (Maître et al. 1986).

Choice of the eluent by analytical TLC for separations on silica gel is quite straightforward (Hostettmann et al. 1980). The R_f values should be kept below 0.5 or else elution by CTLC is too fast. For optimizing the choice of complicated solvent systems, the "PRISMA" technique has also been applied (Nyiredy et al. 1988). This is especially useful for ternary and quaternary mixtures but less so for binary systems, the most frequent CTLC eluents.

The first two reports on applications of centrifugal TLC involved the separation of synthetic intermediates (Derguini et al. 1979) and the separation of various natural products (Hostettmann et al. 1980). Derguini et al. (1979) were able to separate the isomers **1** and **2**, intermediates in the synthesis of retinoids, with a 1-mm silica gel plate. Recovery from the Chromatotron was 90%, compared with 80% for preparative TLC.

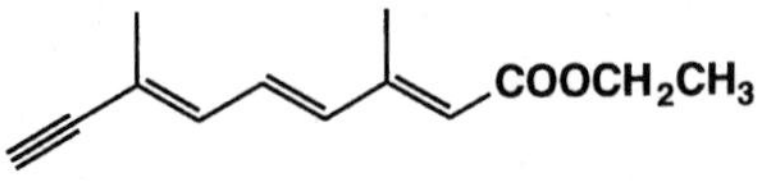

1

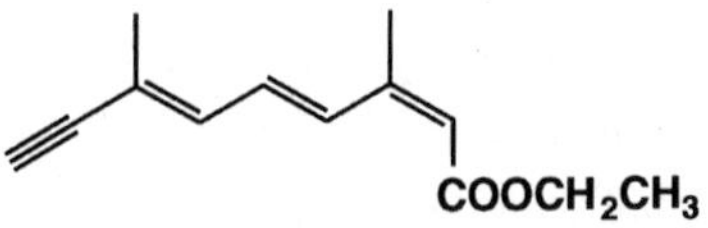

2

The xanthones bellidifolin (**3**) and desmethylbellidifolin (**4**) obtained after acid hydrolysis of a methanol extract of *Gentiana strictiflora* (Gentianaceae) were separated in less than 30 min using chloroform with increasing amounts of methanol as solvent. The corresponding separation by polyamide open-column chromatography required at least 12 h (Hostettmann et al. 1980).

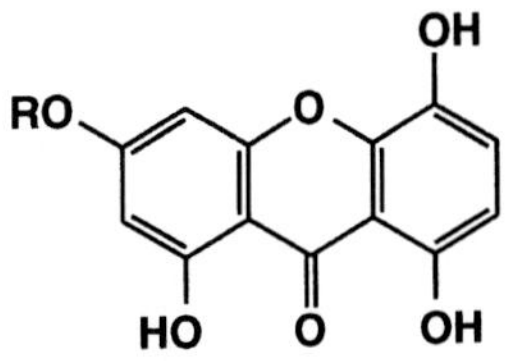

3 R = CH_3
4 R = H

Further applications of CTLC are given in Table 3.2. As can be seen, these cover a wide range of substance classes and polarities. Many CTLC separations are performed on plates prepared with silica gel PF_{254}, which is a support already containing calcium sulphate binder and sold by E. Merck as catalogue no. 7749. The solvent combination hexane-ethyl acetate is to be found in a fair number of applications: its toxicity is low and it is not too volatile. Centrifugal TLC is sometimes used as an intermediate purification step in these examples but it is also of value for the isolation of pure substances (e.g. Marston et al. 1984).

In the search for neo-clerodane insect antifeedants and antifungal agents, *Scutellaria albida* (Lamiaceae) aerial parts have been investigated. An acetone extract (220 g) was chromatographed on silica gel (1 kg; Merck 7734) *deactivated* with 10% water. Elution with a petrol ether-ethyl acetate gradient gave a petrol ether-EtOAc (3:2) fraction which was purified by CTLC (silica gel plate, dichloromethane-methanol 49:1) to give five neo-clerodane diterpenes, including clerodin (**5**; 200 mg) (Bruno et al. 1996).

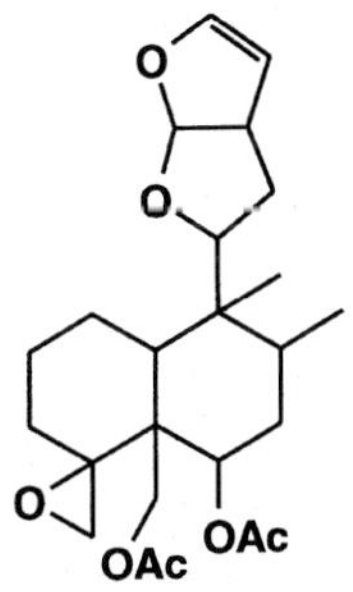

5

The separation of diterpene alkaloids has been successfully performed on aluminium oxide $60GF_{254}$ (Merck) thin layers (Desai et al. 1985). To prepare the Chromatotron rotor, a slurry of 60 g aluminium oxide, 3.5 g calcium sulphate hemihydrate and 65 ml water was spread over the glass surface and scraped to a 1 mm layer, after drying. The plates were pre-washed with hexane and eluted with solvent at 2–4 ml/min. Several diterpene alkaloid mixtures were separated

Table 3.2. Applications of preparative centrifugal TLC

Sample	Sorbent (Thickness)	Sample size	Eluent	Reference
General phenolics				
Dimethylacrylophenones and a chromanone from *Nama hispidum* (Hydrophyllaceae)	Silica gel (1 mm)		nC_6H_{14}-Et_2O 4:1 nC_6H_{14}-$CHCl_3$ 1:1	Roitman and Wollenweber 1993
Phthalide glycoside from *Gentiana pedicellata* (Gentianaceae)	Silica gel		$CHCl_3$-MeOH 9:1	Chulia et al. 1986
Arbutin derivatives from *Gentiana pyrenaica* (Gentianaceae)	Silica gel		$CHCl_3$-MeOH gradient	Garcia et al. 1989
Naphthoquinones				
Juglone derivatives from *Diospyros usambarensis* (Ebenaceae)	Silica gel (2 mm)	100 mg	Toluene-EtOAc 15:1	Marston et al. 1984
Lignans				
Diarylheptanoid glycosides from *Alnus serrulatoides* (Betulaceae)	Silica gel[a] (5 mm)		$CHCl_3$-MeOH-H_2O 85:14:1	Ohta et al. 1984
Lignan from *Justicia pectoralis* (Acanthaceae)	Silica gel		nC_6H_{14}-EtOAc 95:5	Joseph et al. 1988
Peltatin from *Amanoa oblongifolia* (Euphorbiaceae)	Silica gel PF_{254} (2 mm)		Heptane-$CHCl_3$-EtOH 25:25:3	MacRae et al. 1988
Neolignan ketones from *Piper capense* (Piperaceae)	Silica gel		nC_6H_{14}-EtOAc	Green and Wiemer 1991
Flavonoids				
Hydroxyflavanone from *Lonchocarpus minimiflorus* (Leguminosae)	Silica gel		Toluene-EtOAc-HOAc 96:4:1	Mahmoud and Waterman 1985
Kaempferol monoglycosides from *Platanus acerifolia* (Platanaceae)	Silica gel		Benzene-EtCOMe-MeOH 17:2:1, 15:3:2, 13:4:3	Kaouadji 1990

Xanthones				
Xanthones from *Garcinia* species (Guttiferae)	Silica gel		Petrol-EtOAc 96:4 Toluene-EtOAc-HOAc	Ampofo and Waterman 1986
Pterocarpans				
Glyceollins from *Glycine max* (Leguminosae)	Silica gel (4 mm)	700 mg	r.C_6H_{14}-EtOAc 4:1 + 0.1 % MeOH	Giannini et al. 1991
Monoterpenes				
Secoiridoid glucosides from *Gentiana campestris* (Gentianaceae)	Silica gel		$CHCl_3$-MeOH 9:1	Mpondo Mpondo et al. 1990
Sesquiterpenes				
Norbornanol derivative from *Dysoxylum spectabile* (Meliaceae)	Silica gel		r.C_6H_{14}-Et_2O 4:1	Russell et al. 1994
Heliangolide from *Eupatorium quadrangulare* (Asteraceae)	Silica gel (+ 10 % $AgNO_3$)		r.C_6H_{14}-EtOAc 9:1, 75:25, 85:15, 1:1	Hubert et al. 1987
Sesquiterpene lactones from *Centaurea napifolia* (Asteraceae)	Silica gel		CH_2Cl_2-MeOH 99:1, 49:1, 19:1	Bruno et al. 1995
Diterpenes				
Diterpenes from *Stachys mucronata* (Lamiaceae)	Silica gel PF_{254} (1 mm)		r.C_6H_{14}-EtOAc 95:5	Fazio et al. 1994
Ingenane diterpene from *Mabea excelsa* (Euphorbiaceae)	Silica gel (2 mm)		r.C_6H_{14}-Et_2O 9:1 $\rightarrow$ Et_2O $\rightarrow$ Et_2O-Me_2CO 9:1 $\rightarrow$ 5:2 $\rightarrow$ 1:2 $\rightarrow$ Me_2CO	Brooks et al. 1990
Neo-clerodane diterpenes from *Scutellaria albida* (Lamiaceae)	Silica gel		CH_2Cl_2-MeOH 49:1, 19:1 Petrol-EtOAc 1:1	Bruno et al. 1996
Neo-clerodane diterpenes from *Teucrium yemense* (Lamiaceae)	Silica gel PF_{254} (4 mm)	1 g	Toluene-EtOAc-HOAc 65:35:1	Sattar et al. 1995
Quassinoid from *Simaba guianensis* (Simaroubaceae)	Silica gel	380 mg	Toluene-EtOAc 8:2	Cabral et al. 1993

Table 3.2 (continued)

Sample	Sorbent (Thickness)	Sample size	Eluent	Reference
Triterpenes				
Triterpenes from *Artocarpus integrifolia* (Moraceae)	Silica gel PF_{254} (2 mm)		nC_6H_{14}-EtOAc 6:1	Smith-Kielland et al. 1994
Saponins				
Oleanane saponins from *Vigna angularis* (Leguminosae)	Silica gel[a]		$CHCl_3$-MeOH-H_2O 20:3:1 (lower phase) 10:3:1 (lower phase)	Kitagawa et al. 1983
Triterpene saponins from *Bellium bellidioides* (Asteraceae)	Silica gel PF_{254}		$CHCl_3$-MeOH-H_2O 13:3:1 (lower phase)	Schöpke et al. 1996
Steroid alkaloid from *Fritillaria harelinii* (Liliaceae)	Silica gel		$CHCl_3$-MeOH-H_2O 13:7:2 (lower phase)	Zhi-Da et al. 1986
Cardenolides				
Cardenolides from *Asclepias viridis* (Asclepiadaceae)	Silica gel PF_{254} (4 mm)	800 mg	CH_2Cl_2-MeOH 99:1, 97:3, 9:1	Martin et al. 1991
	(2 mm)	500 mg	CH_2Cl_2-MeOH 49:1 $\rightarrow$ 47:3	
Acetogenins				
Acetogenins from *Annona muricata* (Annonaceae)	Silica gel PF_{254}		CH_2Cl_2-EtOAc (3:2)-MeOH gradient $CHCl_3$-MeOH 99.5:0.5 $\rightarrow$ 98:2	Rieser et al. 1996
Alkaloids				
Norditerpene alkaloids from *Delphinium ajacis* (Ranunculaceae)	Al_2O_3 PF_{254} (1 mm)		nC_6H_{14}-Et_2O 75:25, 1:1	Lu et al. 1993
Diterpene alkaloids from *Aconitum crassicaule* (Ranunculaceae)	Al_2O_3 GF_{254}	73 mg	nC_6H_{14}-Et_2O 95:5 $\rightarrow$ 70:30	Wang and Pelletier 1987

Alkaloids from *Zanthoxylum budrunga* (Rutaceae)	Silica gel PF_{254} (1 mm)		$n\text{-}C_6H_{14}\text{-}Et_2O$ mixtures	Joshi et al. 1991
Cyclopeptide alkaloids from *Zizyphus lotus* (Rhamnaceae)	Silica gel PF_{254} (4 mm, 2 mm)		$CHCl_3$-MeOH 99:1, 49:1, 19:1	Ghedira et al. 1993
Ent-kaurene alkaloids from *Anopterus glandulosus* (Escalloniaceae)	Silica gel PF_{254}		Heptane-$CHCl_3$-Me_2CO-Et_2NH 6:5:4:1	Wall et al. 1987
Canthin-6-one alkaloids from *Brucea mollis* (Simaroubaceae)	Silica gel GF_{254} (4 mm)		$CHCl_3$-MeOH 19:1 CH_2Cl_2	Ouyang et al. 1994
Miscellaneous				
Guanidines from *Galega orientalis* (Leguminosae)	Al_2O_3 (2 mm)		$CHCl_3$-MeOH 1:4 → 3:2	Benn et al. 1996
Nitropropanol glycoside from *Astragalus miser* (Leguminosae)	Silica gel		$CHCl_3$-EtOH 4:1 → 2:3	Benn and Majak 1989
Nitropropanol glycosides from *Corynocarpus laevigatus* (Corynocarpaceae)	Silica gel (1 mm)		$CHCl_3$-Me_2CO 9:1 → 3:7 + 1% HCOOH	Majak and Benn 1994
Hydroxycinnamic acid amides from *Iochroma cyaneum* (Solanaceae)	Silica gel (2 mm)		EtOAc-MeOH-H_2O 50:15:2, 50:15:3, 10:3:1	Sattar et al. 1990
Ryanoids from *Ryania* insecticide (Flacourtiaceae)	Silica gel (2 mm)	400 mg	$CHCl_3$-MeOH 20:6 → 6:1 + 2% aq. $MeNH_2$	Jefferies et al. 1991
Sulphide from *Ferula rutabensis* (Apiaceae)	Silica gel PF_{254} (2 mm)	170 mg	nC_6H_{14}-$CHCl_3$-Toluene 2:1:1	Al-Said et al. 1996

[a] Hitachi CLC-5 Chromatograph.

in a single run but the bis-diterpenoid alkaloids staphisine (**6**) and staphidine (**7**), which differ only in a single methoxy group, required a recycling procedure. A 93-mg mixture was applied and the rotor eluted with hexane-acetone 95.5:4.5. After two recycles, a process which took 2.5 h, the two pure compounds were obtained. The same separation by preparative TLC was very difficult and could only be achieved by gradient multiple developments (Desai et al. 1985). Visualization of non-UV-active alkaloids was effected by exposing a small strip of the rotor to iodine vapour. However, monitoring of the separation might have been better carried out by a TLC analysis of fractions collected from the Chromatotron.

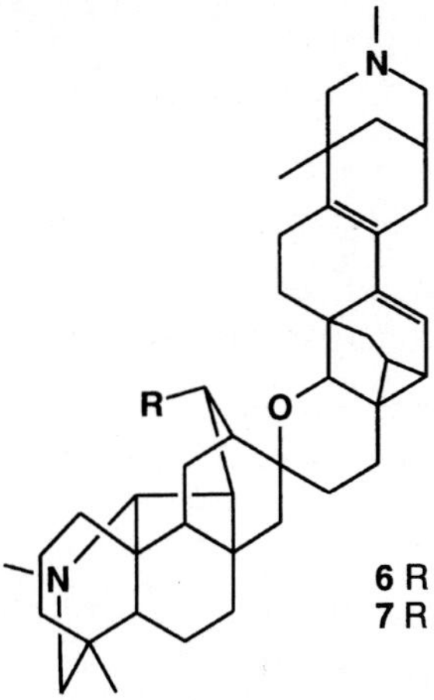

6 R = OCH_3
7 R = H

Other applications of CTLC in the separation of artificial mixtures of alkaloids have been reported by Ferrari and Verotta (1988). The solvent systems contained diethylamine and the rotors were coated with aluminium oxide 60 HF_{254} basic (Merck 1094).

In the characterization of guanidines from fodder galega (*Galega orientalis*, Leguminosae), a step involving CTLC on aluminium oxide plates provided final purification of smirnovine (**8**) and 4-hydroxysmirnovine (**9**). Rotors of 2 mm thickness were prepared with neutral aluminium oxide 60 GF containing 10% calcium sulphate hemihydrate. The two alkaloids were eluted with a chloroform-methanol gradient at a flow rate of 3 ml/min (Benn et al. 1996).

As is evident from the examples described, preparative CTLC is potentially a method of choice for the separation of mixtures containing in the region of

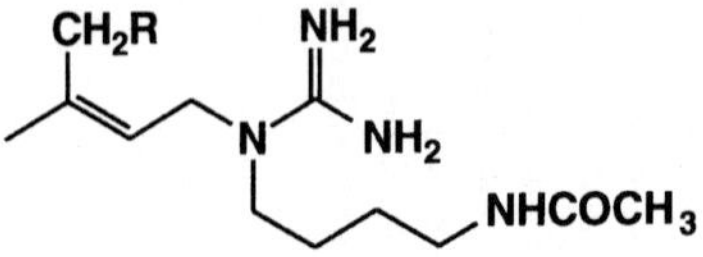

8 R = H
9 R = OH

100 mg sample. Resolution is inferior to that of preparative HPLC but operating conditions are simple and separations are rapid. The major advantage over PTLC is the elution of product without having to scrape the sorbent. Further improvements could be introduced but care has to be taken that these do not detract from the very simplicity of the technique. The method by which eluent is collected could be ameliorated and the use of CTLC would be extended if the number of adsorbents compatible with the glass support plates was increased.

3.3 References

Adolf W, Sorg B, Hergenhahn M, Hecker E (1982) J Nat Prod 45:347

Al-Said MS, Sattar EA, El-Feraly FS, Nahrstedt A, Coen M (1996) Int J Pharmacog 34:189

Ampofo SA, Waterman PG (1986) Phytochemistry 25:2351

Benn MH, Majak W (1989) Phytochemistry 28:2369

Benn MH, Shustov G, Shustova L, Majak W, Bai Y, Fairey NA (1996) J Agric Food Chem 44:2779

Botz L, Nyiredy Sz, Sticher O (1990) J Planar Chromatogr 3:10

Brooks G, Evans AT, Markby DP, Harrison ME, Baldwin MA, Evans FJ (1990) Phytochemistry 29:1615

Bruno M, Fazio C, Paternstro MP, Diaz JG, Herz W (1995) Planta Med 61:374

Bruno M, Piozzi F, Rodriguez B, de la Torre MC, Vassallo N, Servettaz O (1996) Phytochemistry 42:1059

Cabral JA, McChesney JD, Milhous WK (1993) J Nat Prod 56:1954

Caronna G (1955) Chim Ind (Milan) 37:113

Chulia AJ, Garcia J, Mariotte AM (1986) J Nat Prod 49: 514

Dallenbach-Tölke K, Nyiredy Sz, Meier B, Sticher O (1986) J Chromatogr 365:63

Dauphin J, Maugarny M, Berger JA, Dorier C (1960) Bull Soc Chim France 2110

Derguini F, Balogh-Nair V, Nakanishi K (1979) Tetrahedron Lett 4899

Desai HK, Joshi BS, Panu AM, Pelletier SW (1985) J Chromatogr 322:223

Desai HK, Trumbull HK, Pelletier SW (1986) J Chromatogr 366:439

Deyl Z, Rosmus J, Pavlicek M (1964) Chromatographic Reviews 6:19

Erdelmeier CAJ, König GM (1991) Phytochem Anal 2:3

Erdelmeier CAJ, Kinghorn AD, Farnsworth NR (1987) J Chromatogr 389:345

Erdelmeier CAJ, Van Leeuwen PAS, Kinghorn AD (1988) Planta Med 54:71

Fazio C, Passannanti S, Paternostro MP, Arnold NA (1994) Planta Med 60:499

Ferrari M, Verotta L (1988) J Chromatogr 437:328

Finley JW, Krochta JM, Heftmann E (1978) J Chromatogr 157:435

Fullas F, Kim J, Compadre CM, Kinghorn AD (1989) J Chromatogr 464:213

Garcia J, Mpondo Mpondo E, Kaouadji M, Mariotte AM (1989) J Nat Prod 52:858

Ghedira K, Chemli R, Richard B, Nuzillard JM, Zeches M, Le Men-Olivier L (1993) Phytochemistry 32: 1591

Giannini JL, Halvorson JS, Spessard GO (1991) Phytochemistry 30:3233

Green TP, Wiemer DF (1991) Phytochemistry 30:3759

Heftmann E, Krochta JM, Farkas DF, Schwimmer S (1972) J Chromatogr 66:365

Herndon JF, Appert HE, Touchstone JC, Davis CN (1962) Anal Chem 34:1061

Hopf PP (1947) Ind Eng Chem 39:938

Hostettmann K, Hostettmann-Kaldas M, Sticher O (1980) J Chromatogr 202:154

Hubert TD, Okunade AL, Wiemer DF (1987) Phytochemistry 26:1751

Hunter IR, Heftmann E (1983) J Liq Chromatogr 6:281

Jefferies PR, Toia RF, Casida JE (1991) J Nat Prod 54: 1147

Joseph H, Gleye J, Moulis C, Mensah LJ, Roussakis C, Gratas C (1988) J Nat Prod 51:599

Joshi BS, Moore KM, Pelletier SW, Puar MS (1991) Phytochem Anal 2:20

Kitagawa I, Wang HK, Saito M, Yoshikawa M (1983) Chem Pharm Bull 31:674

Kaouadji M (1990) Phytochemistry 29: 2295

Lepoivre A (1972) Bull Soc Chim Belges 81: 213

Lu J, Desai HK, Ross SA, Sayed HM, Pelletier SW (1993) J Nat Prod 56: 2098

MacRae WD, Hudson JB, Towers GHN (1988) J Ethnopharmacol 22: 223

Mahmoud EN, Waterman PG (1985) J Nat Prod 48: 648

Maître JM, Boss G, Testa B, Hostettmann K (1986) J Chromatogr 356: 341

Majak W, Benn M (1994) Phytochemistry 35: 901

Marston A, Msonthi JD, Hostettann K (1984) Planta Med 50: 279

Martin RA, Lynch SP, Schmitz FJ, Pordesimo EO, Toth S, Horton RY (1991) Phytochemistry 30: 3935

McDonald HJ, Bermes EW, Shepherd HG (1957) Chromatographic Methods 2: 1

Mpondo Mpondo E, Garcia J, Chulia AJ (1990) Phytochemistry 29: 1687

Nes WD, Heftmann E, Hunter IR, Walden MK (1980) J Liq Chromatogr 3: 1687

Nyiredy Sz (1990) Anal Chim Acta 236: 83

Nyiredy Sz, Erdelmeier CAJ, Sticher O (1985) J High Res Chromatogr 8: 132

Nyiredy Sz, Erdelmeier CAJ, Sticher O (1986a) In: Kaiser RE (ed) Planar chromatography, vol 1. Alfred Hüthig, Heidelberg, p 119

Nyiredy Sz, Erdelmeier CAJ, Dallenbach-Tölke K, Nyiredy-Mikita K, Sticher O (1986b) J Nat Prod 49: 885

Nyiredy Sz, Dallenbach-Tölke K, Sticher O (1988) J Planar Chromatogr 1: 336

Ohta S, Aoki T, Hirata T, Suga T (1984) J Chem Soc Perkin Trans I 1635

Ouyang Y, Koike K, Ohmoto T (1994) Phytochemistry 36: 1543

Pfander H, Haller F, Leuenberger FJ, Thommen H (1976) Chromatographia 9: 630

Rieser MJ, Gu Z, Fang X, Zeng L, Wood KV, McLaughlin JL (1996) J Nat Prod 59: 100

Roitman JN, Wollenweber E (1993) Phytochemistry 33: 936

Roussis V, Ampofo SA, Wiemer DF (1987) Phytochemistry 26: 2371

Russell GB, Hunt MB, Bowers WS, Blunt JW (1994) Phytochemistry 35: 1455

Sattar EA, Glasl H, Nahrstedt A, Hilal SH, Zaki AY, El-Zalabani SMH (1990) Phytochemistry 29: 3931

Sattar EA, Mossa JS, Muhammad I, El-Feraly FS (1995) Phytochemistry 40: 1737

Schöpke T, Wray V, Nimitz M, Hiller K (1996) Phytochemistry 41: 1399

Sherma J, Fried B (1987) In: Bidlingmeyer BA (ed) Preparative liquid chromatography. Elsevier, Amsterdam, p 105

Shimasaki H, Ueta N (1983) Agric Biol Chem 47: 327

Smith-Kielland I, Malterud KE (1994) Planta Med 60: 196

Snyder LR (1978) J Chromatogr Sci 16: 223

Stahl E (1967) Dünnschicht-Chromatographie: ein Laboratoriums-Handbuch. Springer, Berlin Heidelberg New York

Stahl E (1968) Z Anal Chem 236: 294

Stahl E, Müller J (1982) Chromatographia 15: 493

Székely G (1983) In: Bock R, Fresenius W, Günzler H, Huber W, Tölg G (eds) Analytiker-Taschenbuch. Springer, Berlin Heidelberg New York, p 263

Tanaka T, Kawamura K, Kohda H, Yamasaki K, Tanaka O (1982) Chem Pharm Bull 30: 2421

Terreaux C, Maillard M, Hostettmann K, Lodi G, Hakizamungu E (1994) Phytochem Anal 5: 233

Tyihak E, Mincsovics E (1988) J Planar Chromatogr 1: 6

Tyihak E, Mincsovics E, Kalasz H (1979) J Chromatogr 174: 75

Vieth RF, Sloan HR (1986) J Chromatogr 357: 311

Vuorela H, Dallenbach-Tölke K, Sticher O, Hiltunen R (1988) J Planar Chromatogr 1: 123

Wall ME, Wani MC, Meyer BN, Taylor H (1987) J Nat Prod 50: 1152

Wang F, Pelletier SW (1987) J Nat Prod 50: 33

Yang P, Wang H, Chang W, Che C (1988) Planta Med 54: 349

Zhi-Da M, Jing-Fang Q, Iinuma M, Tanaka T, Mizuno M (1986) Phytochemistry 25: 2008

CHAPTER 4

Special Column Chromatography

Conventional open-column chromatography is universally practised as a result of its simplicity of operation. If silica gel is the support, 30 mg sample loadings per g of 50–200 µm support are feasible (Verzele and Geeraert 1980) but this very high capacity is only possible when the substances to be separated differ greatly in their R_f values. Loadings of 10 mg sample per g of support are more common. Alternatively, in the filtration mode, silica gel chromatography can be performed under overloaded conditions, e.g. in the filtration of hydrocarbon terpenes and oxidized terpenes from essential oils (Verzele and Geeraert 1980), where 1 g of essential oil can be applied per 10 g of silica gel. Deactivation of the silica gel sorbent may be necessary to avoid decomposition of the sample on the column (e.g. Bruno et al. 1996).

The limitations of classical open-column chromatography are as follows:

- slow separations
- irreversible adsorption of solutes
- incompatibility with small granulometry particles

In an attempt to overcome some of these disadvantages, alternative approaches to preparative chromatography have been attempted. Flash chromatography (see Chap. 5) is one of these and, in addition, two further methods will be described here: dry-column chromatography and vacuum liquid chromatography.

4.1 Dry-Column Chromatography

This method requires the filling of a chromatography column with *dry* packing material. The sample is added as a concentrated solution or, better, dried onto a small amount of adsorbent before introduction. The solvent is allowed to move down the column by capillary action until the solvent front nearly reaches the bottom (Fig. 4.1). The solvent flow is stopped and the bands on the column removed by extrusion, slicing or digging out. They are then extracted by a suitable solvent.

There is effectively no liquid flow down the column and no channelling; zone separation is sharp (Loev and Snader 1965). Dry-column chromatography

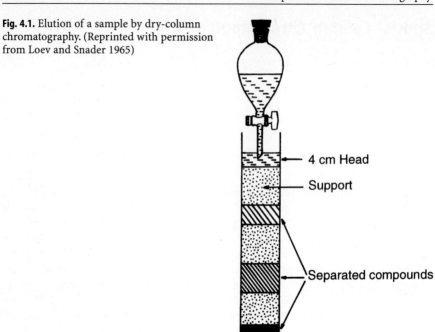

Fig. 4.1. Elution of a sample by dry-column chromatography. (Reprinted with permission from Loev and Snader 1965)

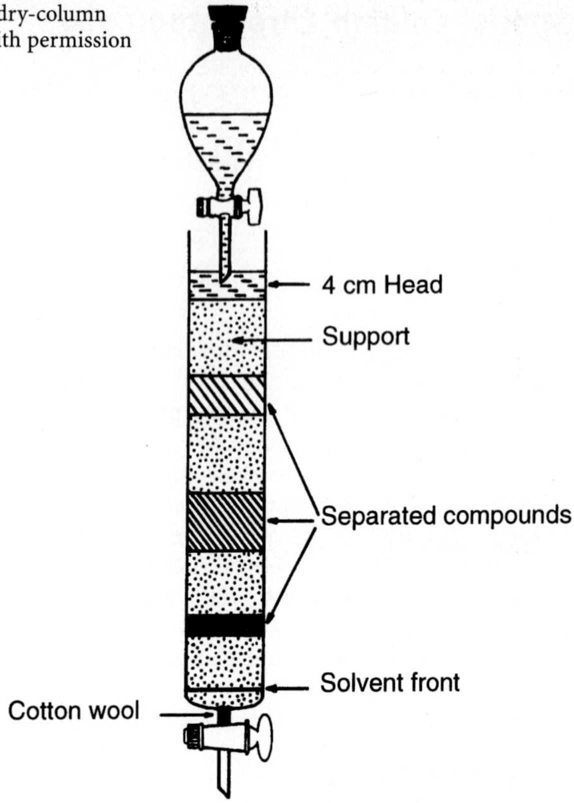

is also rapid (run times typically 15–30 min) and very little solvent is required.

When alumina (and, to a certain extent, silica gel) is used, separations can be extrapolated directly from analytical TLC plates by choosing the same adsorbent in the column. As a result, the method is a variant of preparative TLC, with the same resolution. The loading factors are naturally, on the other hand, much higher. Since the ratio of sample to adsorbent on the TLC plates is approximately 1:500, ratios of 1:300 and 1:500 have been employed for the column work. However, a ratio of 1:100 is possible for easily separable mixtures (Loev and Goodman 1967). Preliminary studies to find the best solvent system are performed on TLC plates and the separation is then transposed to the dry column, taking care to deactive the adsorbent suitably (Loev and Goodman 1967). When separations on silica gel are required, normal column chromatographic grades are used, with the addition of ca. 15% water for deactivation. Eluents composed of solvent mixtures may not always give the resolution of analytical TLC. In this case it is recommended to presaturate the dry-column adsorbent with about 10% mobile phase before packing the column (Engelbrecht and Weinberger 1977).

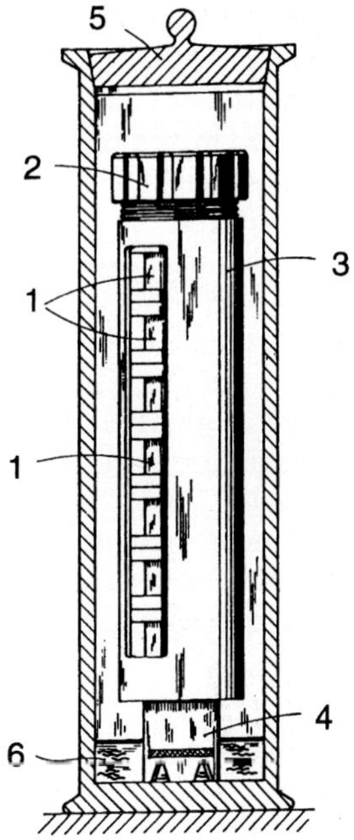

Fig. 4.2. Apparatus for PHS dry-column chromatography. 1: glass or quartz segments; 2: screw top; 3: stainless steel mantle; 4: sample chamber with glass frit; 5: developing chamber; 6: eluent

The easiest way of removing support from the chromatography column after development is to use plastic columns (e.g. Nylon). The column or tube can be cut with a sharp knife into sections corresponding to the migrated bands and the separated compounds are extracted and filtered (Loev and Goodman 1967). Another advantage of a Nylon column is that colourless bands can be observed with a UV lamp to guide sectioning.

However, it is difficult to pack a Nylon tube evenly and tightly with sorbent. For this reason, extrusion or successive removal of zones with a spatula from a glass column may be preferred.

With the aim of simplifying the removal of separated zones, J.T. Baker have introduced the "Preparative High Resolution Segment" (PHS) system. This consists of a number of glass or quartz segments, with ground glass joints, which are held on top of each other in a steel mantle (Fig. 4.2). Elution is in the ascending mode. When the separation is complete, the segments are simply slid apart and the contents are easily removed. The recommended chromatographic supports are 40-µm phases and 20% Kieselguhr can be added to speed up the elution. Different diameters and lengths of columns are available.

Dry-column chromatography is claimed to give better resolution than conventional open-column chromatography and columns of alumina 2 m long have been used to separate 50-g samples! (Loev and Goodman 1967). However, applications are not numerous. More common is *elution* dry-column chromatography. In this variant, a mixture is loaded onto a column of dry packing material and then eluted from the column by the mobile phase.

Beveridge et al. (1988) describe a wet-packing method which is easier to perform, gives more reproducible results and is compatible with alkaline solvent systems. The column was prepared by pouring a slurry of TLC silica gel 60 GF_{254} (Merck) mixed with a column volume of solvent into a glass tube with a cotton cloth attached at the end to retain the stationary phase. The silica was allowed to settle and the solvent run off until there was just sufficient to cover the support. Sample adsorbed onto a small amount of silica gel was added and elution performed until non-retained material reached the end of the column. At this stage, the cotton cloth was removed and the contents of the column extruded with a plunger. The components were detected under UV light and the column sliced as required. Preliminary TLC investigations are directly translatable to the column but slow moving systems such as propanol or butanol are not applicable.

A development of dry-column chromatography, called *vacuum dry-column chromatography*, in which vacuum from a water aspirator is applied to draw the solvent down the column, has been described (Leopold 1982).

4.1.1 Applications

Dry-column chromatography has proved useful for the rapid initial fractionation of antitumour-active plant extracts (Hokanson and Matyunas 1981). Complex mixtures were separated into ten or more fractions so that activity could be localised by bioassay. As an example, an ethanol extract of *Euphorbia cyparissias* (Euphorbiaceae) was chromatographed on a Nylon column (600 × 25 mm) containing Woelm silica gel (175 g). The extract was partitioned between chloroform and water, and 525 mg of the chloroform fraction was introduced with Celite, as a dry powder. Development was with chloroform-diethyl ether 95:5 but, contrary to normal dry-column chromatography, the yellow pigments travelling with the solvent front were allowed to elute before the development was stopped. The column was cut into 15 bands which were extracted with diethyl ether-methanol. Biological assay (P-388 lymphocytic leukaemia in vivo assay and 9KB nasopharynx carcinoma in vitro assay) were performed on the different fractions. After obtaining the relevant biological data, larger scale gradient elution column chromatography was employed to isolate increased quantities of the active constituents.

4.1.1.1 *Essential Oils*

In spite of the high resolution attainable on capillary GC instruments, complex mixtures such as essential oils cannot always be separated in one step. A prefractionation step becomes necessary to split the hydrocarbons from the oxygenated terpenoids and this function has often been filled by conventional column chromatography. The drawbacks of this method include high solvent consumption, dilution of the individual components, poor reproducibility, artefact formation and, of course, the time factor. An alternative but fast method is dry-column chromatography, which has been used by Kubeczka (1973, 1985) for the separation of essential oils into fractions suitable for GC analysis. Apolar components were separated on Woelm silica gel (deactivation with 7% water was essential to prevent rearrangements of terpene hydrocarbons) by elution first with n-pentane (giving fraction 1) and then benzene or isopropyl chloride (giving fraction 2). After all the benzene had run through the column, the support was cut into three bands containing the more polar constituents. Each of these was extracted with diethyl ether-methanol 8:2, thus providing fractions 3–5. All the fractions (1–5) were injected one by one into the gas chromatograph, giving chromatograms which were considerably easier to interpret. In addition, information on the relative polarities of the fractions gained by elution on silica gel was helpful in the identification of the peaks.

For preparative work, the fractions from dry-column chromatography were subjected to further column chromatography, preparative GC, HPLC etc. (Kubeczka 1985).

4.1.1.2
Sesquiterpenes

The isolation of sesquiterpenes from plants is normally achieved by successive chromatographic steps of the non-polar extracts at low or medium pressure and/or separation by preparative TLC and recrystallisation for compounds present in large quantities. Alternatively, RP-8 or RP-18 columns are employed, with mixtures of methanol and water as eluents. These strategies inevitably lead to some irreversible adsorption of the compounds of interest. Jares and Pomilio (1989) have described a procedure involving dry-column chromatography and a final semi-preparative HPLC step which they claim gives better yields. Thus, the petroleum ether extract of *Senecio crassiflorus* (Asteraceae) was chromatographed on a dry silica gel H (7.5 × 7 cm) column using, sequentially, hexane, hexane-ethyl acetate (99:1, 95:5, 90:10, 80:20, 70:30, 60:40, 50:50), ethyl acetate and methanol (3 × 50 ml of each) as eluents. After semi-preparative HPLC on RP-18, six sesquiterpenes, including 10-oxo-iso-dauc-3-en-15-al (**1**), were obtained from the hexane-ethyl acetate 90:10 eluate.

The germacrane ester 8-*O*-angeloylshiromodiol (**2**) was isolated from the roots of *Thapsia villosa* (Apiaceae) using dry-column chromatography (silica gel; hexane-ethyl acetate 7:3) for final purification (Teresa et al. 1985).

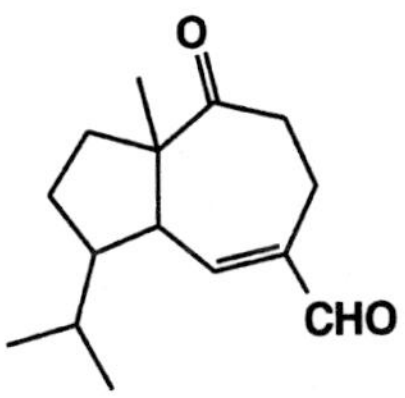

1

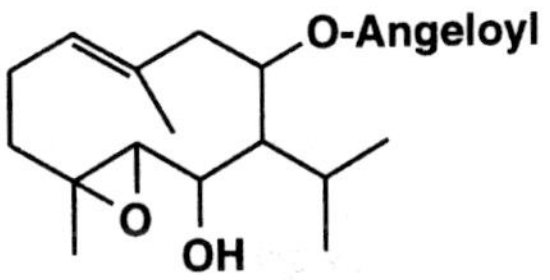

2

4.1.1.3
Diterpenes

Following the discovery of the antitumour activity of paclitaxel, numerous studies of the genus *Taxus* have been undertaken in an effort to find analogous bioactive taxanes. The Chinese species *T. yunnanensis* is one of the plants which has been investigated. Leaves and stems of the plant were extracted with ethanol. After solvent partition, the dichloromethane-soluble fraction was subjected to initial vacuum liquid chromatography. Subsequent chromatography, first on a dry column (silica gel 80–200 mesh; CH_2Cl_2-MeOH 40:1) and then on conventional open columns gave two rearranged abeotaxanes (taxin B (**3**) and 2-deacetyltaxin B (**4**)) (Yue et al. 1995).

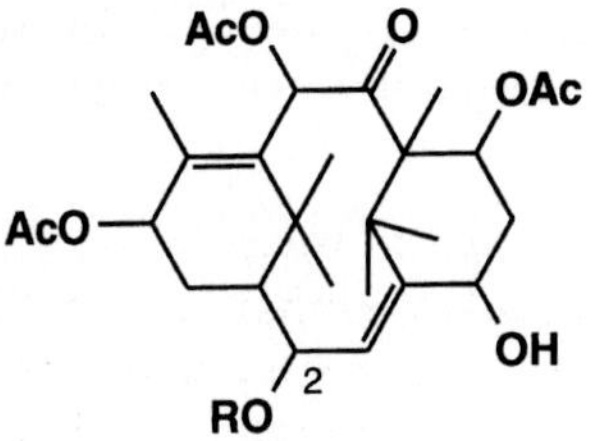

3 R = Ac
4 R = H

The pure *neo*-clerodane diterpenes ajugapitin (**5**) and dihydroajugapitin (**6**) were isolated directly from an ethanol extract of *Ajuga chamaepitys* (Lamiaceae) by dry-column chromatography on 63–200 μm Kieselgel 60 silica gel deactivated with 15% water (Hernandez et al. 1982).

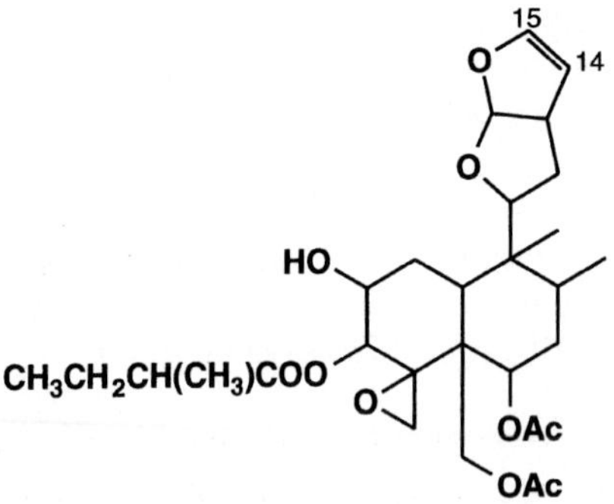

5
6 14,15-Dihydro

A mixture of silica gel (63–200 μm) and silver nitrate (5:1) has been applied to the separation of labdane diterpene methyl esters (Calabuig et al. 1981). An ethanol extract of *Cistus symphytifolius* (Cistaceae) was first chromatographed on a silica gel open column. One diterpene fraction was methylated and separated into its components by dry-column chromatography on the silver nitrate-containing support (petrol ether-benzene 4:1).

4.1.1.4
Cucurbitacins

During the isolation of cytotoxic 11-deoxycucurbitacin from *Desfontainia spinosa* (Desfontainiaceae), a dry column of silica gel was used, developed with ethyl acetate-water 9:1 and hexane-dichloromethane 7:3. Final purification followed with centrifugal and preparative TLC (Amonkar et al. 1985).

4.1.1.5
Lignans

Both lignans (Li et al. 1985) and neolignans (Moro et al. 1987) have been obtained by a combination of dry-column chromatography and preparative TLC.

4.1.1.6
Depsides

Salvianolic acid B (7) is a depside which has been isolated from the roots of *Salvia miltiorrhiza* (Lamiaceae), used in Chinese traditional medicine. A 95% ethanol extract of the roots was partitioned with several solvents and the resulting fraction (45 g) was separated by dry-column chromatography on 2.3 kg silica gel ($CHCl_3$-MeOH-HCOOH 85:15:1). The column was cut into 16 sections which were eluted with warm ethanol. Sections 13–14 were purified on Sephadex LH-20 (MeOH) to give 2.06 g of salvianolic acid B (Ai and Li 1988).

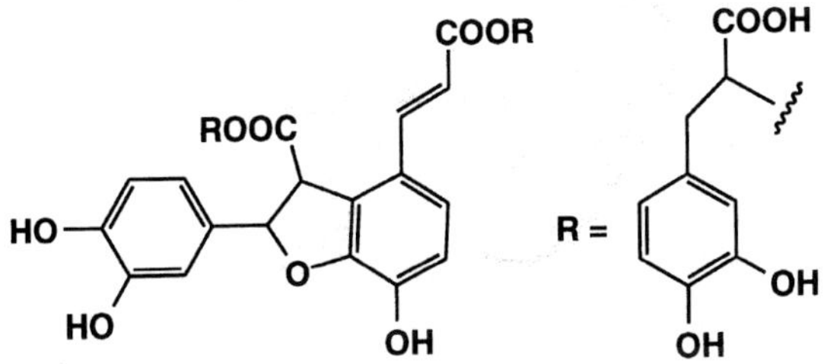

7

4.2
Vacuum Liquid Chromatography

Vacuum liquid chromatography (VLC) seems to have its origins in Australia. In 1977 a short description of the method for the isolation of a cembrenoid diterpene from an Australian soft coral was published (Coll et al. 1977), although the name was first coined by Targett et al. in 1979.

The technique can be considered as preparative TLC run as a column, with a vacuum provided to speed up eluent flow rates. It differs from flash chromatography in that the column is allowed to run dry after each fraction is collected. This is similar to preparative TLC because plates can be dried after a run and then re-eluted.

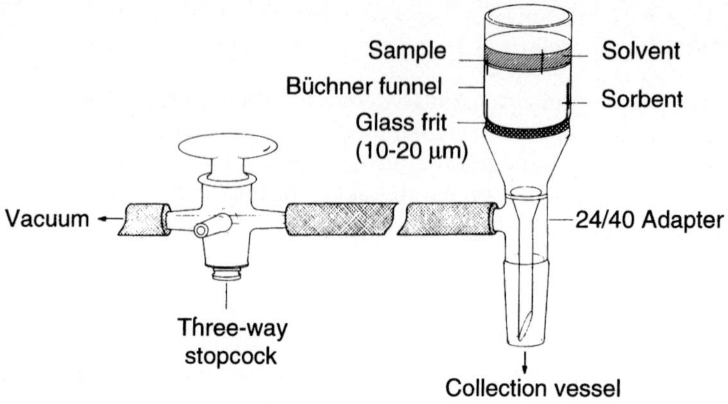

Fig. 4.3. Laboratory VLC apparatus. (Reprinted with permission from Pelletier et al. 1986)

Although Coll and Bowden (1986) describe a very simple VLC apparatus for small-scale laboratory applications and a variant for larger scale separations, the most practical set-up is that shown in Fig. 4.3 (Pelletier et al. 1986).

A short column or a Büchner filter funnel fitted with glass frit (10–20 µm, porosity D or porosity 2) is dry-packed with sorbent (10–40 µm of TLC grade, e.g. Merck 60H or 60G silica gel). The sorbent is allowed to settle by gentle tapping under gravity. Then vacuum is applied via the three-way stopcock and the sorbent compressed to a hard layer by pressing with a rubber stopper and tapping. The vacuum is released, solvent of low polarity is poured quickly onto the surface of the adsorbent and then vacuum is re-applied. When the eluent is through, the column is sucked dry and is ready for loading. The sample in a suitable solvent is applied directly to the top of the column and is drawn gently into the packing by applying the vacuum. Alternatively, the sample is preadsorbed on silica gel, aluminium oxide or Celite. The column is developed with appropriate solvent mixtures, starting with solvent of low polarity and gradually increasing the polarity, pulling the column dry between each fraction collected (this helps to avoid channelling).

Fractions are collected in a round-bottomed flask or in a suitable separatory funnel. The use of a separatory funnel avoids the problem of changing the flask for each fraction.

In contrast to methods which use pressure applied at the top of the column to increase flow rates, manipulations on the VLC column (solvent changes etc.) are easy because the head of the column is at atmospheric pressure.

In general, the height of sorbent should not exceed 5 cm. For small scale operations (< 100 mg), a column of 0.5 – 1.0 cm i.d. and 4 cm height is appropriate; for 0.5 – 1.0 g, the dimensions 2.5 × 4 cm are suitable; for 1 – 10 g, 5 × 5 cm; larger amounts are best separated on a 250-ml sintered glass filter funnel packed to a height of 5 cm (Coll and Bowden 1986).

Wherever possible, the mixture to be separated is added to the silica column in light petroleum (if this is not possible a solid introduction should be performed). Increasing amounts of a more polar solvent are added to each successive solvent fraction. Early increases in polarity are kept small (1%, 2%, 3% etc.) and then the increments can be increased (5%, 10%, 20% etc.). Usually 20–25 fractions will remove all components (Coll and Bowden 1986).

Targett introduced a modification to avoid channelling (Targett et al. 1979). This system was designed to operate under conditions of continuous vacuum and used a longer column to increase resolution. However, the simplicity of the original method is lost in this particular version.

4.2.1 Applications

The value of vacuum liquid chromatography as an alternative to flash chromatography (Chap. 5) for the fractionation of plant extracts prior to HPLC and other sophisticated separation steps will be shown here. In the last few years VLC has been increasingly used in the field of natural products because of its simplicity of operation. Separations of up to 30 g of extract are possible (Coll and Bowden 1986).

Different chromatographic supports have been employed in VLC: silica gel (both normal- and reversed-phase), Al_2O_3, CN, diol and polyamide. The most popular eluent is petroleum ether with increasing proportions of ethyl acetate (e.g. the separation of antimicrobial abietane diterpenoids from *Plectranthus elegans* (Lamiaceae)) (Dellar et al. 1996).

The techniques of flash chromatography and VLC have been evaluated for the separation of the pungent principles of ginger (*Zingiber officinale*, Zingiberaceae). Freeze-dried ginger rhizome powder was extracted with acetone and, after various liquid-liquid partition steps, a stock solution was prepared. For flash chromatography, silica gel 60 (40–63 μm, Merck 9385, 9.5 g) was packed into a 15 cm long column. A charge of ca. 200 mg ginger extract was separated with the solvent toluene-methanol 80:5 (chosen from TLC investigations which showed the desired compounds to have an R_f value difference of more than 0.15). However, the method did not separate the gingerols from the shogaols (Fig. 4.4).

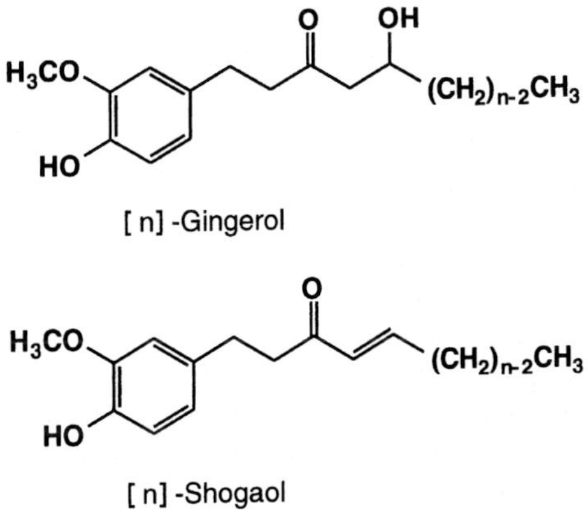

Fig. 4.4. Structures of gingerols and shogaols

For VLC, a 100-ml Büchner flask with an inside diameter of 3.5 cm was filled with silica gel 60 (the same as for the flash experiment, 9.5 g) to a height of 3 cm. Ginger stock solution was mixed with silica gel 60, evaporated and spread evenly over the top of the silica bed (total charge: ca. 300 mg of extract). A filter paper with the same diameter as the inside diameter of the funnel was placed on top of the sample and hexane (25 ml) was added slowly. The solvent was allowed to penetrate the sorbent bed and then sucked through to give the first fraction. A further 20 aliquots of 25 ml solvent mixtures with increasing polarities (hexane-diethyl ether → diethyl ether-methanol) were used for elution. By this means, there was virtually complete separation of the gingerols from the shogaols (Zarate et al. 1992). This result shows the surprising power of VLC: a high separation capacity with a small amount of sorbent.

From his work on alkaloids, Pelletier also claims that the method is superior to flash and dry-column chromatography (Pelletier et al. 1985).

4.2.1.1
Alkaloids

Diterpene alkaloids are easily and rapidly separated by VLC (Pelletier et al. 1985). As an illustration, the alkaloid mixture (1 g) from the pH 9 and pH 12 fraction of an *Aconitum columbianum* (Ranunculaceae) extract was vacuum liquid chromatographed on 65 g *aluminium oxide* 60 (H basic, TLC grade, type E, Merck). Elution with toluene-chloroform 1:4 gave 265 mg of talatizamine (**8**) and elution with chloroform-methanol 19:1 gave 247 mg of cammaconine (**9**) (Pelletier et al. 1985).

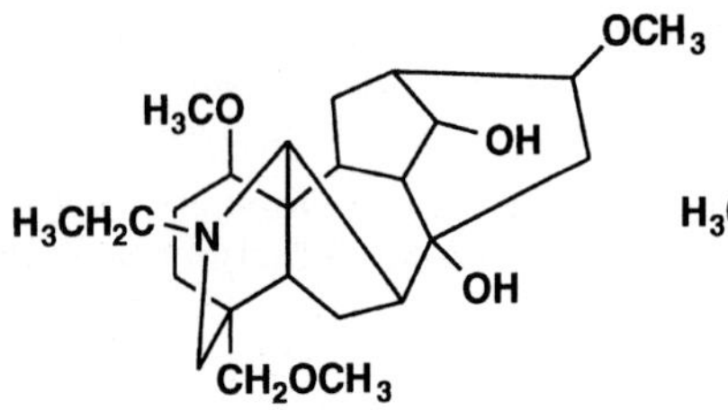

8

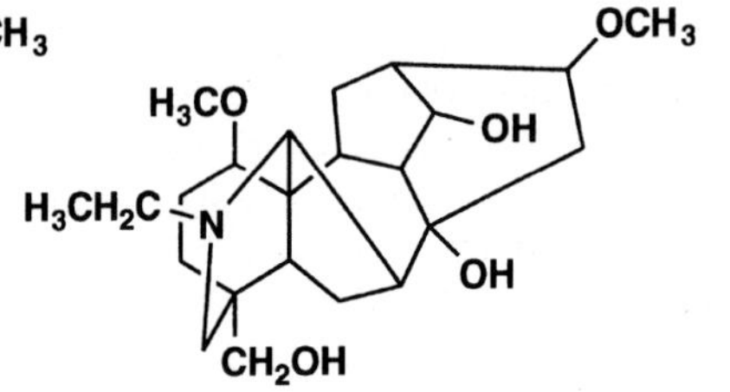

9

It was found that purification of commercial aconitine was quicker, cheaper and more convenient by VLC than by preparative TLC, the most important factor obviously being the ability to elute a sample by VLC without having to scrape a plate (Pelletier et al. 1986).

Other separations of diterpene alkaloids by VLC on Al_2O_3 have been reported (Pelletier and Badawi 1987; Joshi et al. 1990; Liang et al. 1991). In these examples, a combination of VLC and CTLC was employed.

A cytotoxic guanidine alkaloid has been isolated from the Mediterranean sponge *Anchinoe paupertas* (Anchinoidae) by a procedure involving VLC on *diol*, with dichloromethane-methanol (10:0 → 5:5) (Bouaicha et al. 1994).

Fig. 4.5. Chemical structures of phorbol diesters

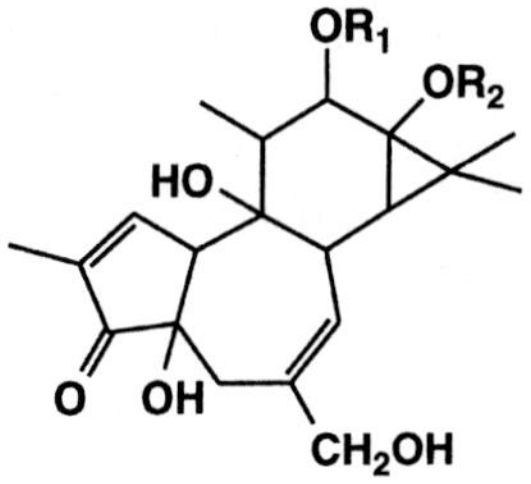

R_1	R_2	
long	short	A-series of phorbol diesters
short	long	B-series of phorbol diesters
short	short	short-chain phorbol diesters

4.2.1.2
Diterpenes

The seed oil of *Croton tiglium* (Euphorbiaceae) contains phorbol-12,13-diesters (Fig. 4.5) which are known as irritants and tumour promoters. These diesters have been purified over silica gel 60 (15–40 μm, Merck) with a modified VLC apparatus which allows a constant solvent flow rate and vacuum. A pre-fractionated batch of croton oil was chromatographed by VLC with the solvent dichloromethane-acetone 3:1. The B series of phorbol long-chain diesters eluted first, followed by the A series and finally the short-chain diesters (Pieters and Vlietinck 1989).

4.2.1.3
Isobenzofuranones

Root preparations of *Polygonum multiflorum* (Polygonaceae) are used in Chinese traditional medicine for cardiovascular complaints. As the mechanism of action may involve the regulation of calcium ions in cells, a search was undertaken for constituents which might affect the calcium pump Ca^{2+}-ATPase from erythrocytes (Grech et al. 1994). A methanol extract of the roots was partitioned between water and ethyl acetate. The ethyl acetate part was chromatographed by VLC over a 2.5 cm depth bed of TLC grade *reversed-phase* silanized silica gel 60H from Merck (no. 7761) (in a Büchner funnel of diameter 13.7 cm, with a porosity 3 glass frit). The sorbent was first washed with methanol (300 ml) and 100 ml portions of water (+ 1% HOAc)-methanol (1:3, 1:1, 3:1, 9:1). The sample (11.09 g) was dissolved in water (+ 1% HOAc)-methanol 9:1 (100 ml) and spread over the sorbent. Elution was then carried out with a step gradient of water (+1% HOAc)-methanol 9:1 → 1:1. Fractions from the 50% and 55% methanol eluate were active against the enzyme and were evaporated to dryness (660 mg). This material was then chromatographed over a second reversed-phase short column of the same dimensions, after introduction in water-acetonitrile 9:1 (80 ml). Elution commenced with water-acetonitrile 9:1

and employed a step gradient of water and acetonitrile. Bioactive fractions eluted with 30%, 35%, 40% and 45% acetonitrile and were combined. Final purification over a C-18 semi-preparative HPLC column gave *trans*- (**10**) and *cis*- (**11**) 6,7-dihydroxyligustilides, both exhibiting inhibitory activity against Ca^{2+}-ATPase (Grech et al. 1994).

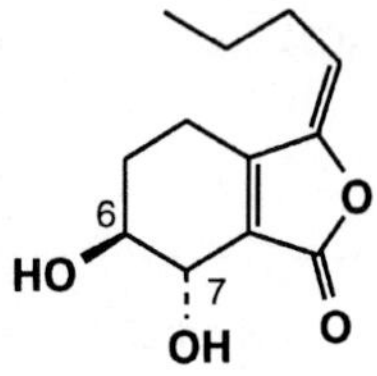

10

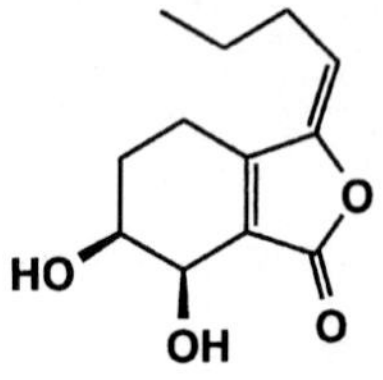

11

4.2.1.4
Triterpene Glycosides

The molluscicidal activity of the Nigerian plant *Dialium guineense* (Leguminosae) is due to the presence of oleanolic acid saponins. These were isolated from the fruit by a procedure which first involved flash chromatography on silica gel and then VLC on LiChroprep RP-18 with methanol-water 7:4. Three monodesmosidic saponins, all with strong molluscicidal activity, were thus obtained (Odukoya et al. 1996).

4.2.1.5
Limonoids

Azadirone (**12**) has been isolated from the unripened seed kernels of *Trichilia havanensis* (Meliaceae). A hexane extract of the kernels was partitioned between hexane and methanol-water 19:1. The aqueous methanol solution was extracted with ethyl acetate and VLC of this latter extract over silica gel deactivated with 5% H_2O (elution with petrol-ethyl acetate 19:1) gave the limonoid directly (Arenas and Rodriguez-Hahn 1990).

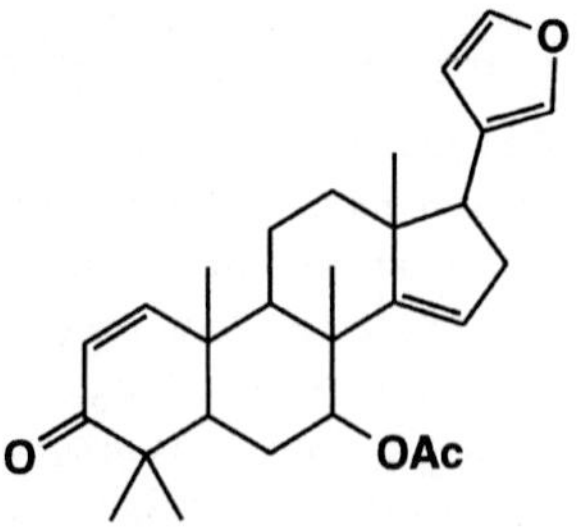

12

4.2.1.6
Xanthones

Five xanthones were isolated from *Garcinia cowa* (Guttiferae) by a procedure involving VLC as the key separation step. A methanol extract of the latex was separated into eight fractions by VLC on silica gel (80 g), eluting sequentially with hexane, hexane-benzene mixtures, benzene and dichloromethane. The xanthones were obtained either by direct crystallization from the fractions or by subsequent centrifugal TLC. Cowanol (**13**) showed moderate antimicrobial activity against *Staphylococcus aureus* (Na Pattalung et al. 1994).

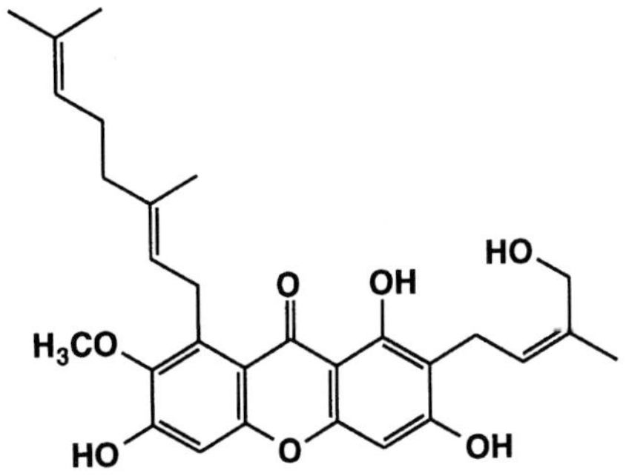

13

VLC on silica gel has likewise been employed for the separation of xanthones from the root bark of *Garcinia subelliptica* (Guttiferae), in this instance with n-hexane-ethyl acetate as eluent (Iinuma et al. 1995).

4.2.1.7
Iridoids

The application of VLC to the separation of various iridoids has been reported by Handjieva et al. (1991a,b). For this purpose, either neutral alumina 60G (5–40 µm, Merck 1090), TLC silica gel 60H (15 µm, Merck 11695) or *Florisil* (10–63 µm, Merck 12519) and Büchner filter funnels of different diameters (according to sample size) were used. Some small modifications to the method of Coll and Bowden (1986) were made: heating of the sorbent and use of a plastic foam layer on top of the support, for example.

For example, a dichloromethane extract (490 mg) from the dried roots of *Centranthus ruber* (Valerianaceae) was applied to 26 g Florisil (magnesium silicate) and eluted with petroleum ether (50 ml), followed by dichloromethane-acetone-ethyl acetate (100:0.1:01 → 100:0.3:0.3 → 100:0.5:0.5). Fractions eluting from the last solvent mixture were chromatographed on 4 g silica gel with dichloromethane-acetone-ethyl acetate 100:0.1:0.1 to give pure valepotriates IVHD-valtrate (**14**, 28 mg) and AHD-valtrate (**15**, 8 mg). Comparison of VLC and preparative TLC for the separation of valepotriates from *Valeriana offi-*

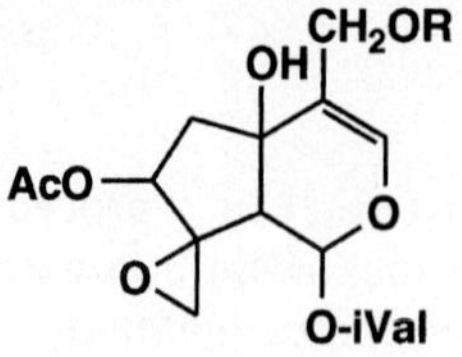

14 R = (α-*O*-iVal)-iVal
15 R = (α-OAc)-iVal

cinalis (Valerianaceae) showed the following advantages for VLC: five times shorter separation, three times less solvent, sorbent could be re-used six to seven times. Furthermore, VLC gave much less degradation of valepotriates when compared with open-column chromatography.

The secoiridoid glucosides gentiopicroside (**16**) and swertiamarin (**17**) were obtained in a pure state from fresh roots of *Gentiana punctata* (Gentianaceae) after VLC. A methanol extract of the roots was partitioned with water, chloroform and then butanol. The butanol extract was separated by VLC on 40 g silica gel with chloroform-methanol (6:1 → 1:1) to give in a single step the pure secoiridoid glucosides (Handjieva et al. 1991a).

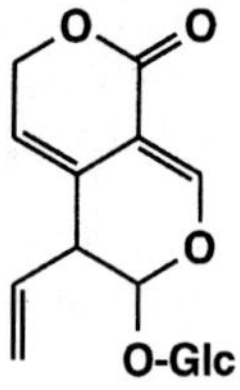

16

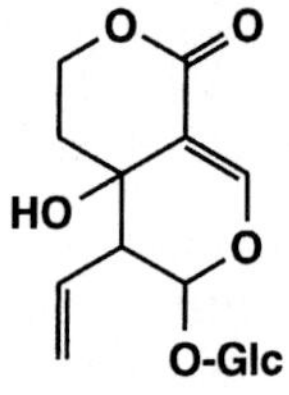

17

Further separations of iridoid glycosides by VLC include reports by Boros et al. (1990, 1991), Akunyili et al. (1991) and Justice et al. (1992). In the application reported by Boros et al. (1990), "mini-VLC" on disposable Pasteur pipettes was employed for the separation of small quantities of material.

4.2.1.8
Flavonoid Glycosides

In the isolation of a new flavonol glycoside (kaempferol-3-(2,3-diacetoxy-4-*p*-coumaroyl)rhamnoside) from the leaves of *Myrica gale* (Myricaceae), the initial purification step was VLC over *polyamide*-TLC 6 AC (Macherey-Nagel), eluting with 50-ml aliquots of methanol-water of decreasing polarity (methanol-water 1:1 → methanol 100%) (Carlton et al. 1990).

4.2.1.9
Acetylenic Alcohols

VLC over a *cyano*-bonded phase (40 µm) was used for the isolation of cytotoxic acetylenic alcohols (e.g. **18**) from the marine sponge *Cribrochalina vasculum* (Niphatidae) (Hallock et al. 1995). The crude sponge extract was eluted in VLC with a gradient of *t*-butyl methyl ether-hexanes; reversed-phase HPLC was necessary for final separation of the long-chain alcohols.

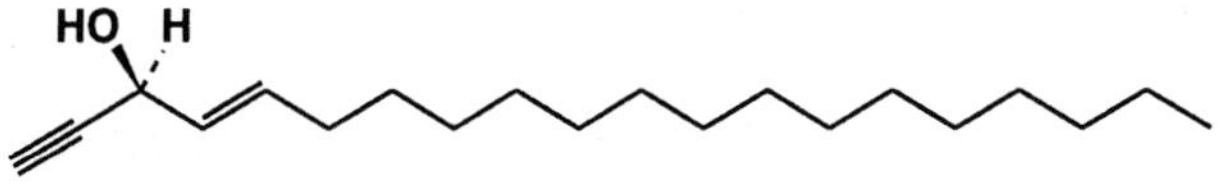

18

4.2.1.10
Polyacetylenes

A cytotoxic polyacetylene, (–)-17-hydroxy-9,11,13,15-octadecatetraynoic acid (**19**) has been isolated from *Minquartia guianensis* (Olacaceae), a tree from Ecuador. The stem bark was extracted with solvents of increasing polarity and the chloroform extract was found to have activity against P-388 murine lymphocytic leukaemia cells. This extract was first chromatographed by open-column chromatography and the active fraction (13 g) subjected to VLC on an 8 × 40 cm column of 70–230 mesh silica gel (1 kg). After elution with chloroform-ethyl acetate-methanol (18:1:1) and final crystallization, the active polyacetylene (**19**) was obtained (Marles et al. 1989).

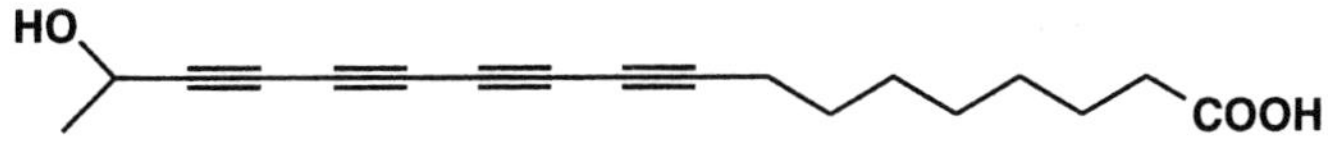

19

4.2.1.11
Kawalactones

Extraction of the leaves of *Cryptocarya kurzii* (Lauraceae) with ethanol gave a residue which was cytotoxic to KB cells. The compound responsible for this activity, a new kawa-type lactone, kurzilactone (**20**), was isolated by a combination of VLC, preparative TLC and reversed-phase HPLC. Silica gel 60H (70–230 mesh) was used for VLC, with dichloromethane-methanol (100:0, 99:1, 99:2, 19:1, 9:1) as eluent (Fu et al. 1993).

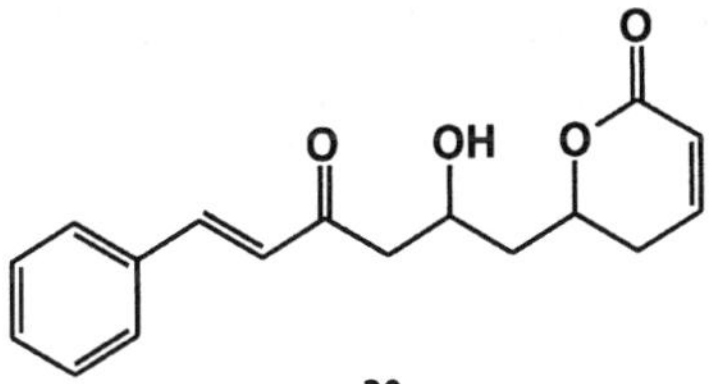

20

4.3 References

Ai C, Li L (1988) J Nat Prod 51:145
Akunyili DN, Houghton PJ, Raman A (1991) J Ethnopharmacol 35:173
Amonkar AA, McCloud TG, Chang CJ, Saenz-Renauld JA, Cassady JM (1985) Phytochemistry 24:1803
Arenas C, Rodriguez-Hahn L (1990) Phytochemistry 29:2953
Beveridge DJ, Harrison DM, McGrath RM (1988) J Chromatogr 450:443
Boros CA, Stermitz FR, Harris GH (1990) J Nat Prod 53:72
Boros CA, Marshall DR, Caterino CR, Stermitz FR (1991) J Nat Prod 54:506
Bouaicha N, Amade P, Puel D, Roussakis C (1994) J Nat Prod 57:1455
Bruno M, Piozzi F, Rodriguez B, de la Torre MC, Vassallo N, Servettaz O (1996) Phytochemistry 42:1059
Calabuig MT, Cortes M, Francisco CG, Hernandez R, Suarez E (1981) Phytochemistry 20:2255
Carlton RR, Gray AI, Lavaud C, Massiot G, Waterman PG (1990) Phytochemistry 29:2369
Coll JC, Bowden BF (1986) J Nat Prod 49:934
Coll JC, Hawes GB, Liyanage N, Oberhänsli W, Wells RJ (1977) Aust J Chem 30:1305
Dellar JE, Cole MD, Waterman PG (1996) Phytochemistry 41:735
Engelbrecht BP, Weinberger KA (1977) Am Lab 9:71
Fu X, Sevenet T, Hamid A, Hadi A, Remy F, Pais M (1993) Phytochemistry 33:1272
Grech JN, Li Q, Roufogalis BD, Duke CC (1994) J Nat Prod 57:1682
Hallock YF, Cardellina JH, Balaschak MS, Alexander MR, Prather TR, Shoemaker RH, Boyd MR (1995) J Nat Prod 58:1801
Handjieva N, Saadi H, Popov S, Baranovska I (1991a) Phytochem Anal 2:130
Handjieva N, Spassov S, Bodurova G, Saadi H, Popov S, Pureb O, Zamjansan J (1991b) Phytochemistry 30:1317
Hernandez A, Pascual C, Sanz J, Rodriguez B (1982) Phytochemistry 21:2909
Hokanson GC, Matyunas NJ (1981) J Pharm Sci 70:329
Iinuma M, Tosa H, Tanaka T, Asai F, Shimano R (1995) Phytochemistry 38:247
Jares EA, Pomilio AB (1989) J High Res Chromatogr 12:565
Joshi BS, Desai HK, Pelletier SW, Snyder JK, Zhang X, Chen S (1990) Phytochemistry 29:357
Justice MR, Baker SR, Stermitz FR (1992) Phytochemistry 31:2021
Kubeczka KH (1973) Chromatographia 6:106
Kubeczka KH (1985) In: Vlietinck AJ, Dommisse RA (eds) Advances in medicinal plant research. Wissenschaftliche Verlagsgesellschaft, Stuttgart, p 197
Leopold EJ (1982) J Org Chem 47:4592
Li L, Hung X, Rui T (1985) Planta Med 51:297
Liang X, Ross SA, Sohni YR, Sayed HM, Desai HK, Joshi BS, Pelletier SW (1991) J Nat Prod 54:1283
Loev B, Goodman MM (1967) Chem Ind. 2026
Loev B, Snader KM (1965) Chem Ind 15
Marles RJ, Farnsworth NR, Neill DA (1989) J Nat Prod 52:261
Moro JC, Fernandes JB, Vieira PC, Yoshida M, Gottlieb OR, Gottlieb HE (1987) Phytochemistry 26:269
Na Pattalung P, Thongtheeraparp W, Wiriyachitra P, Taylor WC (1994) Planta Med 60:365
Odukoya OA, Houghton PJ, Adelusi A, Omogbai EKI, Sanderson L, Whitfield PJ (1996) J Nat Prod 59:632
Pelletier SW, Badawi MM (1987) J Nat Prod 50:381

Pelletier SW, Joshi BS, Desai HK (1985) In: Vlietinck AJ, Dommisse RA (eds) Advances in medicinal plant research. Wissenschaftliche Verlagsgesellschaft, Stuttgart, p 153
Pelletier SW, Chokshi HP, Desai HK (1986) J Nat Prod 49:892
Pieters LAC, Vlietinck AJ (1989) J Nat Prod 52:186
Targett NM, Kilcoyne JP, Green B (1979) J Org Chem 44:4962
Teresa J, Moran JR, Hernandez JM, Grande M (1985) Phytochemistry 24:1779
Verzele M, Geeraert E (1980) Chromatographia 18:559
Yue Q, Fang Q, Liang X, He C, Jing X (1995) Planta Med 61:375
Zarate R, Yeoman S, Yeoman MM (1992) J Chromatogr 609:407

CHAPTER 5

Preparative Pressure Liquid Chromatography

The purpose of this chapter is not to provide an introduction to HPLC, amply covered in a number of excellent books (e.g. Snyder and Kirkland 1979; Simpson 1982; Johnson and Stevenson 1978; Henschen et al. 1985; Poole and Poole 1991; McMaster 1994; Scott 1995) but to concentrate on the preparative aspects of HPLC and related techniques. At the same time, examples are provided which can be used as the basis for the solution of a particular separation problem. For a more extensive description of preparative LC, the reader is referred to other authors (Verzele and Dewaele 1986; Bidlingmeyer 1987; Ganetsos and Barker 1993; Unger 1994; Porsch 1994).

The term "pressure liquid chromatography" is introduced here to include any method involving the application of pressure to a chromatography column, as distinct from gravity-driven column chromatographic separations. As a result, all categories ranging from flash liquid chromatography (2 bar) to preparative HPLC (100 bar), and sample sizes from mg to kg can be included under this heading.

Conventional chromatography is often not efficient enough to resolve gram quantities of closely related chemical structures. On the other hand, pressure liquid chromatographic techniques such as semi-preparative HPLC with smaller particle sorbents can accomplish far more difficult separations because of correspondingly higher separation factors (α).

The difference between preparative and analytical chromatographies is that whereas analytical chromatography (for separation, identification and determination) does not rely on the recovery of a sample, preparative chromatography is a *purification* process and aims at the *isolation* of a pure substance from a mixture. The high levels of sample load often associated with preparative LC usually require special chromatographic apparatus and appropriate operating conditions – items which will be discussed here.

The isolation of macromolecules is often performed by HPLC but this particular aspect is developed in detail in Chap. 8.

5.1 Basic Principles

Pressure liquid chromatography hinges on *speed* and *resolution* for its effectiveness. In preparative pressure liquid chromatography, another factor, namely *load*, is important (Fig. 5.1).

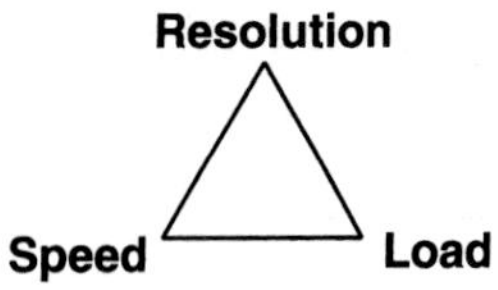

Fig. 5.1. Inter-relation of separation parameters for preparative pressure liquid chromatography

Optimisation of one separation parameter affects other separation parameters. Thus, increasing the flow rate results in a decrease of the resolution. Resolution is also impaired if the load is too high, column overloading arising from either an excessive sample volume or excessive sample mass (Golshan-Shirazi and Guiochon 1994). The loading capacity depends in turn on variables such as column radius, column length, particle diameter and packing density of the support.

For occasional separations, the aim is to isolate a particular compound in a suitable amount, with a certain degree of purity and a certain recovery ratio. As for production-scale chromatography, the amount of a component that can be separated per unit time is an essential criterion. *Throughput*, the weight of solute isolated in a given period of time, depends on parameters such as column dimensions and eluent flow-rate, and is often increased at the expense of purity (Unger and Janzen 1986). In other words, resolution is not always the primordial factor in preparative liquid chromatography (Guiochon and Katti 1987). However, preparative LC should give the product of interest as quickly and cheaply as possible.

The application of pressure can have one of the two rôles a) or b) or a combination of both:

a) increasing the pressure on a column filled with a packing material of a given granulometry increases the rate of flow of eluent, leading to a faster separation;
b) columns eluted with solvent under pressure can accommodate finer granulometry packing material and thus give higher resolutions.

The great advantage of a shorter elution time is the minimisation of decomposition which may take place during lengthy open-column separations of sensitive compounds.

In the context of this book, the term "preparative" can include amounts of isolated product ranging from µg to kg and includes any separation which is not solely used for analytical ends. The quantity of pure material required depends on the purpose for which it is to be employed:

- µg to mg quantities for spectroscopic identification purposes, as in the isolation of natural products;
- mg quantities for further identification purposes, chemical reactions and limited biological testing;
- g quantities for further biological testing, synthetic or semi-synthetic work and isolation of reference compounds.

In addition, the preparative techniques described here are useful for obtaining ultrapure standards, labelled compounds, trace impurities, decomposition products of pharmaceuticals and metabolites, as well as higher molecular weight biological materials, such as peptides.

There is a considerable divergence of opinion on the terms used for preparative pressure liquid chromatography. Snyder and Kirkland (1979), when referring to high-pressure liquid chromatography, for example, differentiate column throughput into *analytical* scale (1–2 mg for a totally porous support; 15 min separation of 2 compounds with resolution $R_s = 1.25$ and >99% purity of recovered fractions), *semi-preparative* scale (100–200 mg) and *preparative* scale (0.5–1.0 g). They also include *scaling-up* of an analytical separation under the semi-preparative order of magnitude. The global term *large-scale* LC is used for both preparative and scale-up operations. To these terms can be added *process* chromatography, which covers hundreds of grams to kilograms of product. Herbert (1991) defines preparative chromatography as the separation of gram quantities of sample on columns with i.d. 25–150 mm, while process chromatography concerns applications on columns with i.d. > 150 mm. Hence there are important distinctions to be made among the various quantities of material isolated under the so-called "preparative" conditions defined for this chapter.

Different methods of performing pressure liquid chromatography are available and will be described in this chapter. Rough application limits for these variants are shown in Fig. 5.2. Of course these ranges must be treated with caution, especially if overloading occurs.

Choice of the preparative system for a separation problem must not only take into account the size of the sample but also the nature of the separation to be undertaken. Consideration of these factors and the right choice of conditions, such as column dimensions, stationary phase, pressure and eluent will lead to the required separation. To avoid the empiricism often involved in the search for preparative separation conditions, attempts have been made to calculate these parameters by theoretical methods (Hupe and Lauer 1981; Cretier and Rocca 1982; Knox and Pyper 1986; Guiochon and Katti 1987). A theoretical con-

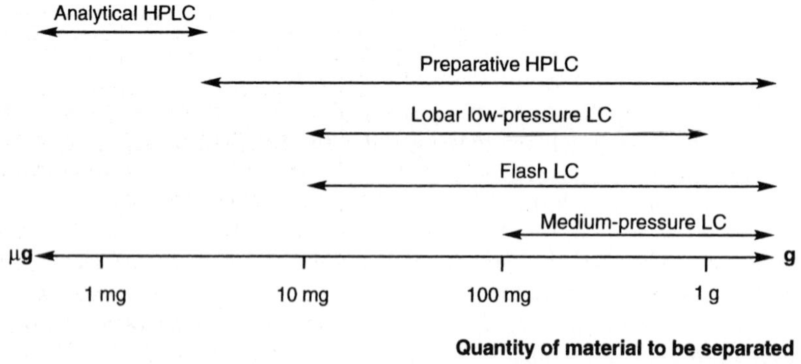

Fig. 5.2. Approximate division of pressure liquid chromatography methods according to the preparative scale required

cept has also been introduced to treat scale-up in preparative chromatography (Heuer et al. 1996). However, the phenomenon of *overloading*, often found in preparative separations, makes the prediction of chromatographic parameters hazardous. Furthermore, the optimisation of a separation must also take into account whether a *linear* or *non-linear* elution process is involved (Gareil and Rosset 1982a).

Sometimes it is necessary to complete a separation in two steps: a coarse separation on a low-efficiency column and then a second high efficiency separation.

5.1.1 Method Development and Optimisation

The following sequence can be adopted for the development of a preparative pressure LC separation:

1) Basic choice of LC system. In certain cases, TLC analysis of the sample can be used as a first indication of the correct operating conditions (Haywood and Munro 1980) – silica gel plates for normal-phase columns and silylated silica gel plates for reversed-phase columns. When using this method, though, it should be borne in mind that the surface areas of silica gels employed for TLC are about twice those used in column packings and that the R_f should be ≤ 0.3. The direct transposition from a TLC analysis to a preparative column separation has been discussed by, e.g. Loev and Goodman (1967), Bidlingmeyer (1987).
2) Separation of a small quantity of sample on an analytical LC column. Once the appropriate mobile phase has been discovered, the TLC separation should be extrapolated to an analytical LC column containing the packing to be used in the preparative column. A preliminary analytical search for the right conditions saves time, sample and solvent.
3) Optimisation of the analytical LC separation. One should try for small capacity factors (k') because:
   - the transformation to a preparative system is usually accompanied by a loss in separation efficiency;
   - a small capacity factor implies a quick separation and small peak volumes.

   A good analytical LC separation is usually a prerequisite for a successful preparative operation. Once an analytical LC separation has been found, the resolution is normally improved beyond that required for analytical LC, to accommodate overloading of the column during the separation (Fig. 5.3). One way of achieving this, in the case of reversed-phase columns, is to increase the water content of the solvent system.
4) Extrapolation to preparative LC apparatus (Gareil and Rosset 1982b; Cox and Snyder 1988). In direct scale-up runs from analytical LC, the applied pressure will be about one third of that used in the analytical separation (Johnson and Stevenson 1978). The procedure TLC – analytical HPLC – preparative HPLC for the development of a separation is well illustrated by

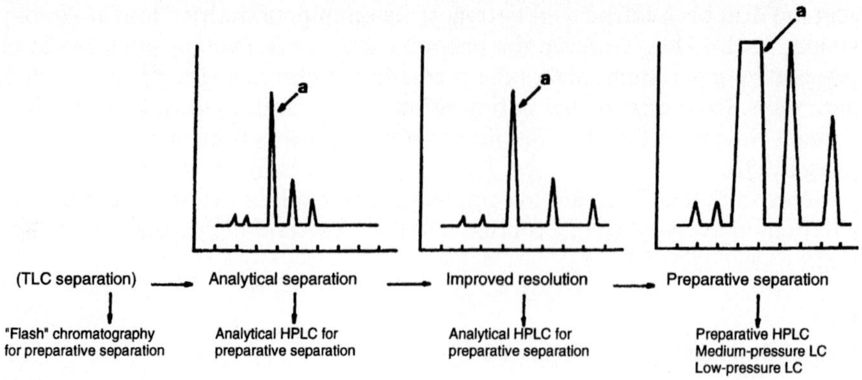

Fig. 5.3. Development of conditions for preparative pressure LC

Bidlingmeyer (1987). There is a tendency nowadays to produce preparative columns which contain versions of analytical-size packings so that direct scale-up from analytical to preparative chromatography is possible without drastically modifying mobile phase conditions. This is exemplified by the Symmetry range of columns (which contain spherical 100-Å packing materials) from Waters.

5) Performance of separation.
6) Recovery of purified material.
7) Analysis of substances obtained. When the separation has been carried out (e.g. peak *a* in Fig. 5.3), the pure product is analysed by TLC or analytical HPLC.

After the separation, the columns can be washed as follows (Simpson 1982):

normal phase: acetone → water → methanol → acetone-THF → dichloromethane
reversed-phase: methanol-THF 1:1.

5.1.2 Columns

The heart of any preparative pressure chromatographic separation is, naturally, the column. The dimensions of the column depend on the quantity of sample to be separated and vice versa (Hupe and Lauer 1981; Verzele et al. 1988). Some typical column diameters and sample loadings are shown in Table 5.1.

Other parameters of the column are of vital importance for useful separations (Fig. 5.4). The terms α (selectivity) and k' (capacity factor) are the major controlling influences on resolution (R_s), whereas the effect of efficiency (plate count, N) is less. It is essential to maximise the α value (selectivity or relative retention).

Increasing the diameter of columns means that larger samples may be injected and the throughput can be increased. Increasing the length (the *recycling* technique effectively also increases the column length) means that the inject-

Table 5.1. Typical sample loadings in preparative liquid chromatography with porous supports

Separation \ Column i.d.	4.6 mm	10 mm	25 mm
Difficult - (α < 1.2)	1 mg	20 mg	100 mg
Easy - (α > 1.2)	10 mg	200 mg	1 - 5 g

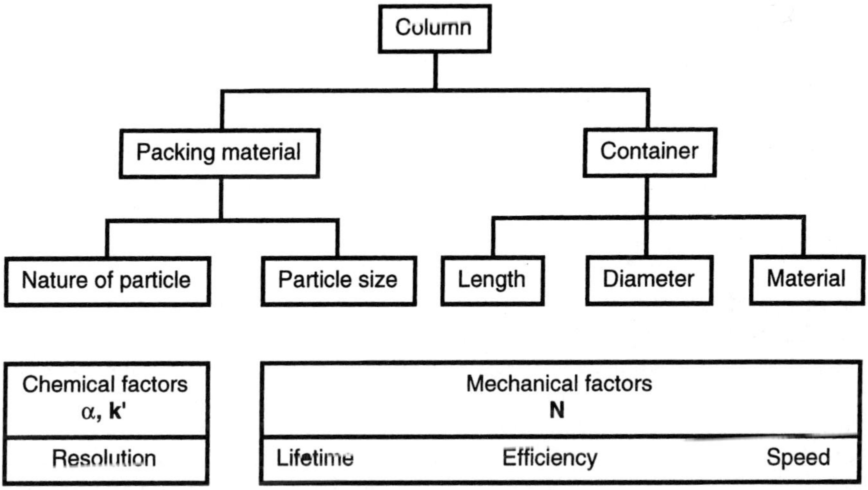

Fig. 5.4. The components of a column and their contribution to chromatographic performance

able quantity and the resolution can be augmented, but with an accompanying increase in back pressure. The throughput in this case, however, is unchanged except in certain circumstances (Gareil and Rosset 1982b). Shorter, larger diameter columns are filled with smaller granulometry particles and the longer columns with larger particles. In the chemical literature, there are proponents of both the large diameter (De Jong et al. 1978) and the long column (Scott and Kucera 1976) approaches. Smaller particles (and large diameter columns) are generally favoured for more difficult separations ($\alpha < 1.2$) (Jones 1988b).

Increasing the diameter of the column cannot be continued indefinitely, however, because the wall thickness must be correspondingly increased to withstand the pressure.

For high-pressure preparative scale chromatography, large-bore columns made of stainless steel are used which can withstand pressures of up to 300 bar. Glass columns are cheaper but have a lower pressure limit and are more fragile.

5.1.3
Stationary Phases

General discussions on liquid-solid (adsorption), partition (liquid-liquid and bonded-phase), gel and ion-exchange chromatographies, and their corresponding stationary phases can be found in any standard text on liquid chromatography (e.g. Johnson and Stevenson 1978; Snyder and Kirkland 1979). A good review on the packing materials used in preparative pressure liquid chromatography has been given by Unger and Janzen (1986). In this article, the authors list commercially available adsorbents with particle sizes (d_p) greater than 20 µm. It must be emphasised, however, that this market is continually evolving and that new stationary phases are appearing all the time.

Suitable column packings must be decided by taking into consideration not only their chemistries but also *all* the following variables:

- particle size (not forgetting pore volume and pore size)
- column length
- operating pressure

and suitable separations are obtained by a judicious choice of a combination of these. Whereas increasing the length of the column, to augment separation efficiency, implies higher operating pressures, changing the diameter of the column has little effect on permeability; column capacity is, therefore, best increased by increasing the internal diameter, within the limits of column packing techniques and mechanical strength of the walls. However, a consequence of increasing the size of the columns is that they become more expensive and require a much higher solvent consumption.

Spherical particles are preferred to irregularly shaped particles (Unger and Janzen 1986) because they have a higher mechanical stability, generate less fines, give better reproducibility of column packing and increased column permeability. Their cost is higher but they have a longer lifetime.

5.1.3.1
Particle Size

A wide range of particle sizes is commercially available, and while those in the range 40–63 µm are easily dry-packed into columns, give a high throughput and are moderately priced, the optimum size for efficient separations is approximately 15 µm (Verzele et al. 1988). Decreasing the particle size increases the plate height and separation efficiency but at the same time causes packing difficulties (for small particles it is difficult to pack columns greater than 50 cm in length), leads to higher working pressures and increases the cost considerably. However, there are situations when the product is rare, expensive or difficult to purify and the use of high-efficiency 10 μm diameter particles may be justified. Nevertheless, the trend is towards smaller particle, higher pressure systems, as illustrated by chromatography of a selection of natural products on a 50×21.4 mm column (3 μm; 100 Å RP-18 particles) (Franke and Verillon 1988).

5.1.3.2
Silica-Based Materials

Macroporous silica gel is the most important adsorbent for liquid-solid chromatography and is also the material used to prepare most bonded phase packings. According to the method of preparation, it can vary in porosity, surface area and particle shape (irregular or spherical). For analytical HPLC, 5–10 μm particle sizes are most often used, and for preparative LC, 10–40 μm and larger sizes are common.

The majority of preparative pressure LC separations are still carried out on unmodified silica gel, mainly due to its lower cost, application of a broad range of solvents as eluents, easy removal of solvents after fractionation and high flow rates. However, bonded phases (especially reversed-phase) are increasingly being used, these leading to a diminished risk of sample decomposition and less irreversible adsorption, among other advantages.

Treated silica gel phases have been employed to effect useful separations. In one instance, a low-pressure chromatography column of Adsorbosil silica gel was washed with water before purifying vomitoxin (a trichothecene mycotoxin). Dichloromethane was first pumped through the column to remove extraneous material and then water was used to elute the trichothecene (Witt et al. 1985). It would appear that liquid-liquid partitioning was involved, rather than the usual adsorption mechanism. Another method is to treat the silica gel with a *buffer*. The procedure involves washing the support with a suitable buffer, followed by a thorough drying of the modified material. Elution must then be performed with a solvent which does not dissolve the buffer on the silica gel. Impressive separations of polar solutes have been achieved with minimal tailing (Schwarzenbach 1985). Although very little preparative work has yet been performed with these modified columns, their potential for the separation of acids, bases, other polar compounds and isomers is considerable.

Silver nitrate-coated silica gel. Argentation chromatography is of practical significance for the separation of compounds containing olefinic double bonds, such as geometrical isomers (van Beek and Subrtova 1995). Preparative applications of silver nitrate impregnated supports are not frequent but methods for the preparation of these columns have been described (Heath and Sonnet 1980; Morita et al. 1983). Li et al. (1995) prepared silica impregnated with 10% silver nitrate for flash applications by simply grinding 200–300 mesh silica gel with an aqueous solution of silver nitrate, followed by drying in an oven at 150 °C. Separations of steroid and triterpene mixtures were performed by using 50 g of adsorbent for each gram of mixture. A method for preparing coated silica gel TLC plates was also reported (Li et al. 1995).

A mixture (536 mg) of the sesquiterpenes hinesol and β-eudesmol was separated on a 300×22 mm column with n-hexane-ethyl acetate 4:1 (Morita et al. 1983). Itokawa's group has carried out preparative work on silver nitrate-coated silica gel (n-hexane-ethyl acetate) for the separation of eight sesquiterpenes from *Alpinia japonica* (Zingiberaceae) (Itokawa et al. 1985), and Kieselgel 60 (40–63 μm) impregnated with 10% silver nitrate has been used for the separation of diterpenes from coffee beans (Lam et al. 1985). Flash chromatography on silver nitrate impregnated silica gel has been applied to the isolation of various sesquiterpenes from the essential oil of *Thuja occidentalis* (Cupressaceae) (Weyerstahl et al. 1996). Wood oil (25 g) was subjected to two flash chromatography steps on silica gel to furnish several known sesquiterpenoids. A further isolation step by flash chromatography on silver nitrate impregnated silica gel (ICN silica 32–63 μm; elution with petroleum ether-diethyl ether) allowed the isolation of 12 mg of the new guaiadiene derivative 1βH,5αH,7βH-guaia-3,10(14)-dien-11-ol (**1**).

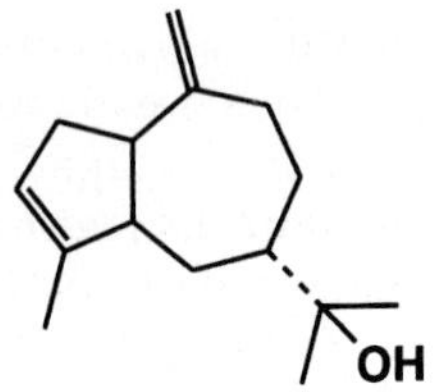

1

The same procedure was applied to obtain vassoura oil sesquiterpenes from the leaves of *Baccharis dracunculifolia* (Asteraceae) (Weyerstahl et al. 1995).

5.1.3.3
Ion Exchangers

Highly oxygenated and dehydrated abietanoic diterpene plant pigments such as royleanones and coleons (e.g. quinone methide **2**) are difficult to separate by classical chromatography because they exhibit severe tailing. Although buffered silica gel (see above) works better, elution of the surface coating is a problem. In the search for a support containing chemically bonded acidic groups, it was found that strong cation exchangers in the protonated form were effective for the separation of these diterpenes. Columns were filled with Partisil 10 SCX (benzene sulphonic acid type; as the mechanism of separation is predominantly

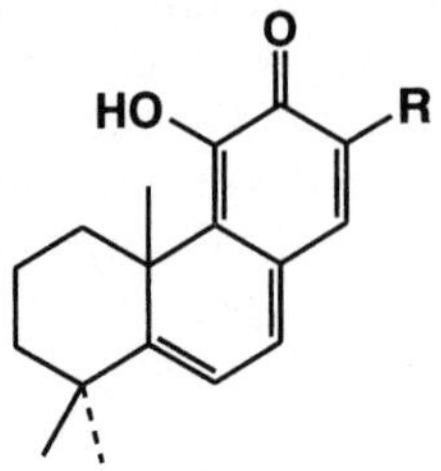

2

adsorption, chemically bonded silica with available free silanol groups is required) and eluted with hexane-dichloromethane and hexane-dichloromethane-methanol mixtures. By this means, eight parviflorones (quinone methides of general structure **2**) from *Plectranthus parviflorus* and *P. strigosus* (Lamiacae) were separated with hexane-dichloromethane-chloroform-methanol 100:100:150:3 as eluent (Rüedi 1985).

In an extension of this work, argentation chromatography on ion exchangers has also found application for the separation of isomeric alkenes, terpenes and fatty acids. Separations can most conveniently be performed with methanol but mixtures with tetrahydrofuran and *tert*-butyl methyl ether have increased selectivity for work on essential oils, for example (Evershed et al. 1982; van Beek and Subrtova 1995).

A comparison of this technique with centrifugal partition chromatography and non-aqueous reversed-phase HPLC for the separation of triterpene acetates has been published (Abbott et al. 1989).

5.1.3.4
Paired-Ion Chromatography

This technique involves the addition of a counter-ion to an ionized compound which is to be chromatographed, forming an ion pair which behaves as though it is non-ionic. The principle can be applied to normal- or reversed-phase supports and has the effect of improving elution behaviour and sharpening up peaks. Paired-ion LC is especially suitable for alkaloids, e.g. the separation of pyrrolizidine alkaloids from comfrey (Huizing et al. 1981). Several quaternary alkaloids from *Zanthoxylum usambarense* and *Z. chalybeum* (Rutaceae) have been isolated, including the new isoquinoline alkaloid **3**, using a mixture of sodium perchlorate solution and perchloric acid in acetonitrile on a Cosmosil 5C18-AR 250×20 mm column. Stems of *Z. usambarense* were extracted with methanol, tartaric acid was added and fat-soluble material was removed with diethyl ether. The aqueous layer was basified and extracted with diethyl ether and chloroform. Sodium perchlorate was added to the remaining aqueous layer and this was extracted with 1,2-dichloroethane. The aqueous phase was acidified with perchloric acid and extracted with 1,2-dichloroethane. The organic washings were combined and dissolved in a mixture of DMSO and 0.5 mol/l sodium perchlorate (1:1) and then chromatographed on the HPLC column. A gradient of mobile phase A:B 80:20 to 60:40 over 40 min was used, where A = 0.2 mol/l sodium perchlorate-60% perchloric acid 1000:0.2 and B = acetonitrile. Usambanoline (**3**) was obtained in 0.00039% yield (Kato et al. 1996).

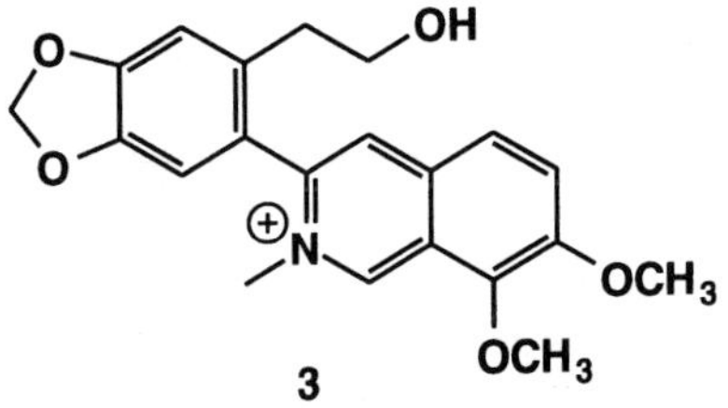

3

5.1.3.5
Polymeric Columns

The advantage of polymeric supports over normal- or bonded-phase silica gel is that a pH range of solvent from 1 to 13 can be tolerated. In addition, they have large capacities and, in some cases, selectivities superior to those of silica-based supports (Smith 1984). Porous, non-polar polystyrene-divinylbenzene copolymers (such as Amberlite XAD-2, XAD-4, Hamilton PRP and Polymer Laboratories PLRP-S) are available. Pietrzyk's group (Pietrzyk and Stodola 1981; Pietrzyk et al. 1982) have evaluated Amberlite XAD-4 and reported separations of amino-acids and peptides. A number of styrene-divinylbenzene copolymers were compared for their abilities to separate polymyxin antibiotics: Diaion CHP-3C, Amberlite XAD-2, Hitachi gel 3011, Jasco HP-01 (Kimura et al. 1981). Relatively low flow-rates were employed (ca. 1 ml/min) but good preparative separations were obtained with Hitachi gel 3010 and Amberlite XAD-2. The separation of polyhydroxylated sterols from a sponge has been investigated on Hamilton PRP-1 (West and Cardellina 1991). Buta (1984) examined the HPLC separation of plant phenolic acids and flavonoids on a Hamilton PRP-1 styrene-divinylbenzene (10 μm) column eluted with acetonitrile in dilute formic acid and came to the conclusion that results were reproducible and comparable to those obtained with an octadecylsilyl column. Final purification of a fumonisin fungal toxin from *Fusarium moniliforme* was achieved by HPLC on a PLRP-S column with an acetonitrile-water-TFA gradient (Musser et al. 1996).

A reversed-phase, highly porous polymer (Mitsubishi Chemical Industries Diaion HP-20) finds a great deal of use for fractionation of polar compounds. For example, the isolation of saponins, especially in Japan, often involves a preliminary passage over such a polymer, eluted with water-methanol, water-acetonitrile or water-acetone gradients (Hostettmann and Marston 1995). MCI CHP 20P polymer with 70% aqueous methanol was efficaceous for the separation of dammarane saponins from ginseng roots (Matsuura et al. 1984).

Benson and Woo (1984) and Poole and Poole (1991) deal with aspects of polymeric columns in more detail.

5.1.3.6
Cyclodextrins

Although these are mainly employed in the separation of enantiomers, diastereomers and structural isomers (Chap. 9), cyclodextrin-bonded stationary phases do have other limited (and expensive!) applications. They have similar properties to diol columns in the normal-phase mode (Ward and Armstrong 1986). In the isolation of polyhydroxy sterols from the sponge *Dysidea etheria*, the final step was chromatography over an Astec β-cyclodextrin column (250×4.6 mm) with acetonitrile-water (1:1 or 2:1 or 4:1 or 5:1, depending on the sterol) (West and Cardellina 1991). The cyclodextrin column allowed the resolution of analogues with slightly different side chains.

5.1.3.7
Gel Permeation Columns (see Chap. 8)

Theoretically these rely on size exclusion effects for their separating powers but for their application to small molecules, solvent-solute-support interactions are important. LC gels (for the purification of proteins, for example) have a strengthened structure which ensures stable hydrodynamic properties and prevents large volume variations upon pH or ionic strength changes; they are often based on copolymers (chapter 8). An excellent review which deals with all aspects of preparative gel permeation chromatography has been published (Lesec 1985). Semi-rigid gels, which include Styragel and TSK gels, have become very popular and they are also available in smaller (e.g. 10 μm) particle sizes. Rigid gels, such as Porasil or μPorasil are generally made from glass or silica and consequently should not be employed at a pH higher than 7.5. These porous materials are especially suitable for the separation of water-soluble substances.

5.1.4
Column Packing Methods

Depending on the preparative LC method to be used, the particle size of the support and the dimensions of the column, different column packing methods can be employed (Haywood and Munro 1980; Gareil and Rosset 1982b; Prusiewicz et al. 1982; Verzele et al. 1988; Unger 1994). The efficiency of packing depends on the particle size. A decrease in particle size leads to increased difficulty of packing and slurry methods are, as a rule, necessary for particles smaller than about 20–30 μm. However, for spherical silica gel, dry packing has been possible with 20 μm particles. Wetting with ether even allows packing with 10 μm silica gel (Verzele and Geeraert 1980). This particular aspect of preparative LC, which is very important for the separation method concerned, is treated under the relevant subdivisions of preparative pressure LC, although for HPLC the columns are often purchased ready-filled with the required packing material. A brief résumé of the most common packing methods is given below.

5.1.4.1
Dry Packing

The so-called "tap-and-fill" technique is suitable for particles larger than about 20–30 μm (Cox 1990). As an exception, MPLC columns can successfully be packed with 15 μm silica, using alternating vacuum and nitrogen pressure (Zogg et al. 1989a).

5.1.4.2
Sedimentation

The two disadvantages of this method are: a) the time required; b) the formation of aggregates. The latter problem can be solved (deflocculation) by choice of a suitable packing solvent. For example, acetone has been used to pack 16 μm

C_{18} silica 250×21 mm columns. Methanol-water (1:1) was employed to consolidate the packed bed (Wang et al. 1990a). For 50 mm diameter columns, methanol alone was the slurry solvent for 15 μm silica and C_{18} (Porsch 1994).

5.1.4.3
Slurry Packing

The basic objective of slurry packing is to force a relatively dilute slurry into the column at high speed in order to minimise void formation. Slurry is dispaced into the column under pressure and the accumulating bed grows by filtration ("high pressure filtration"). The high pressure slurry packing technique becomes increasingly complex with column diameters greater than 20 mm, at pressures of 30–40 bar.

5.1.4.4
Variable-Geometry Columns

For optimising the packing of larger preparative LC columns with small particles, bed *compression* can be used. A slurry (or, exceptionally, dry packing material) is introduced into the column which is then compressed – the packing is physically squeezed together. This can be achieved by two methods: radial compression and axial compression (Fig. 5.5).

Radial compression (Sarker et al. 1996) is obtained by gas or liquid pressure on a flexible sheet against the column wall inside the metal column. This technology is most suitable, therefore, for cartridges. It allows dry packing of particles as small as 15 μm. Radial compression was developed by Little et al. (1976) in the mid-seventies and is extensively used in the Waters patented RCM systems. Provided one has access to the right cartridge holder, a variety of cartridge dimensions are possible, ranging from 8 to 40 mm i.d. and a maximim length of 100 mm. With extension kits, it is possible to assemble a column of up

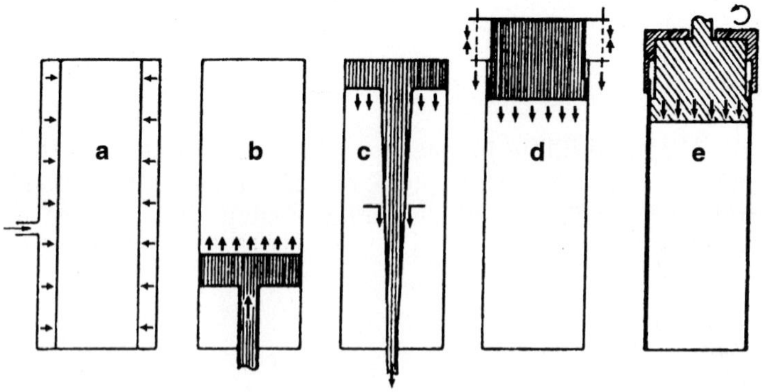

Fig. 5.5a–e. Different column compression techniques: **a** radial; **b** axial; **c** combined axial and radial; **d** axial with flange; **e** axial with outer nut. (Reprinted with permission from Porsch 1994)

to 300 mm in length. Ready-made cartridges can be purchased containing Nova-Pak, Bondapak, Delta-Pak and Porasil packings. Several areas of application have been described on Waters PrepPAK cartridges, with different supports (Hostettmann et al. 1986).

Axial compression can be exerted from the top of the column, the bottom of the column or both together. The piston can be actuated either mechanically or hydraulically. The required amount of concentrated slurry is compressed until the piston stops. The column is then ready for use, although the compressed bed continues to shrink during elution (Porsch 1994). Here a moving piston is useful because constant pressure is maintained on the packed bed during operation ("dynamic axial compression") (Colin et al. 1990). This ensures that packing material remains consistent, uniform and void free. Dynamic axial compression columns are probably the most popular pressurized columns and are available up to 80 cm i.d. A feature of certain axial compression columns is that the packing can be forced out of the column by the piston at the end of the run.

Combined radial and axial compression (Fig. 5.5c) is obtained by pressing a wedge-shaped shaft into the column, as developed by Separations Technology (SepTech) in 1987 ("annular expansion").

Axial compression was developed by Godbille and Devaux (1974) and in the first applications (with the Jobin-Yvon system; for examples see Hostettmann et al. 1986) the pressure applied to the bed was of the order 5–20 bar. The original patent of Roussel-Uclaf is presently being exploited by the firm Prochrom (for example in the Prochrom 150, which has a 15 cm i.d. axially compressed column). Other companies marketing axially compressed columns include Amicon (N-Pack-300 and others), Axxial (formerly Jobin-Yvon), Modcol, CEDI (pre-packed cartridges of plastic or steel with 25–150 mm diameter and 50–300 mm length) and E. Merck (Prepbar) (Unger 1994). With the larger columns, samples of 50–100 g can be separated per injection.

In 1985, Rainin developed a column (*dynam*ic *ax*ially compressed or "Dynamax" system) fitted at both ends with a screw cap which can compress the packing material (manually) and eliminate the voids which progressively appear during elution (Fig. 5.5e). Macherey-Nagel produce a variant of this system with a screw cap at one end only ("Vario-Prep" columns). These are available in diameters from 10 mm to 80 mm and up to 250 mm in length.

5.1.5 Sample Introduction

A number of factors have to be considered before injecting a sample onto a preparative column:

- sample preparation,
- sample solvent,
- sample mass,
- sample volume,
- injection mechanism,
- column and fitting.

If possible, the mobile phase should be chosen as the sample solvent. Care must be taken to ensure that the solubility of the sample in the mobile phase does not limit the quantity to be injected. On the one hand, if the sample volume is too great, separation efficiency decreases (Verzele and Dewaele 1986) and on the other hand, if the sample is too concentrated, precipitation on the top of the column may occur (N.B. a column acts as a sample concentrator, notably when the sample is introduced in a solvent with a lower elution power than the mobile phase). Despite this, small volumes of concentrated sample solutions in the mobile phase are advantageous for a higher production (Guiochon and Katti 1987). Scaling up a successful analytical separation may also cause problems associated with solubility of the sample. This is especially true of reversed-phase HPLC, when the compounds under investigation are poorly soluble in aqueous eluents. For example, coumarins from *Musineon divaricatum* (Apiaceae) could not be isolated by reversed-phase HPLC because of their poor solubility in the solvent suitable for their separation (Swager and Cardellina 1985).

It may sometimes be possible to introduce the sample in a solvent different from the mobile phase, although care has to be taken with this approach. The crude extract of *Aristolochia clematitis* roots, for example, is poorly soluble in methanol-water systems. In order to obtain aristolochic acids from this extract, the sample was dissolved in methanol-THF 1:1 before reversed-phase HPLC (Makuch et al. 1992).

The technique of sample introduction is a very important aspect of a successful LC separation and great care needs to be taken in order to ensure a correct distribution of the sample over the top of the column. Modifications of the top of the column have, for example, been carried out to improve dispersal of the sample onto the column (De Jong et al. 1978). The usual methods of introduction (Haywood and Munro 1980) are:

- syringe injection,
- stopped-flow injection,
- injection via valves (with or without loop),
- injection via main pump,
- injection via auxiliary pump,
- solid introduction.

Solid introduction is a technique for circumventing low sample solubility. Either dry powdered sample can be mixed with or pre-adsorbed to the column packing material and introduced onto the top of the column or the mixture can be packed into a small precolumn and connected to the main column (Miller et al. 1989). For extremely insoluble samples (<5 mg/ml), between five and ten parts packing to one part sample is needed. Applications involving the precolumn method have been reported for samples ranging from 0.1 to 1900 g in weight, with between two and five parts packing to one part sample (Miller et al. 1989).

5.1.6 Pumps

Due to the larger particle size packings used in preparative pressure chromatography, the permeability of the columns is generally greater than analytical columns. Consequently, lower pressures are required for solvent delivery (10–100 ml/min) and correspondingly less powerful pumps can be employed. Reciprocating or pneumatic amplifier pumps are quite adequate. Factors such as precision and complete pulse damping, essential to the accuracy of analytical separations, are not overwhelmingly important.

The exception is preparative HPLC on large-bore columns packed with small diameter particles. Here, of course, stronger pumps are necessary. In certain cases, pressures up to 150 bar are possible and membrane pumps are to be recommended.

For smaller diameter preparative columns, analytical LC pumps with maximum flow rates of around 10–15 ml/min are suitable.

5.1.7 Detectors

Most commercially available detectors are constructed for analytical work and, due to saturation, are unsuitable for preparative (especially MPLC) applications. However, specially designed detectors with cells for preparative separations do exist and one example is provided by the GOW-MAC 80–800 LC-UV detector, in which the eluate passes as a thin film over a quartz cell (Leutert and von Arx 1984). With this detector, flow rates of 500 ml/min are possible. Alternatively, detectors with variable cell lengths can be incorporated (manufactured by Knauer, for example). In general, a UV detector with a 0.05 mm cell length will tolerate flow rates of up to 200 ml/min.

Detectors for preparative pressure LC should be capable of accommodating the high flow rates usually employed. However, a loss in sensitivity at these rates is easily tolerated because concentrations of material eluting from the column are high. In fact, concentrations of material in the eluate are often so high that the detector is overloaded. One way of getting round this is to detect at a wavelength where the solute absorbs feebly ("de-tuning"). Another alternative is to use a bypass splitter, in which a small amount of eluate from the column is channelled through an analytical detector (Simpson 1982; von Arx et al. 1982).

Methods of detection for pressure LC are limited and the two most commonly used techniques – ultraviolet and refractive index detection – both have their limitations. Differential refractometers are very sensitive to temperature changes, are not ideal for small quantities and cannot handle gradients. However, they provide a universal means of detection. Another method which may become more popular for non-UV active solutes is evaporative light scattering detection (ELSD). Since ELSD responds to non-volatile analytes, it can be used under gradient conditions. It is necessary, on the other hand, to incorporate a splitting arrangement because the detection method requires heat to evaporate the solvent. The separated products are thus likely to be degraded unless they are collected separately.

The high concentrations of solute mean that TLC monitoring of fractions is also relatively straightforward so when other methods of detection are inadequate, recourse to TLC is possible.

5.1.8
Mobile Phases

The purity of the solvents used to make up the eluent is especially important in preparative LC because evaporation of large volumes of mobile phase from fractions containing pure product can produce significant amounts of impurities when the solvent is only contaminated with traces of non-volatile matter. As some preparative LC methods require large volumes of solvents, a compromise has to be reached between solvent purity (and cost!) and quantity employed.

Non-volatile additives to the mobile phase (e.g. ion-pair chromatography, displacement chromatography) cause problems in product recovery (see next section). However, volatile buffers provide a method of improving separations without concomitant difficulties in removing additives. For example, the steroid alkaloid veratridine (4) has been purified from crude veratrine (mixture of al-

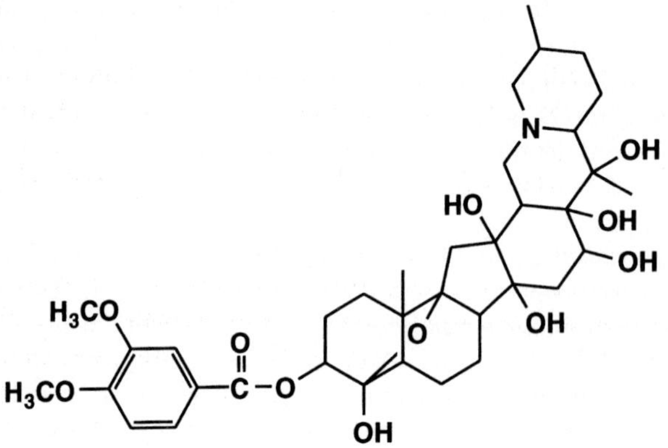

4

kaloids) by HPLC on a Whatman Partisil 10 μm M9/C_8 (500×9.4 mm) column with methanol-0.1 mol/l ammonium acetate (pH 5.5) (60:40) as mobile phase. The required fraction was collected and simply lyophilized to obtain veratridine of 98.4% purity (Reed et al. 1986).

Similarly, mycinamicin macrolide antibiotics have been isolated on a semi-preparative scale using octadecylsilyl columns and acetonitrile-0.01 mol/l ammonium acetate pH 4.0 buffer (40:60) (Mierzwa et al. 1985). An acetonitrile-ammonium acetate eluent was also used in the HPLC separation of cyclic peptides from *Microcystis aeroginosa* cyanobacteria (Edwards et al. 1996).

Peak sharpening can be achieved by running a gradient or even by a simple stepwise change of mobile phase composition (Porsch 1994).

5.1.9
Collection of Separated Material

In preparative-scale work, large volumes of eluted solvent have to be handled and suitable fraction collectors are required. If possible, a means of recycling used solvent should be employed. This is naturally facilitated if mixtures are avoided.

When reversed-phase or polymeric sorbents are used to effect separations, the recovery of sample from aqueous solvents may sometimes be difficult. One solution is to evaporate the organic part, leaving an aqueous dispersion. The solute can then be extracted by an organic solvent such as toluene or chloroform. An illustration is the n-hexane extraction of urushiol constituents after reversed-phase and TSK-gel chromatography with CH_3CN-H_2O-HOAc mixtures (Du et al. 1984). Bruce et al. (1990) performed preparative HPLC with an acetonitrile-tetrahydrofuran-water-acetic acid solvent and removed the acetonitrile and THF under reduced pressure before neutralising the acid with sodium bicarbonate. The aqueous solution was extracted with chloroform to obtain the products. Another approach, especially when using buffer systems, is to rechromatograph the purified product. In this manner, pure chlorogenic acids from coffee beans were chromatographed on C-18 supports to remove H_3PO_4 (eluent: H_2O) after separation with a gradient of 10 mmol/l H_3PO_4 in methanol (Morishita et al. 1984).

Removal of H_3PO_4 from purified *Quercus robur* (Fagaceae) ellagitannins was achieved by gel filtration over Sephadex LH-20, first eluting with water to remove the acid and then with methanol to desorb the tannins (Scalbert et al. 1990). Desalting can also be performed over Sephadex G-10 or Bio-Gel P-2, for example.

For methanol-water solutions, the collected fraction can be diluted 5x with water and then pumped back onto the column. To elute the compound, pure methanol is used. In this way, desalting of cyclic peptides was achieved by first washing (with water) the material re-introduced on the column and then eluting the required product with methanol (Edwards et al. 1996). An alternative is to pass the collected sample over Sephadex LH-20. By this means, H_3PO_4 contained in an HPLC eluent can be removed by gel filtration with water, as in the separation of flavonoid glycosides from *Rosa* species (Seto et al. 1992). The purified products are eluted with methanol after the acid has been washed out.

The latest trend is towards automation of fraction collection and injection, allowing continuous operation and repetitive separations.

5.1.10
Shave and Recycle Chromatography

When two or more closely-running major components are required, and the selectivity of the chromatographic system is not sufficient to separate the mixture, the technique of recycle chromatography is employed (Charton et al. 1994) (Fig. 5.6). After optimising the analytical separation conditions, preparative scale work is undertaken and pure *a* and *b* can be collected by taking ("shav-

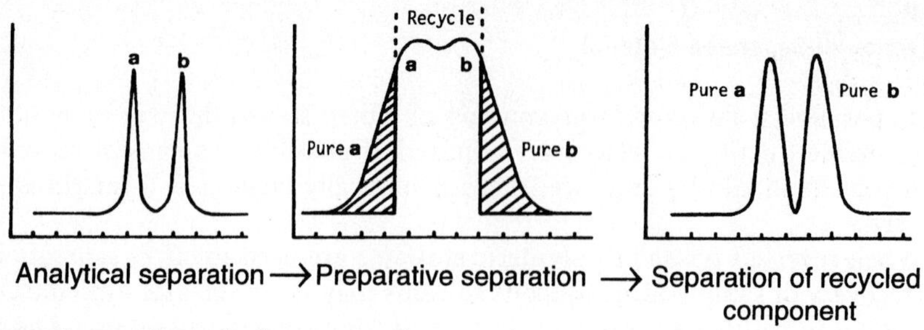

Fig. 5.6. Shave and recycle chromatography

ing") the leading edge and trailing edge of the respective peaks. On re-injecting the mixture of *a* and *b* (recycled component), further pure products are isolated. Repetition of the recycling may be required if a portion of *a* and *b* is unresolved after the first recycling run (Coq et al. 1981). In short, recycling has a similar effect to increasing column length.

Band-spreading, a problem caused by extra-column factors such as the volume of solvent in the pump, must be kept to a minimum (Haywood and Munro 1980) and the first eluted peak in the recycled portion must not overlap with the last peak in the original separation (total band width must be less than the recycle volume).

Two basic systems exist for recycle chromatography (Haywood and Munro 1980):

- closed loop;
- alternate column recycling (see also Sect. 5.1.12).

The first approach involves connecting the detector outlet from the column to the pump inlet via a multi-port valve and the second approach makes use of two columns, such that the required fraction is passed directly from the first column to the second column. If necessary, the sample cut can be transferred back to the first column (by means of valves) from the second column before the pure material is collected. The advantage of the alternate pumping recycling is that the sample only passes once through the pump.

Mixtures of isomeric labdadienes and mixtures of isomeric labdatrienes have been resolved by the peak shaving and recycle technique (Mohanraj and Herz 1981) even though the HPLC chromatogram showed only one peak for all three isomers in each case. There was a 98% recovery of starting material and, since a one-component solvent system (n-hexane) was used, the solvent could also be recycled.

Other examples are given in Sect. 5.2.4.

5.1.11 Column Overloading and Heart Cutting

Increased capacity for preparative separations can be achieved by overloading the column (mass or concentration overload, as opposed to volume overload).

As the separation is no longer being performed under linear conditions, optimum parameters are no longer predictable from data obtained under analytical conditions. The technique of "heart cutting" may have to be practised to avoid contamination by minor components which are present in the front and back zones (Fig. 5.7).

Necessarily a certain amount of *a* is lost as a mixed fraction by this method but the throughput is correspondingly higher. In practice, successively higher sample masses are loaded onto the column until the correct degree of overloading is reached and heart cutting results in a product uncontaminated with faster- or slower-running impurities. For silica-based media, mass overload begins with sample amounts surpassing approximately 1 mg per gram packing material (Herbert 1991).

To achieve maximum loading, the sample should be dissolved, if possible, in a solvent weaker than the mobile phase.

An example of heart cutting can be found in the separation of the isoflavones genistin and daidzin from soybean (Farmakalidis and Murphy 1984).

Overloading is also useful for the separation of minor components in a mixture (Fig. 5.8). A cut of the overloaded column gives a fraction enriched in compound *a*, which can then be isolated in the pure state by recycling.

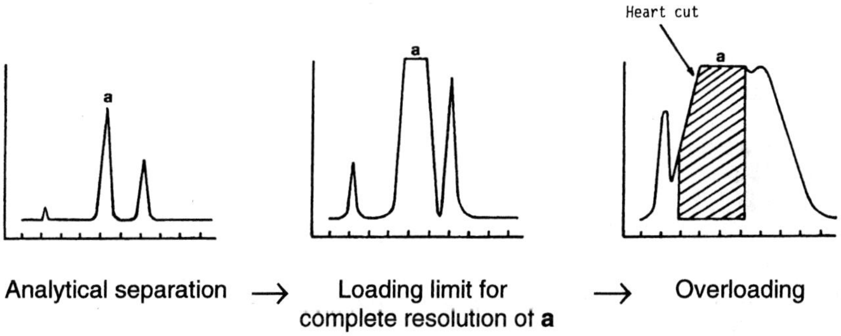

Fig. 5.7. Overloading of chromatography columns and heart cutting

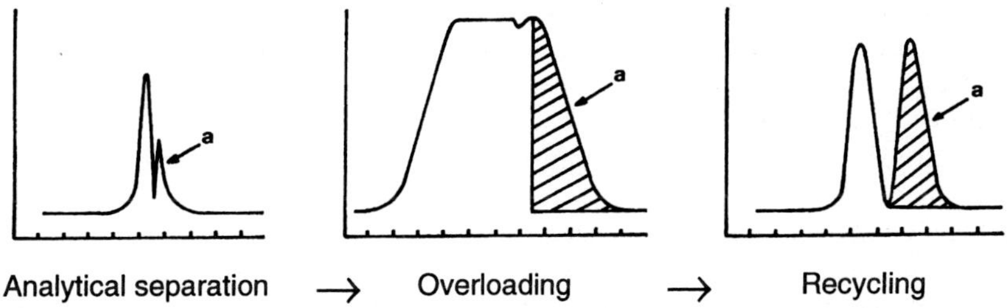

Fig. 5.8. Separation of a minor component by overloading and recycle chromatography

5.1.12
Column Switching

The term column switching includes all techniques by which the direction of flow of the mobile phase is changed so that effluent from one column is passed to a second column (Ramsteiner 1988).

It is important for several reasons:

- resolution and selectivity are improved,
- on-line sample clean-up is possible,
- partially pure fractions from a first separation can be brought to 100% purity by chromatographing on subsequent columns (includes recycling),
- trace amounts of sample can be enriched,
- times required for, e.g. washing may be reduced.

The transfer of a fraction from one column to a second column is achieved with switching valves and while separation is taking place on the second column the first column can be simultaneously cleaned with solvent (Ramsteiner 1988). This method is very useful when overloading the column because the heart cut can immediately be passed to the next column for further purification. By this means, the sweetener stevioside (5) was directly isolated from an aqueous me-

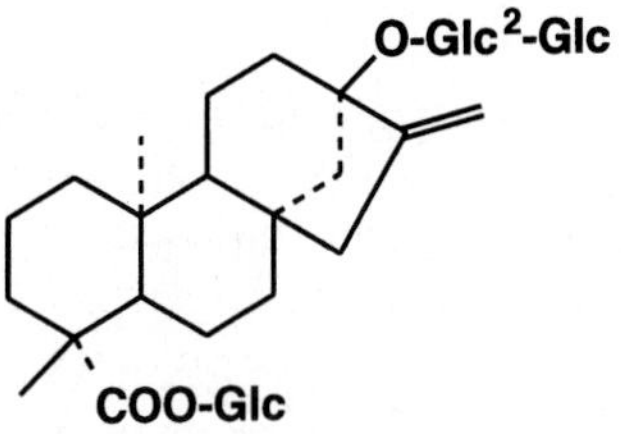

5

thanol extract of *Stevia rebaudiana* (Asteraceae) leaves (Little and Stahel 1984). The dried leaves were macerated with methanol-water 80:20 and, after evaporation of the solvent, the extract was injected directly onto a 10-μm Spherisorb ODS column (250×10 mm) and eluted with methanol-water 7:3. The heart cut containing stevioside was directed onto an identical second column by a switching valve and eluted with the same solvent. At the same time, the first column was flushed with methanol, thus saving a considerable amount of preparation time for the next run. Even with gross overloading (throughput 30–50 mg per injection), pure stevioside was obtained after passage through the second column.

A combination of column switching and recycling has been used in the purification of 25-hydroxyvitamin D_2 from rabbit plasma (Okano et al. 1984). The material to be separated was passed through a silica gel column (1) (Fig. 5.9). The eluate was led to the drain via ports 5′ and 4′ of a 6-port valve and, when necessary, to silica gel column 2 via ports 5′ and 6′. Repeated recycling of the vitamin fraction between the two columns gave a progressively purer sample.

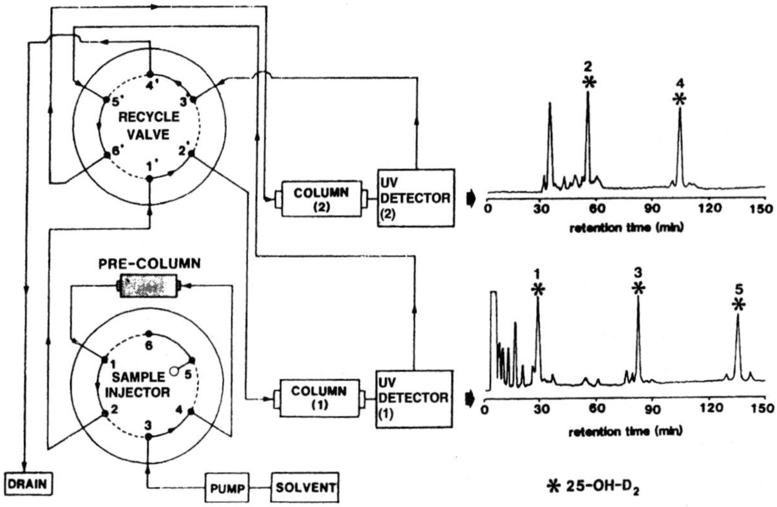

Fig. 5.9. Flow diagram of the recycle HPLC system and chromatograms for the separation of 25-hydroxyvitamin D_2 from rabbit plasma. (Reprinted with permission from Okano et al. 1984)

5.1.13 Peak Magnitude

Peak size depends not only on the relative mass contribution of the individual components in the sample but also on their optical properties. A major weakly UV-absorbing component of a mixture may, for example, be completely eclipsed by a very minor but strongly UV-absorbing component and the important constituent may be missed during collection of fractions corresponding to the peaks. An example in which peak size does not reflect the importance of the compound in question is given in Fig. 5.10.

The problem of peak detection becomes important when, in order to avoid overloading the detector at a certain wavelength, absorbance is monitored at another wavelength at which impurities may no longer be observable.

5.2 Different Preparative Pressure Liquid Chromatographic Methods

Since column (and sample) sizes vary so much, it is better to classify the different preparative techniques according to the pressure employed for the separation:

- flash chromatography (ca. 2 bar/30 psi);
- low-pressure LC (<5 bar/75 psi);
- medium-pressure LC (5–20 bar/75–300 psi);
- high-pressure LC (>20 bar/300 psi).

There is a considerable overlap between low-pressure, medium-pressure and high-pressure LC, and it is only for convenience that the three classes are treated separately.

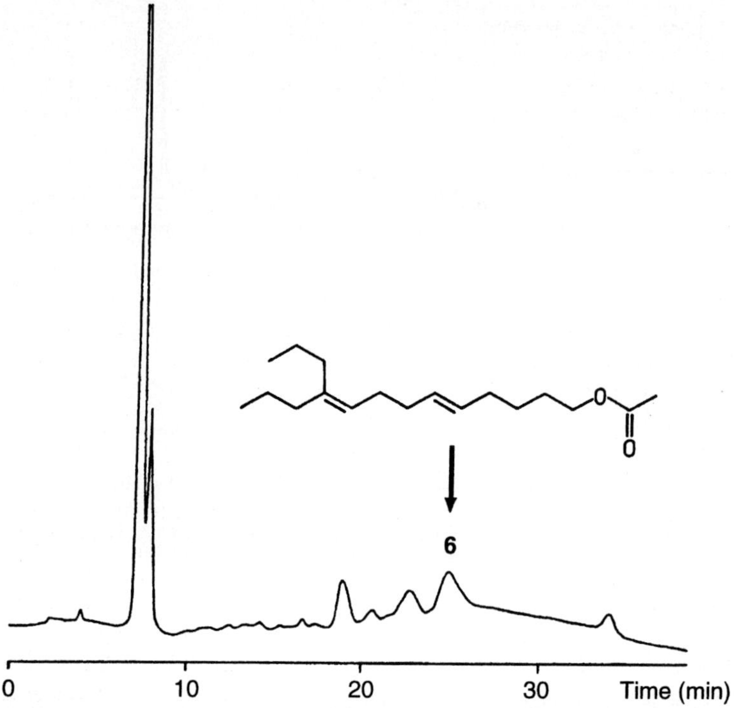

Fig. 5.10. Preparative separation of a pheromone. Sample: 1 ml solution containing 220 mg solute. Column: 7 µm silica gel, 250×21.2 mm. Mobile phase: dichloromethane-hexane 1:1, 14 ml/min. Detector: refractive index. (Reprinted with permission from Meyer 1984)

The sizes of the columns and particles used for separations depend on each individual separation problem. For difficult separations ($\alpha \sim 1$) with small samples, small (5–10 µm) particles should be used. Longer columns with larger particles will give equivalent separation efficiencies. In easier separations ($\alpha \gg 1$), columns with larger particles and higher sample loads are advantageous. For these easier separations, lower pressures may therefore be possible, and if the selectivity is really high, flash chromatography is the obvious solution.

5.2.1 Flash Chromatography

The concept of flash chromatography is exceptionally simple; preparative separations employing this variant of conventional column chromatography are very easy to perform, using readily available and cheap laboratory glassware. For these reasons, flash chromatography is very popular with researchers who are confronted with straightforward separation problems.

The performance of flash chromatography is lower than that of MPLC systems (which have a similar loading capacity) but considerations of simplicity and economy often dominate and make it a method of choice in many cases.

The fore-runner of flash chromatography was a very similar technique known as *short column chromatography* (Hunt and Rigby 1967), in which mixtures were separated on Merck Kieselgel G (the fine grade of silica gel used in preparative TLC) by the application of pressure from a nitrogen cylinder. Typical column dimensions reported for this version of column chromatography were 3–10 cm diameter and 7–15 cm length.

Details of flash chromatography were first published in 1978 (Still et al. 1978) and patented in 1981 (U.S. Patent 4,293,422) as an attempt to cut down on open column chromatography operation times. These long elution times have two main disadvantages: the risk of decomposition of sensitive compounds and the tailing of bands, factors which often cause problems in the separation of natural products. An example of the necessity of flash chromatography comes from work on antifeedant limonoids. These compounds gradually decomposed on silica gel open columns and could only be obtained in the pure state by a combination of flash chromatography and semi-preparative HPLC (Nakatani et al. 1994).

The setup of a typical flash chromatography column is shown in Fig. 5.11. A glass column of suitable length, fitted with an exit tap, is either dry-filled or slurry filled with suitable packing material. Dry-filling gives better packing but requires the passage of a large amount of solvent before the support is fully moistened. It is advisable to introduce a layer of sand at the top of the sorbent. Enough space should be left above the sorbent to allow for repeated filling of the column with eluent. Alternatively, a dropping funnel can be attached to the top of the column to act as a solvent reservoir. The sample is introduced and the column connected to an air inlet fitted with a needle valve to control the compressed air supply. By this means, pressures of ca. 1 bar above atmospheric pressure can be reached, in order to elute the sample.

Depending on the size of the column, samples anywhere in the range 0.01–10.0 g can be separated in as little as 15 min.

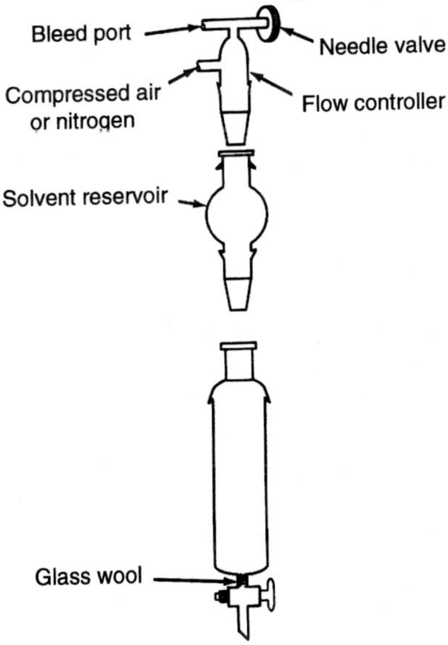

Fig. 5.11. The basic design of flash chromatography apparatus

Various solutions to the original problem of storing a large volume of eluent have been found, including a low-cost modification using a separatory funnel. This is not directly attached to the top of the separation column and thus the risk of breakage at the joint is avoided (Leung et al. 1986). Another variant introduces a column with plungers at the top and bottom. By this means, an increase in reproducibility of separations is claimed (Reichenbächer et al. 1991).

A technique called "*TLC mesh column chromatography*" using 10–15 µm silica gel operates under very similar conditions to flash chromatography and is claimed to give superior resolution (Taber 1982).

Some flash systems are commercially available:

- *Aldrich* has a line of columns, flow controllers and solvent reservoirs;
- *Baker* market columns, flow controllers, reservoirs with ball joints and a whole range of stationary phases;
- the *Eyela* "flash" chromatograph is produced by the Tokyo Rikakikai company but this uses a reciprocating glass plunger pump for delivering solvent and resembles a low-pressure LC system;
- the *Biotage* (Charlottesville, VA, USA) radially compressed cartridge system allows separations of 1–250 g sample per run at maximum flow rates of 250 ml/min (up to 7 bar). This comes in two versions: the Flash 75 system has a compression module which accepts 7.5 cm diameter cartridges (pre-packed with 200, 400 or 800 g KP-Sil 32–63 µm 60 Å silica gel); the Flash 150 system accepts 15 cm diameter cartridges (with 2.5 or 5 kg silica gel). Cartridges are connected to solvent reservoirs (4–60 l) and the solvent flow is driven by gas from compressed air cylinders.

5.2.1.1
Silica Gel as Adsorbent

The most widely used stationary phase in flash chromatography is silica gel. Merck LiChroprep Si 60 25–40 µm, 40–63 µm or 63–200 µm can be employed and J.T. Baker has a 40-mm silica gel specially for this particular technique. YMC also produces dedicated spherical stationary phases. The original work by Still et al. (1978) recommends 40–63 µm for best resolution. With 63–200 µm silica gel, flow rates of 50 ml/min are perfectly reasonable. Since flow rates are so high and separation times short, manual fraction collection in 100-ml flasks is often performed. Analysis of fractions is then carried out by TLC.

Compounds having ΔRf ~ 0.10 on TLC can be separated in quantities of 1 g, on 50 mm diameter columns (Still et al. 1978). If less resolution is required, up to 10 g (of crude plant extracts, for example) may be successfully separated on these columns. In general, the amount of sample is proportional to the cross-sectional area of the column (Table 5.2).

Final purification of natural products is sometimes performed by flash chromatography on silica gel; more often, though, crude extracts or mixtures are pre-purified by this method before other techniques with higher resolution are employed. In other words, flash chromatography provides a rapid preliminary fractionation of complex mixtures. These mixtures can equally well derive from synthetic work, flash chromatography providing an ideal means of purifying

Table 5.2. Relationship between column diameter and sample size (depth of sorbent: ca. 15 cm) with 40–63 μm silica gel (Still et al. 1978)

Column diameter (mm)	Eluent volume (ml)	Sample loading (mg)		Typical fraction size (ml)
		ΔRf ≥ 0.2	ΔRf ≥ 0.1	
10	100	100	40	5
20	200	400	160	10
30	400	900	360	20
40	600	1600	600	30
50	1000	2500	1000	50

intermediates, as shown in recent work on the total synthesis of the ansamycin antibiotics (+)-trienomycins A and F from *Streptomyces* (Smith et al. 1996).

Dry-column flash chromatography, as the name suggests, is a combination of dry-column chromatography (Chap. 4) and flash chromatography. The column is filled with dry packing material and the sample is then added – most frequently dried onto a small amount of adsorbent. Solvent is introduced and pushed down the column by application of air or nitrogen pressure. The method has been employed for the separation of monoterpenes and a dibenzo-α-pyrone from *Anthocleista djalonensis* (Loganiaceae). Chromatography was performed on silica gel, with petrol ether-diethyl ether mixtures (Onocha et al. 1995). A combination of dry-column flash chromatography on silica gel and C-18 reversed-phase silica gel, with a final semi-preparative HPLC step yielded cycloartane triterpenoids and their hydroperoxides (e.g. 7) from *Tillandsia recurvata*

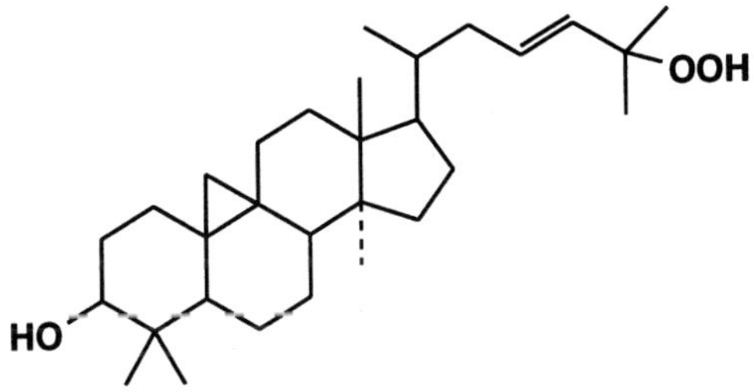

7

(Bromeliaceae). In this instance, silica gel was used for the first dry-column flash chromatography step (eluting with hexane, mixtures of hexane-dichloromethane and then finally dichloromethane). The fraction eluted with dichloromethane was fractionated by dry-column chromatography on C-18, with mixtures of methanol-water and methanol-dichloromethane of decreasing polarity, to obtain the hydroperoxide-rich fractions (Cabrera and Seldes, 1995).

Examples of flash chromatography in the separation of synthetic products and the isolation of natural products appear with increasing frequency in the literature. In fact, the technique is now so routine that column dimensions and flow rates are rarely published – the solvent alone is usually mentioned. A few selected applications are collected in Table 5.3.

Table 5.3. Applications of flash chromatography in the separation of natural products

Substances separated	Granulometry of silica gel	Column dimensions	Mobile phase	Reference
Silica gel stationary phase				
Alkaloids from *Corydalis remota* (Fumariaceae)			Cyclohexane-EtOAc-diethylamine 15:1:0.5 → 6:1:0.4 gradient	Fu et al. 1988
Pyrrolizidine alkaloids from *Senecio chrysocoma* (Asteraceae)			$CHCl_3$-MeOH-NH_3 85:13:2	Grue and Liddell 1993
Piperidine alkaloids from tunicate *Pseudodistoma kanoko*		45×2.2 cm	$CHCl_3$-nBuOH-HOAc-H_2O 1.5:6:1:1	Ishibashi et al. 1987
Alkaloid from fungus *Penicillium verrucosum*		15×5 cm	Hexane-EtOAc gradient CH_3CN-CH_2Cl_2 1:10 Hexane-EtOAc-CH_3CN 20:3:2	Hodge et al. 1988
Phenolic glycosides from *Salix lasiandra* (Salicaceae)		4 cm i.d.	$CHCl_3$-MeOH 7:93 → 20:80	Reichardt et al. 1992
Sinapic acid esters from *Polygala virgata* (Polygalaceae)			$CHCl_3$-MeOH-H_2O 80:20:2 → 60:40:10	Bashir et al. 1993
Methylated flavones from *Helichrysum nitens* (Asteraceae)			Petrol ether-EtOAc gradient	Tomas-Barberan et al. 1988
Flavones from *Centaurea species* (Asteraceae)	40–63 μm		Petrol ether-Et_2O gradient	Christensen et al. 1991
Isoflavones from *Ulex europaeus* (Leguminosae)			CH_2Cl_2-MeOH 1:1, 9:1 Petrol ether-Et_2O	Russell et al. 1990
Prenylated flavanone from *Petalostemum purpureum* (Leguminosae)	40–63 μm	120 g support	Toluene-EtOAc-HOAc gradient	Hufford et al. 1993
Xanthones from *Garcinia livingstonei* (Guttiferae)			$CHCl_3$-MeOH 100:1	Sordat-Diserens et al. 1992b
Stilbenes from *Picea abies* (Pinaceae)			$CHCl_3$-EtOH 9:1, EtOH Petrol ether-EtOH 1:1	Mannila et al. 1993
Lignans from *Forsythia intermedia* (Oleaceae)	40–63 μm	15×2.2 cm	CH_2Cl_2-MeOH 25:1	Rahman et al. 1990

Neolignans from *Ocotea veraguensis* (Lauraceae)			Toluene-EtOAc 3:2 → 2:3	Dodson et al. 1987
Diprenylquinones from Ascidian *Aplidium* sp.	20 - 50 μm		Hexane-Et_2O gradient	Guella et al. 1987
4-Hydroxy-2-cyclopentenone from *Passiflora tetrandra* (Passifloraceae)	35 - 70 μm (150 Å)	20 g support	$CHCl_3$ → $CHCl_3$-MeOH 4:1	Perry et al. 1991b
Norisoprenoid glycosides from *Rubus idaeus* (Rosaceae)	32 - 63 μm		$CHCl_3$-MeOH-H_2O 80:20:1	Pabst et al. 1992
Iridoid glycosides from *Castilleja wightii* (Scrophulariaceae)			$CHCl_3$-MeOH 7:3 → 1:1	Belofsky and Stermitz 1988
Sesquiterpene ethers from *Baccharis dracunculifolia* (Asteraceae)	32 - 63 μm		Petrol ether-Et_2O gradient	Weyerstahl et al. 1995
Diterpenes from *Psiadia arabica* (Asteraceae)			$CHCl_3$-MeOH 1:1, 9:1	El-Domiaty et al. 1993
Triterpene acid from *Combretum* (Combretaceae)			Petrol ether-EtOAc → EtOAc-EtOH	Rogers and Subramony 1988
Saponins from *Dolichos kilimandscharicus* (Leguminosae)	63 - 200 μm	60×3 cm	$CHCl_3$-MeOH-H_2O 50:10:1	Marston et al. 1988
Tetranortriterpenoid from *Azadirachta indica* (Meliaceae)	40 μm	18×7 cm	CH_2Cl_2, Et_2O	Yamasaki et al. 1988
Limonoids from *Dysoxylum richii* (Meliaceae)	40 - 63 μm		Hexane-EtOAc gradient	Jogia and Andersen 1989
Cucurbitacins from *Picrorhiza kurroa* (Scrophulariaceae)	40 - 63 μm		$CHCl_3$-MeOH 23:0.1, 9:1, 4:1	Stuppner and Müller 1993
Cardenolides from *Erysimum cheiranthoides* (Brassicaceae)	32 - 63 μm		Water-sat. EtOAc-MeOH 99.5:0.5, 99:1, 98:2	Sachdev-Gupta et al. 1993
Polyacetylene from *Heteromorpha trifoliata* (Apiaceae)	63 - 200 μm		$CHCl_3$-MeOH 99:1 → 96:4	Villegas et al. 1988
Acetogenins from *Annona muricata* (Annonaceae)	40 μm		CH_2Cl_2-MeOH 100:0 → 75:25	Rieser et al. 1996

Table 5.3 (continued)

Substances separated	Granulometry of silica gel	Column dimensions	Mobile phase	Reference
Reversed-phase flash				
Phenolic glycosides from *Populus tremuloides* (Salicaceae)	40 µm (C-18)		Acetone-H_2O 42:58	Clausen et al. 1989
Phenolic glycosides from *Salix lasiandra* (Salicaceae)	(C-18)		Acetone-H_2O 85:15, 50:50, 90:10	Reichardt et al. 1992
Flavonoid glucuronides from *Malva sylvestris* (Malvaceae)	40 µm (C-18)	45×2.5 cm	MeOH-H_2O gradient	Billeter et al. 1991
Isoflavones from *Ulex europaeus* (Leguminosae)	(C-18)	7 cm i.d.	MeOH-H_2O gradient	Russell et al. 1990
Naphthoquinone from brown alga *Landsburgia quercifolia*	40–63 µm (C-18)	20 g support	MeOH-H_2O 3:2 → CH_2Cl_2 gradient	Perry et al. 1991 a
Lignans from *Libocedrus plumosa* (Cupressaceae)	40–63 µm (C-18)	50 g support	MeOH-H_2O-CH_2Cl_2 gradient	Perry and Foster 1994
Norditerpene lactones from *Ileostylus micranthus* (Loranthaceae)	40–63 µm (C-18)		MeOH-H_2O 75:25	Bloor and Molloy 1991
Saponins from *Glycine max* (Leguminosae)			H_2O → MeOH	Curl et al. 1988
Tetranortriterpenoid from *Azadirachta indica* (Meliaceae)	40 µm (C-18)	18×7 cm	MeOH-H_2O 4:1	Yamasaki et al. 1988
Trinor-eremophilane from marine deuteromycete *Dendryphiella salina*	40–63 µm (C-18)		MeOH-H_2O gradient	Guerriero et al. 1988 a
Capsaicinoids from *Capsicum* (Solanaceae)	40 µm (C-18)	40×2 cm	MeOH-H_2O 50:50 → 80:20	Krajewska and Powers 1986
Fatty acid from black coral *Leiopathes* sp.	40–63 µm (C-18)		MeOH-H_2O 9:1, 9:2	Guerriero et al. 1988 b
Bryostatin 1 from bryozoan *Bugula neritina* (Bugulidae)	50 µm (C-18)	25×10 cm	MeOH-H_2O → MeOH	Schaufelberger et al. 1991
Zwiebelanes (organosulphur) from *Allium cepa* (Liliaceae)	(C-18)		MeOH	Bayer et al. 1989 b

5.2.1.2
Bonded Phases As Adsorbents

The technique of flash chromatography is by no means confined to silica gel as support; other adsorbents, including bonded phases, are of potential use.

Kühler and Lindsten (1983) presented a simple reversed-phase flash chromatography system consisting of a pump (higher pressures, between 2 and 13 bar, are involved than in regular flash chromatography) connected to a column filled with octadecyl-modified silica. The best resolution and narrowest peaks were obtained with 15–40 µm phase (40–63 µm particles also gave good results). For a 125×34 mm column containing 70 g sorbent, up to 500 mg sample was applied. Flow rates of ca. 70 ml/min gave complete separations in 5–15 min.

Extensive use of reversed-phase flash chromatography as the first purification step has been made in the isolation of polar (often water soluble) constituents from marine organisms (Blunt et al. 1987). The authors load extracts as an aqueous slurry or powder (with phase) onto slurry-packed columns. Elution is first with water and then a steep methanol-water to methanol gradient, finally terminating with dichloromethane. A typical procedure is as follows: extract (10 g) is dissolved in dichloromethane-methanol-water (250 ml) and reversed-phase material (5 g) is added. Concentration under reduced pressure gives an aqueous slurry (50 ml) which is added to the top of a column (25 mm i.d.) containing slurry-packed support (50 g). This system is claimed to handle up to 20 g crude extract per 100 g of packing material (Blunt et al. 1987). As an example, a charge of 10.7 g crude sponge extract was chromatographed on an RP-18 column eluted with water and methanol-water mixtures. Final purification by LPLC gave 1,3,7-trimethylguanine (Perry et al. 1987).

Reversed-phase flash chromatography has also been used in the isolation of an antibacterial α-bromoenone from the Mediterranean sponge *Aplysina cavernicola* (D'Ambrosio et al. 1985). An extract of the sponge was flash chromatographed on a 60×60 mm column of 25–40 µm LiChrosorb RP-18 with a H_2O-MeOH gradient. The α bromoenone eluted with H_2O-MeOH 95:5 and was purified to homogeneity by semi-preparative HPLC.

An insecticidal tetranortriterpenoid, salannin (**8**), has been isolated from neem seeds (*Azadirachta indica*, Meliaceae) by a procedure involving two flash chromatographic steps and two HPLC final purification steps. The seeds (29 kg)

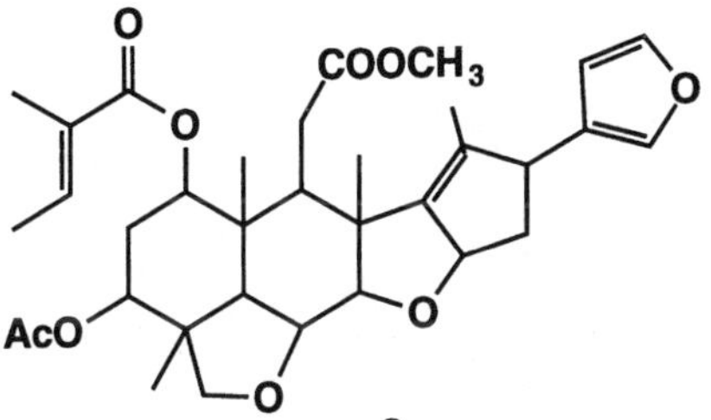

8

were extracted with hexane. After evaporation of the hexane, 8.1 kg of oil resulted. This was flash chromatographed in 1 kg batches (diluted with dichloromethane) over silica gel, first with dichloromethane and then with diethyl ether to displace the salannin. The second flash chromatography step was over a column of octadecylsilica with methanol-water 4:1 as eluent (20 ml/min), yielding 34.3 g crude salannin. The two final HPLC runs (first with a silica gel column and then with a phenyl column) gave 9.6 g of salannin (>99% pure) (Fig. 5.12) (Yamasaki et al. 1988). A similar procedure was used for the purification of azadirachtin from neem seeds (Yamasaki et al. 1986).

Other separations involving reversed-phase flash are shown in Table 5.3.

Microcystins (hepatotoxic cyclic peptides) have been purified from cyanobacteria by a method involving flash chromatography on a Biotage 75S cartridge system. As stationary phase, either spherical Hyperprep C_{18} (30 μm, 120 Å) or irregular Bondapak C_{18} (37–55 μm, 125 Å) was packed into 9×7.5 cm i.d. cartridges. Aqueous extract containing the microcystins was applied at a flow rate of 100 ml/min and elution was by a step gradient of 0–100% methanol in 10% increments. Final purification was by HPLC (Edwards et al. 1996).

A rapid method for the initial purification of molecules from biological fluids on a 400 μm octadecylsilyl support has been reported (Wainer et al. 1985). A glass column (527×33 mm) was packed with 100 g of support and urine samples were eluted with a stepwise MeCN-H_2O gradient at 2–3 bar nitrogen pressure (15 ml/min). Fractions containing Na^+, K^+-ATPase inhibitors were then collected, ready for further purification of the inhibiting factors. Effectively, a concentration effect was being achieved.

Chiral stationary phases in flash chromatography are described in Chap. 9.

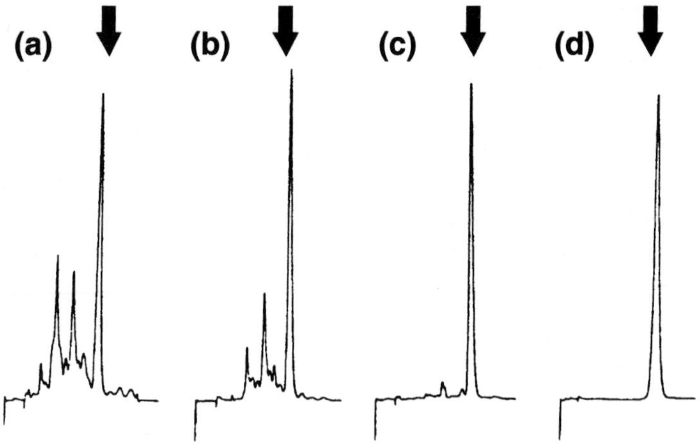

Fig. 5.12 a–d. Analytical HPLC (phenyl column; CH_3CN-H_2O 2:3, 1.5 ml/min) of salannin (*solid arrows*) after: **a** silica gel flash chromatography; **b** reversed-phase flash chromatography; **c** silica gel HPLC; **d** phenyl HPLC. (Reprinted with permission from Yamasaki et al. 1988)

5.2.1.3
Aluminium Oxide As Adsorbent

In work on alkaloids, aluminium oxide has successfully been applied in flash chromatography. The alkaloid (–)-argyrolobine was isolated from *Lupinus argenteus* (Leguminosae) after flash chromatography on basic aluminium oxide (Merck, Type T) by elution with mixtures of diethyl ether and methanol (Arslanian et al. 1990). Similarly, the quaternary benzophenanthridine alkaloids chelerythrine (**9**), sanguinarine (**10**) and chelirubine (**11**) have been obtained from *Dicranostigma lactucoides* (Papaveraceae), using aluminium oxide for TLC (5–50 μm, BDH), prepared by stirring with hydrochloric acid (0.3 mol/l; 1:2 w/w adsorbent-acid), filtering after 24 h and drying at room temperature. A 560×38 mm column was used, eluting at 5–10 ml/min with toluene-methanol mixtures, under argon pressure (Dostal et al. 1992).

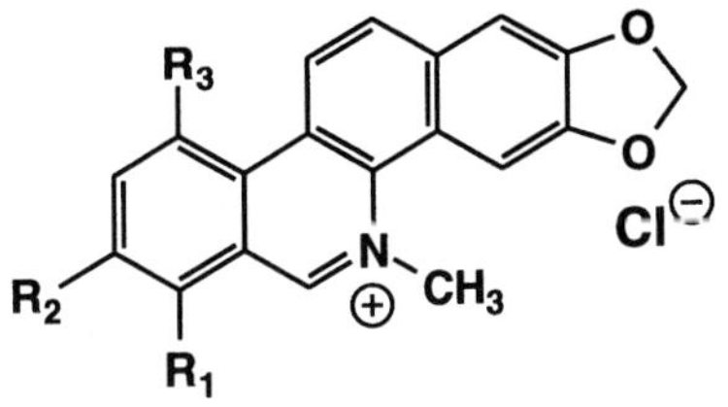

9 $R_1 = R_2 = OCH_3$, $R_3 = H$
10 $R_1 + R_2 = OCH_2O$, $R_3 = H$
11 $R_1 + R_2 = OCH_2O$, $R_3 = OCH_3$

5.2.1.4
Polyamide As Adsorbent

A report of the separation of flavonoid aglycones and two phenolic acids by flash chromatography with column-grade polyamide has appeared. Chloroform, chloroform-methanol and methanol were the elution solvents and the source plant was *Indigofera heterantha* (Leguminosae) (Hasan et al. 1989).

Similarly, Polyamide-6 (Serva, 100–300 μm) in a 10×6 cm frit filter was used in the isolation of two esterified diterpene alcohols from *Sarcodictyon roseum*, a marine stolonifer. A water-acetonitrile gradient provided the eluent (D'Ambrosio et al. 1987).

5.2.2
Low-Pressure LC (LPLC)

While flash chromatography requires a minimum of equipment (the most expensive part is usually the needle valve!) and is characterised by its simplicity, low-pressure LC systems are a little more complex and include a pump.

Apart from home-made columns, the most popular preparative low-pressure LC system is the Lobar range (E. Merck). A series of ready-filled columns is avai-

Table 5.4. Sizes and separation capacities of Lobar columns

Size	Packing material	Length (mm)	Inner diameter (mm)	Outer diameter (mm)	Charge
A	LiChroprep Si60	240	10	13	Up to 0.2 g per 0.3–1 ml
A	LiChroprep RP-8	240	10	13	Up to 0.2 g
A	LiChroprep RP-18	240	10	13	Up to 0.2 g
B	LiChroprep Si60	310	25	28	Up to 1 g per 1–5 ml
B	LiChroprep Diol	310	25	28	Up to 1 g per 1–5 ml
B	LiChroprep CN	310	25	28	Up to 1 g per 1–5 ml
B	LiChroprep NH_2	310	25	28	Up to 1 g
B	LiChroprep RP-8	310	25	28	Up to 1 g
B	LiChroprep RP-18	310	25	28	Up to 1 g
C	LiChroprep Si60	440	37	42	Up to 3 g per 2–10 ml
C	LiChroprep RP-8	440	37	42	Up to 3 g
C	LiChroprep RP-18	440	37	42	Up to 3 g

lable and, since these are made of glass, the support is visible. Table 5.4 gives an idea of the sizes of the columns, together with loading factors. The packing material is sealed into the column with a glass frit and the connection to the pump is provided by a metal cannula attached to a PTFE ring, the whole being held in place by a screw cap (Fig. 5.13).

The columns are available with different stationary phases and are designed to be re-usable, although the lifetime of the reversed-phase columns is generally much longer than that of the normal-phase columns. Particle size is large (40–60 μm) and this is the reason why such high flow rates can be obtained with the low pressures. The silica columns use the same material as that for Merck TLC plates. Injection is by way of a syringe/septum system or by a six-port valve.

Pumps suitable for providing the pressure (up to 6 bar/90 psi) are available from Chemie and Filter GmbH (Heidelberg, Germany), Fluid Metering Inc. (Oyster Bay, NY, USA), Pharmacia (Uppsala, Sweden), Orlita (Giessen, Germany), Lewa (Leonberg, Germany) or Milton Roy (Riviera Beach, FL, USA). Certain HPLC pumps are compatible and even separations under hydrostatic pressure are possible. With a pump, it is always advisable to connect a manometer in order to avoid overpressuring, which might rupture glass columns.

Lobar columns can accommodate separations of gram quantities with resolutions sometimes approaching those of HPLC and, in any case, for greater resolution or larger charges, columns can be coupled in series. When performing separations on reversed-phase columns, one way of minimising the water content (or avoiding water altogether) is to connect several columns (up to six have been reported) in series. In this way, pure methanol can be used as eluent and this is easier to evaporate than methanol-water mixtures (Henke and Rülke 1987).

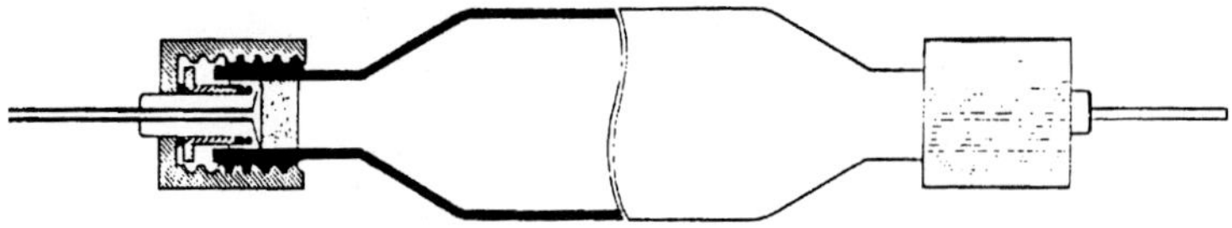

Fig. 5.13. A Lobar pre-packed column

The simplicity of the set-up for Lobar separations is the reason why these columns are so widely used, although it would be even more of an advantage if the spent columns could be refilled.

Applications of Lobar columns in the literature are usually not very detailed – information about the column size, for example, is often lacking. However, this low-pressure technique is frequently employed and some representative examples can be found in Table 5.5.

Lobar columns are generally used in combination with other separation methods and may form intermediate or final steps of a purification. Because of the pumping arrangement, isocratic separations are normally run. However, gradient runs are not unusual, e.g. in the separation of xanthones from *Gentiana lactea* (Gentianaceae) with a methanol-water gradient (Schaufelberger and Hostettmann 1988).

A tetracyclic antifungal polyphenol (**12**) was obtained from the root bark of *Bauhinia rufescens* (Leguminosae) after flash chromatography and LPLC on a Lobar Diol column (toluene-ethyl acetate 9:1). The dihydro derivative **13** required chromatography on a Lobar RP-8 column (methanol-water 7:3) (Maillard et al. 1991).

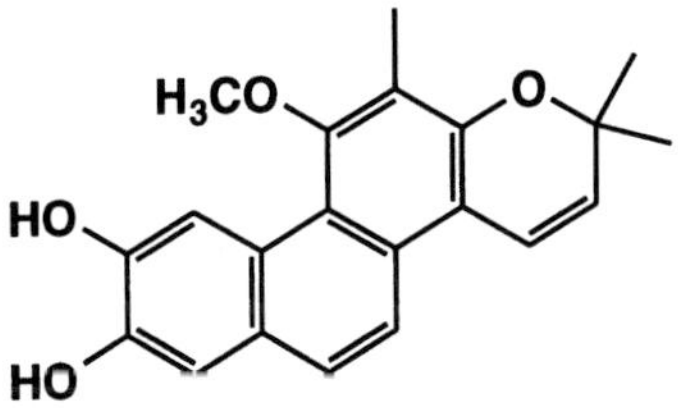

12

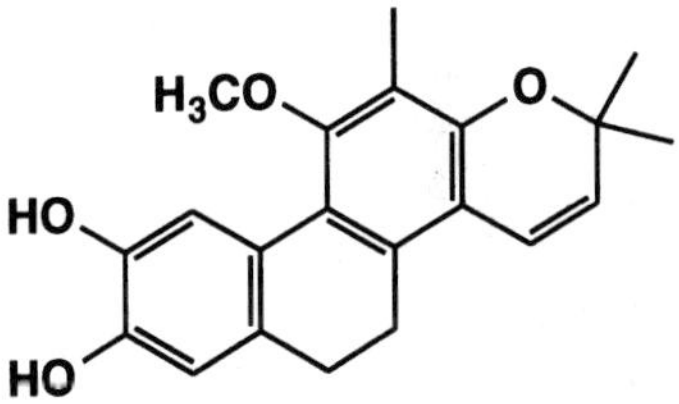

13

The isolation of five antibacterial and molluscicidal 6-alkenylsalicylic acids from *Spondias mombin* (Anacardiaceae) involved a final Lobar step on a B-size RP-18 column. A gradient of MeOH and 5% HCOOH from 85:15 to 99:1 was used, with the following elution profile: 0–15 min 85% MeOH, 15–80 min linear to 90%, 80–120 min 90% MeOH, 120–240 min linear to 99% MeOH and 240–300 min 99% MeOH (Corthout et al. 1994).

Glutamyl peptides have been obtained from garlic (*Allium sativum*, Liliaceae) bulbs by a combination of LPLC and semi-preparative HPLC. LPLC was performed on an 800×26 mm column filled with RP-18 support (RSil RP-18 HL 15–35 μm). As solvent, a gradient starting with water-1% aqueous acetic acid and ending with 40% acidified methanol was used (Mütsch-Eckner et al. 1992).

Table 5.5. Applications of Lobar LC columns in the isolation of natural products

Substance class	Column size	Support	Mobile phase	Reference
Alkaloids	B	Diol	Hexane-Et_3N 200:1	
			Hexane-$CHCl_3$-Et_3N 80:20:1	Tokuyama et al. 1987
Indole alkaloid		RP-8	MeOH-0.02 M NH_4OAc 3:2	Angenot et al. 1990
Phenolic acids	B	RP-18	MeOH-5% HCOOH 85:15 → 99:1	Corthout et al. 1994
Chlorogenic acid derivatives	B	Diol	$CHCl_3$-MeOH 93:7	Wang et al. 1992
Chalcones		RP-8	MeOH-H_2O 17:3	Pistelli et al. 1996
Chalcone glycosides	C	RP-8	CH_3CN-H_2O 1:9	Miething and Speicher-Brinker 1989
Chromones	B	SiO_2	$CHCl_3$-nC_6H_{14}-MeOH 80:8:0.3	Décosterd et al. 1987b
Furanocoumarins	B	RP-8	MeOH-H_2O 5:5 → 7:3	Wawrzynowicz and Waksmundzka-Hajnos 1990
Flavonols	B	Diol	$CHCl_3$-MeOH-HOAc 950:50:1	Wang et al. 1989
Flavonol glycosides	B	RP-8	MeOH-H_2O 2:3	Bashir et al. 1991
		RP-8	Acetone-H_2O 2:8 → 3:7	Brasseur and Angenot 1986
Xanthones	B	RP-8	MeOH-H_2O 3:7 → 5:5 gradient	Schaufelberger and Hostettmann 1988
	B	SiO_2	$CHCl_3$-MeOH 20:1	Marston et al. 1993
	B	Diol	Toluene-EtOAc 2:1	Marston et al. 1993
	B	Diol	$CHCl_3$-EtOAc 100:1	Marston et al. 1993
Prenylated xanthones	B	Diol	$CHCl_3$-MeOH 50:1	Sordat-Diserens et al.
	B	RP-18	MeOH-H_2O 4:1	1992a
Pterocarpinoids	B	SiO_2	Petrol ether-EtOAc 5:5, 2:8	Gunzinger et al. 1988
Phloroglucinols		RP-8	MeOH-H_2O 2:1	Ishiguro et al. 1987
Iridoid glycosides	C		MeOH-H_2O 1:25 → 25:75	Damtoft 1992
	B	RP-18	MeOH-H_2O 4:6 → 6:4	Stuppner et al. 1993

Secoiridoid glycosides	B	RP-8	MeOH-H_2O 25:75 → 50:50	Schaufelberger et al. 1987
Sesquiterpenes	B	SiO_2	$CHCl_3$-MeOH 87:13	Kouno et al. 1990
Diterpenes	B	RP-8	MeOH-CH_3CN-H_2O 1:1:5 → 1:1:40	Mishra et al. 1986
	B	RP-18	MeOH-H_2O 95:5	Slimestad et al. 1995
Secosqualene		RP-2	Dioxane-H_2O 10:3	Asari et al. 1989
Withanolides		Diol	Hexane-iPrOH	Alfonso and Kapetanidis 1994
Triterpenes	C	RP-8	MeOH-H_2O 4:1	Pei-Wu et al. 1988
Triterpene saponins		RP-18	MeOH-H_2O 1.1:1	Zhao et al. 1994
		RP-18	CH_3CN-H_2O	Zhao et al. 1994
	B	RP-8	MeOH-H_2O 8:2, 7:3	Hamburger et al. 1992
Polyacetylenes	B	RP-8	MeOH-H_2O 3:1	Wang et al. 1990b
	B	SiO_2	Toluene-EtOAc 85:15	Villegas et al. 1988
Cyclohexylethanols	B	RP-8	H_2O	Potterat et al. 1992

Discodermin A, an antimicrobial peptide from the marine sponge *Discodermia kiiensis*, was purified by a sequence of operations which included a Lobar step, with a size C LiChroprep Si60 column and a $CHCl_3$-MeOH-H_2O solvent mixture (Matsunaga et al. 1985a,b). A final reversed-phase HPLC separation was necessary to obtain the pure tetradecapeptide.

Instead of using a methanol-water mobile phase for RP-8 chromatography, Jansen et al. (1985) prepared a buffer consisting of 0.5% formic acid in water (neutralised with triethylamine) for the separation of three antibiotics from culture broth of the gliding bacterium *Corallococcus coralloides.*

The final purification of two saponins (**14**, **15**) from the fruit pericarp of *Sapindus rarak* (Sapindaceae) by LPLC on Lobar RP-8 columns and semi-preparative HPLC has been reported (Hamburger et al. 1992). These saponins, glycosides of hederagenin, were both molluscicidal.

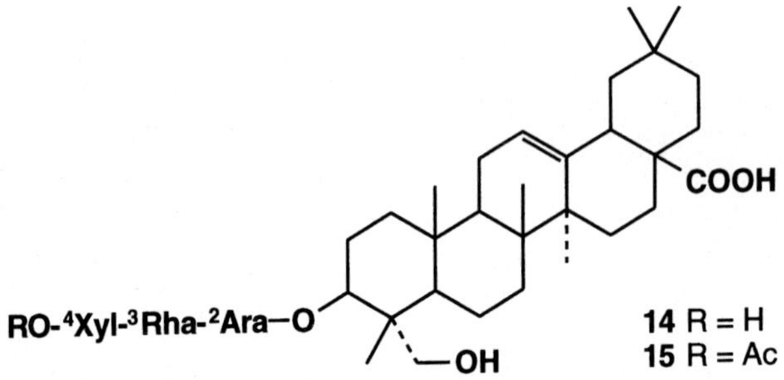

14 R = H
15 R = Ac

Another example of the use of Lobar reversed-phase columns in the separation of triterpene glycosides comes from work on *Mussaenda pubescens* (Rubiaceae), a Chinese plant. The final purification of the saponins was performed on RP-18 columns, with first EtOH-H_2O (1.1:1) and then CH_3CN-H_2O (1:1) as eluent (Zhao et al. 1994).

In the separation of steroid alkaloid glycosides from the roots of *Veratrum patulum* (Liliaceae), a Lobar B RP-18 column was employed. The mobile phase was methanol-pH 3.5 buffer solution (2:1). This buffer solution was prepared by mixing 50 g ammonium acetate with 50 ml water and then adding 280 ml 2 mol/l hydrochloric acid and 1000 ml water (Irsch et al. 1993).

The phthalides **16** and **17** have been isolated from the fungus *Emericella desertorum* (*Aspergillus nidulans* group) via LPLC on 200×10 mm and 200×20 mm glass columns filled with CQ-3 silica gel (30–50 µm, Wako).

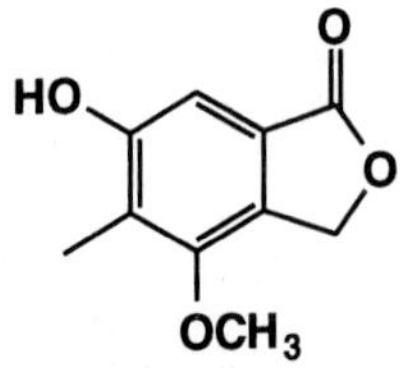

16

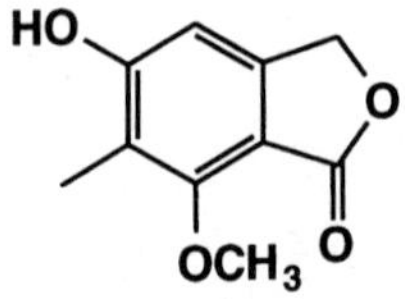

17

Eluent (benzene-acetone 10:1) was pumped with a Chemco Low-Prep pump 81-M-2 (Nozawa et al. 1987).

For the final purificaion of fusarin C (**18**), a bacterial mutagen produced by *Fusarium moniliforme* on corn, a Lobar size B Si60 column was used, with CH_2Cl_2-MeOH 98:2 as solvent (Gaddamidi et al. 1985).

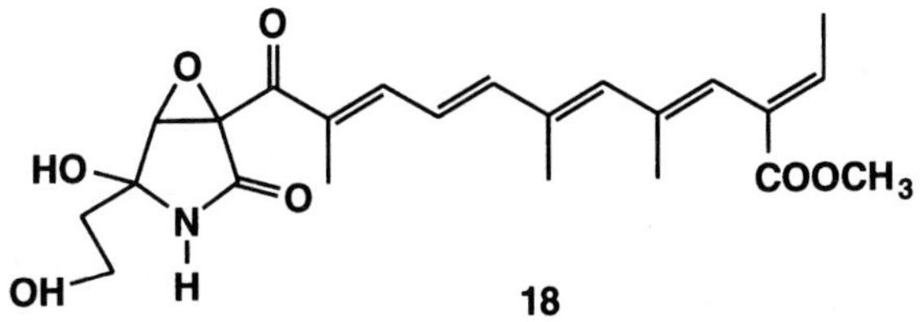

18

Kronlab (Sinsheim, Germany) markets a modular "flash chromatography" system which in fact uses a rotating piston pump and is very similar to a typical Lobar set-up. Kronlab columns are recommended for this system; these can be refilled. Another method described as "flash chromatography" for the purification of reaction mixtures on octadecylsilyl supports has been published (Kühler and Lindsten 1983). A more apt description of this procedure would be LPLC since a solvent pump is also used, giving pressure drops of around 2–13 bar.

A patent for replenishable LPLC columns has been filed. The columns are fitted with an axially movable core element (Poppe et al. 1988).

The use of refillable columns has been described for several LPLC separations (see Calis et al. 1990 and references cited therein). The stationary phase was in these cases Sepralyte C-18 (40 µm). Kubo et al. (1984) have used low-pressure LC on a Pharmacia SR 10/50 column, with LiChroprep RP-18 (particle size 25–40 µm), eluted with methanol-water, for the isolation of the diterpene dihydroxyverrucosane from a liverwort. A preliminary separation was carried out by silica gel open-column LC.

Low-pressure preparative liquid chromatography has been tried with 3-aminopropyl- and 3-(2-aminoethylamino)propyl-siloxane polar bonded phases (Zief et al. 1982a). Since separation conditions could not be predicted by TLC, an analytical column filled with the same material was used to choose suitable elution solvents. Separations of aliphatic and aromatic alcohols, aromatic amines and simple carbohydrate mixtures were attempted with both aqueous and non-aqueous solvents at pressures from 1–4 bar. With $\alpha \geq 1.7$, 200–300 mg quantities of material were isolated in less than 1 h on 300×20 mm columns. The amine bonded phases appear to have about twice the capacity for separations of moderate difficulty, when compared with normal silica (Zief et al. 1982a, b).

A peristaltic pump has been used to provide a suitable flow of eluent through a reversed-phase column for the separation of monoterpene glycosides from free sugars and organic acids of grape juice and wines (Williams et al. 1982).

5.2.3 Medium-Pressure LC (MPLC)

The advent of Merck Lobar pre-filled columns in 1971 marked a big step forward in the methodology of preparative liquid chromatography. Unfortunately, separations of more than 5 g of sample are rarely possible with these columns and, in order to accommodate larger sample loads, recourse is generally taken to medium-pressure LC. This technique involves longer columns (easily filled and re-filled) with large internal diameters and requires higher pressures than low-pressure LC to enable sufficiently high flow-rates. Together with a simple pumping set-up, involving either compressed air or reciprocating pumps, MPLC fulfils the requirement for a simple complementary or supplementary method to open-column chromatography and flash chromatography, with both higher resolution and shorter separation times.

Several different commercially-available systems are accessible for high sample throughput under conditions of moderate pressure and a selection is presented below.

Leutert and von Arx (1984) have described the optimized conditions suitable for MPLC in the Ciba-Geigy (now Novartis) research laboratories: for the separation of 1–100 g lipophilic mixtures within 1–2 h at flow rates of 25–170 ml/min and a pressure maximum of 20 bar. Criteria such as pulse-damping, sample introduction, column filling and detectors were also discussed for these systems and a comparison of MPLC with open-column chromatography and flash chromatography was performed (Table 5.6). Open column and flash chromatography gave the same resolution, whereas MPLC performed much better. MPLC and flash chromatography both led to considerable savings of time. Also notable was the high loading capacity of MPLC columns: up to a 1:25 sample to packing material ratio could be achieved for the test mixture.

In all the MPLC variants mentioned here, the columns are filled by the user. Particle sizes of 25–200 μm are usually advocated (15–25 μm, 25–40 μm or 40–63 μm are the most common materials) and either slurry packing or dry packing is possible. Verzele and Geeraert (1980) have published useful packing methods for these larger-bore columns; since the right way of filling the column is such an essential pre-requisite for good separations, this item cannot be overemphasized. The methods are summarized by Porsch (1994).

Zogg et al. (1989a–c) have also studied MPLC operating conditions, including column packing. Five different methods of packing columns (one slurry and

Table 5.6. Separation of 900 mg test mixture (dimethyl-, diethyl- and dibutyl esters of phthalic acid (1:1:1)) with hexane-ethyl acetate 4:1 (Leutert and von Arx 1984)

	Silica gel (240 g) particle size (μm)	Pressure (bar)	Flow-rate (ml/min)	Resolution (Rs)	Time (min)
Open column	63–200	0	25	1.5	64
Flash	40–63	0.75	140	1.6	9
MPLC	25–40	12	100	3.4	12

four dry packing) were investigated. The most efficient methods were two dry packing procedures, using vacuum and nitrogen overpressure. It was possible to pack the MPLC columns with 5–25 μm TLC silica gel GF_{254} by these means, thus giving excellent separations with test mixtures. A higher packing density of at least 20% was found for the dry packing techniques (Zogg et al. 1989b). One of the practical results to come from the investigation of Zogg et al. (1989c) was that resolution was increased for a long column of small internal diameter when compared with a shorter column of larger internal diameter (with the same amount of stationary phase).

Choice of solvent systems can be very efficiently performed by analytical HPLC. Transposition to MPLC is straightforward and direct (Schaufelberger and Hostettmann 1985).

A search for optimal MPLC conditions on silica gel columns has been described by Nyiredy et al. (1990). Solvent systems are first searched by TLC using, for example, the PRISMA model (Nyiredy et al. 1988). These conditions can either be transposed directly onto MPLC or transferred via an intermediate analytical OPLC step. The advantage of OPLC is that it can be employed as an equilibrated system, whereas TLC is a non-equilibrated system. Thus transposition through OPLC gives a more accurate pilot method.

5.2.3.1
Büchi B-680 A System

Büchi market a complete MPLC system, with a wide range of column combinations, providing a very flexible separating capacity suitable for 100 mg–100 g sample sizes. A piston pump and exchangeable pump heads allow flow rates from 3 to 160 ml/min at a maximum pressure of 40 bar. The pump features a built-in pulse-damper and a manometer with upper and lower pressure limits for pressure cut-off. A gradient former can also be added to the system. Different UV detectors are available, including a variable wavelength monitor (Fig. 5.14).

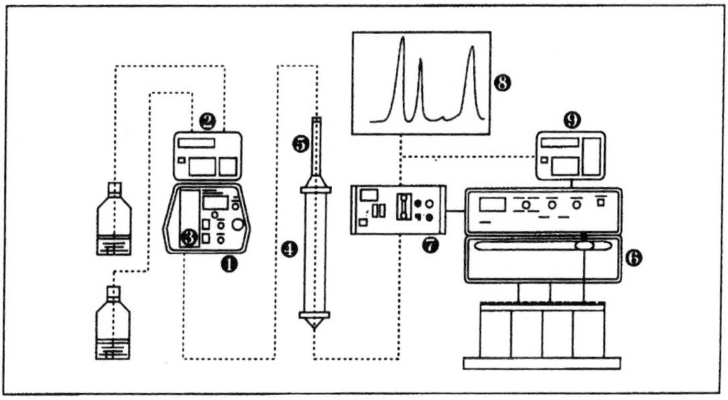

Fig. 5.14. Büchi B-680 A MPLC system. 1: B-688 pump; 2: gradient former; 3: sample introduction; 4: glass column; 5: pre-column; 6: fraction collector; 7: UV detector; 8: recorder; 9: peak detector

The chromatography columns are of strengthened glass with a plastic protective coating, giving a visual control of the separation, and come in sizes varying from 130 to 1880 ml. They can be coupled together very simply by means of flanges to increase the resolving power. A Teflon ring serves to seal the joint between the columns but care must therefore be taken not to include acetone in the solvent systems.

In the case of silica gel, the columns may be filled with dry media or with a slurry of the stationary phase in a suitable solvent. The dry-filling method gives a 20% higher packing density than the slurry technique and is carried out manually by adding support to a recipient screwed to the top of the column. Once the column is filled, the extra recipient is checked to ensure that it contains enough stationary phase to fill another 10% of the column and is then connected to a nitrogen cylinder. Nitrogen is passed into the column at 10 bar (with the column exit open) until the level of the packing material remains constant. The nitrogen valve is then closed and the pressure in the column is allowed to drop completely before conditioning the column (Fig. 5.15). Although the pressure limit on the glass column is only 40 bar, the use of 15 μm particles gives efficiencies and separations similar to those obtained on HPLC columns (Zogg et al. 1989a–c).

For bonded phases, the slurry method is always used for packing.

Samples can be injected through a septum directly onto the column or via a sample loop. It is also possible to perform solid introduction of the sample with the aid of a small Prep-Elut column connected just in front of the main separation column.

If a pre-column (screwed directly onto the preparative column) is used during separations, the contaminated packing material at the top of this column can be removed after each separation. Silica gel supports can be regenerated by washing in the sequence methanol → ethyl acetate → hexane but after a certain time the support should be thrown away. Bonded-phase columns are easier to clean and have a longer working life.

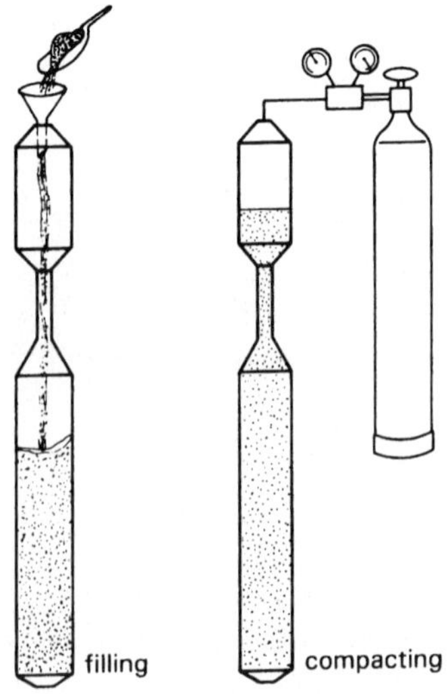

Fig. 5.15. Dry-filling of Büchi MPLC glass columns

5.2.3.2
Labomatic Installations

The early investigations of Leutert and von Arx have influenced the development of the Labomatic MPLC systems. Labomatic, as well as Büchi, now supply a complete range of MPLC equipment, from pumps to columns and even column packing materials. The MD-50/80/100 pump has three interchangeable piston heads, giving flow rates from 0.5 to 156 ml/min up to 40 bar. A pulse-damper is also fitted.

There are 38 different dimensions of columns in the Labomatic range (Labochrom PGC and MPGC), with internal diameters from 9 to 105 mm and lengths from 100 to 1760 mm (volume 12.8–13,000 ml). These columns are made of glass and can withstand pressures of up to 20 bar. They have a smooth inner surface and can be packed with any choice of support. The narrow entrances at the conical ends of the column do however make the packing process difficult. Also available are axial compression columns (the Labochrom AMC range) for larger-scale applications.

Leutert and von Arx (1984) describe the use of five different column sizes, starting with 480×37 mm (for ca. 5 g sample) and extending to 760×105 mm (for ca. 60 g sample). In their separations, Merck LiChroprep Si60 Kieselgel (25–40 μm) was employed to good effect.

Labomatic also market a series of reversed-phase packing materials for MPLC separations. In the preparation of these phases, attention is paid to careful end-capping of the treated silica gel, to give a fully hydrophobic surface. The packings come in 20–45 μm and 35–70 μm particle sizes and are either C_{18} or C_8. These are compatible with the 20 bar pressure limit of the chromatography columns.

Zogg et al. (1989a) have described the dry filling of Labomatic glass columns and the transposition of analytical OPLC conditions to MPLC for the separation of furanocoumarins.

5.2.3.3
Kronlab/Stagroma System

Kronlab (Sinsheim, Germany), in conjunction with Stagroma (Wallisellen, Switzerland) also market a range of MPLC components, including pumps and columns. The columns come in dimensions varying from 220×30 mm to 660×60 mm and can withstand pressures of up to 50 bar.

5.2.3.4
C.I.G. Column System

The Kusano Scientific Instrument Company (Tokyo, Japan) manufactures pre-packed C.I.G. glass columns for medium-pressure applications. These come in different sizes: 100×10 mm, 100×15 mm, 100×22 mm, 300×22 mm and 500×50 mm and are filled with either 10 μm silica gel or 20 μm RP-18 stationary phases. They can resist pressures up to 50 bar. Also available are pre-columns and a choice of two pumps: model KP-7 (up to 100 bar; 1–10 ml/min) and model KPW-20 (up to 350 bar; 0.1 – 20 ml/min).

As the system originates in Japan, most of the applications are from Japanese natural products chemists.

5.2.3.5
Miscellaneous

A number of laboratory separations have been reported in which other MPLC equipment has been used. For example, a very economical system has been designed for the separation of isomeric pyranosides, using silica gel of particle size 40 – 63 μm (Loibner and Seidl 1979). Since the pressure drop in the columns (e.g. 500 × 120 mm) was only around 5 bar when organic solvents were used, the columns could be constructed of Pyrex glass. Dry-filling was performed and pumps (FMI, Oyster Bay, NY, USA) were capable of delivering up to 200 ml/min, although the separations themselves were at flow rates of ca. 40 ml/min.

Hwu et al. (1987) describe the construction of an MPLC system for the separation of up to 20 g sample at a maximum pressure of 15 bar, with flow rates typically around 60 ml/min. The system was fitted with a guard column and silica columns were generally re-packed after 50 separation runs.

5.2.3.6
Applications

Since the last edition of this book, the applications of MPLC have become very numerous. The technique is now found in the pharmaceutical, food and chemical industries; it is widely employed for the separation of natural products as well.

In what must have been one of the first applications of MPLC, the separation of diastereomeric oxazolines was described (Meyers et al. 1979). A piston pump capable of producing a flow rate of 19 ml/min at 7 bar was connected via a four-way valve (for injection of sample) to a 250 × 15 mm pre-column and a main 1000 × 25 mm glass chromatography column. Both columns were filled with 40 – 63 μm silica gel 60 (Merck) and provided separations of, e.g. 4.4 g of **19** and **20**, with a hexane-acetone solvent pair. Monitoring of the eluted material was done by TLC and analytical HPLC.

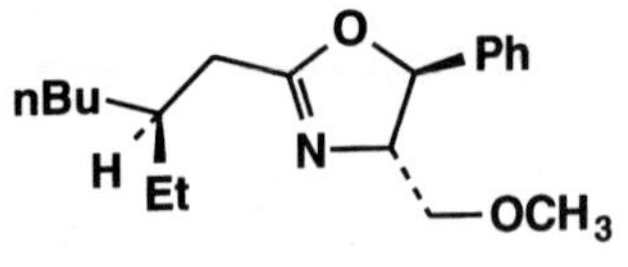

19

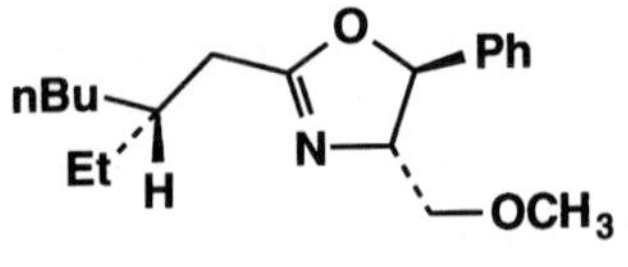

20

The Büchi 680 A system has been used for the separation of secoiridoid glycosides from *Gentiana lactea* (Gentianaceae) on Merck LiChroprep RP-8 with methanol-water as eluent (Schaufelberger and Hostettmann 1985). Conditions for the separation were found by analytical HPLC analysis (Fig. 5.16) and these were trans-

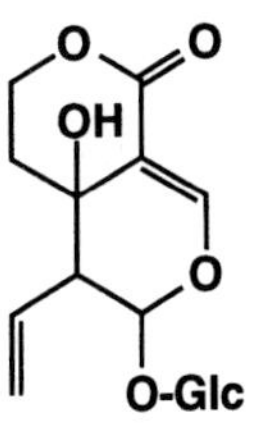

21

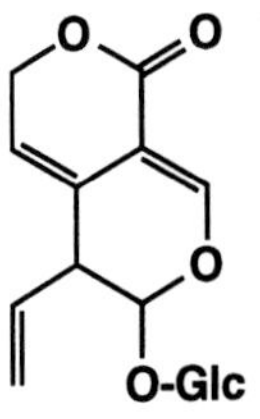

22

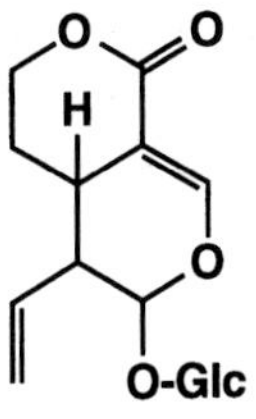

23

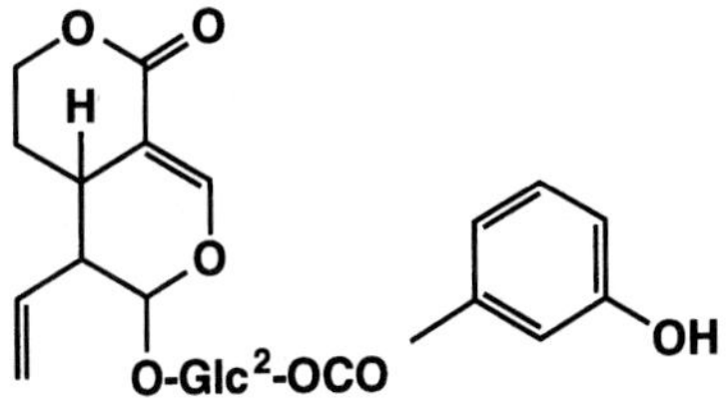

24

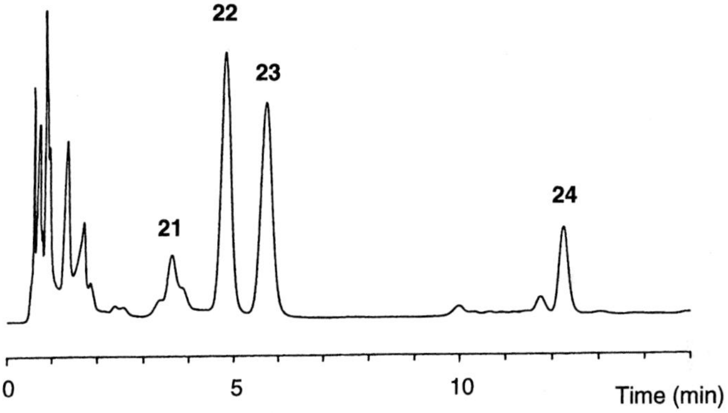

Fig. 5.16. Analytical RP-HPLC separation of secoiridoid glycosides from *Gentiana lactea*. Column: Hypersil RP-8, 5 μm (100×4.6 mm). Eluent: 20% and 30% (8 min after injection) aqueous methanol, flow-rate 1.5 ml/min. Detection: 254 nm. (Reprinted with permission from Schaufelberger and Hostettmann 1985)

posed to MPLC (Fig. 5.17). The resolution in the MPLC run approached that of the analytical system and the separation was complete within 3 h.

Another example of the scale-up of an analytical HPLC separation to MPLC is shown in Fig. 5.18. In this case, the isolation of xanthone glycosides was performed from the methanol extract of *Halenia corniculata*, a member of the Gentianaceae from Mongolia (Rodriguez et al. 1995a). The methanol extract was first chromatographed over Sephadex LH-20 and the glycoside-rich fraction (300 mg) was then purified by MPLC with an RP-18 column, after establishing the separation conditions by analytical HPLC, yielding a total of six xanthone glycosides.

A summary of certain applications of MPLC in the isolation of different classes of natural products is shown in Table 5.7. Most examples are separations

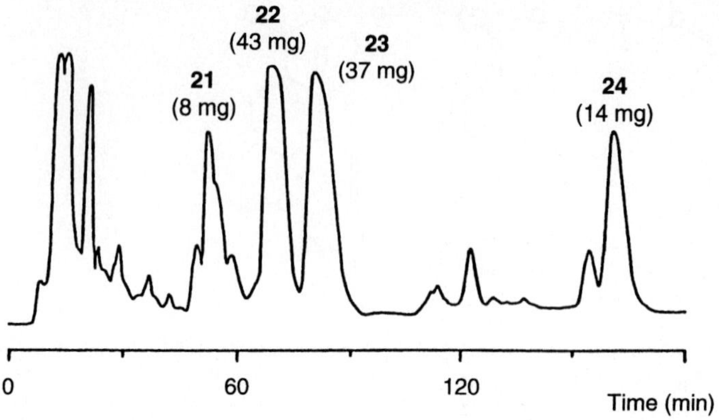

Fig. 5.17. MPLC of secoiridoid glycosides from *Gentiana lactea*. Column: LiChroprep RP-8, 15–25 µm (460×25 mm). Eluent: 20% and 30% (90 min after sample introduction) aqueous methanol, flow rate 18 ml/min. Max. pressure: 36 bar. Detection: 254 nm. Sample: 1.5 g. (Reprinted with permission from Schaufelberger and Hostettmann 1985)

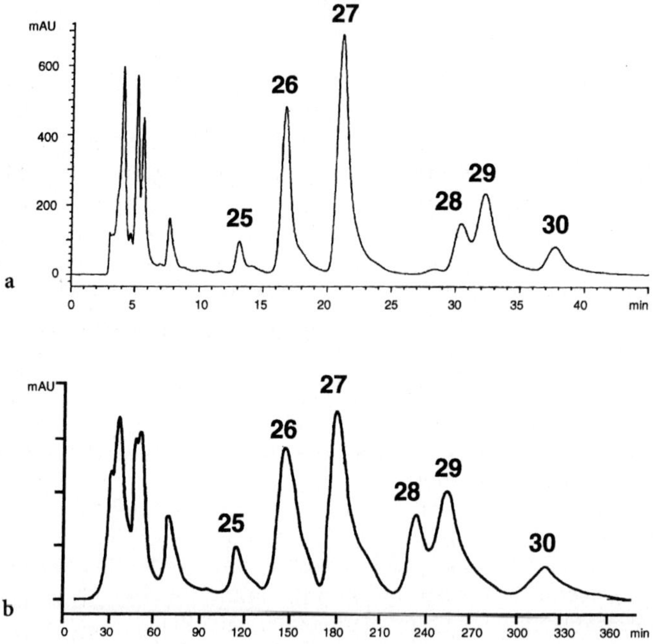

Fig. 5.18 a, b. Transposition of conditions for the MPLC separation of xanthone glycosides from *Halenia corniculata* (Gentianaceae): **a** analytical HPLC on a LiChrosorb 7µm RP-18 (250× 4 mm) column with MeOH-H_2O 40:60; flow rate 1 ml/min; **b** MPLC on LiChrosorb RP-18 (15–25 µm) with MeOH-H_2O 40:60; flow rate 3 ml/min; column dimensions 460×12 mm

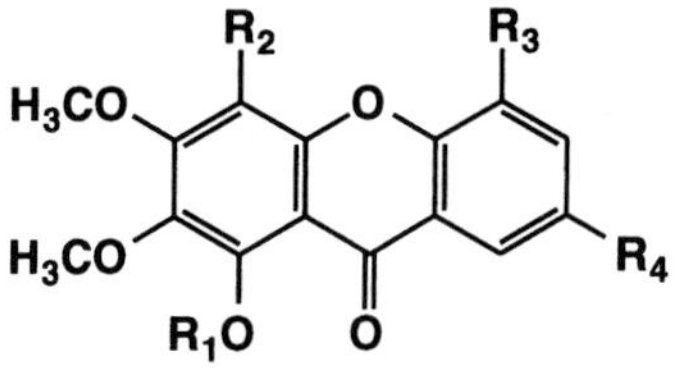

25 R_1=gentiobiosyl, R_2=R_4=H, R_3=OCH_3
26 R_1=primeverosyl, R_2=R_4=H, R_3=OCH_3
27 R_1=primeverosyl, R_2=R_3=OCH_3, R_4=H,
28 R_1=primeverosyl, R_2=R_3=H, R_4=OCH_3
29 R_1=primeverosyl, R_2=R_4=OCH_3, R_3=H
30 R_1=primeverosyl, R_2=R_3=R_4=OCH_3

on silica gel or reversed-phase columns but polyamide and cellulose supports have also been employed.

Fifteen coumarins have been isolated from the roots of *Angelica archangelica* L. ssp. *archangelica* (Apiaceae) by medium pressure techniques alone (Härmälä et al. 1992). Extraction of the dried, ground roots (870 g) was achieved by pumping chloroform through a column containing the plant material. A total of 42 g chloroform extract was obtained. For optimization of MPLC mobile phases, the "PRISMA" method was used. This system was applied to TLC plates for the normal phase mode and to analytical HPLC for the reversed phase mode. Initially, the chloroform extract (6 g) was applied to an MPLC column (460×49 mm), dry-packed by the vacuum/nitrogen overpressure method with TLC grade Kieselgel 60 F_{254} (5–25 μm; Merck) and prepared by washing with 5% methanol in hexane and then with hexane alone. After introduction of the sample, elution (1 ml/min) was with the PRISMA-optimized solvent, i.e. a step gradient of methyl ethyl ketone-chloroform-diethyl ether-n-hexane 6.9:1.0:1.4:90.7 to 48.0:6.8:10.0:35.2. The purified fractions were then submitted to further MPLC to obtain the pure coumarins, either on the same normal phase MPLC system (with an eluent of modified solvent strength) or on the PRISMA-optimized reversed phase system which involved a 460×26 mm column slurry-packed with Spherisorb ODS-2 (16–18 μm; Phase Sep) in THF. The mobile phase in this latter separation was a step gradient of THF-n-propanol-acetonitrile-water 7.5:1.3:1.7:89.5 to 45.1:7.5:10.0:37.4 at a flow rate of 1 ml/min. For the preparative separations, Büchi glass columns were used, with an LC Pump T-414 (Kontron) and 2138 Uvicord S detector (Pharmacia) set at 313 nm.

Extensive application of MPLC was also reported for the isolation of biologically active thiosulphinates (e.g. **31**) and α-sulphinyl-disulphides (e.g. **32**) from onion (*Allium cepa*, Liliaceae). The α,β-unsaturated thiosulphinates exerted antiasthmatic activity and both the thiosulphinates and α-sulphinyl-

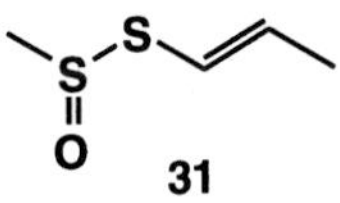

31

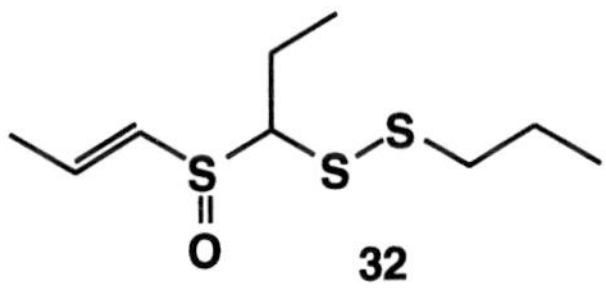

32

Table 5.7. Applications of MPLC in the isolation of natural products

Substance class	Column size (mm)	Support	Mobile phase	Reference
Indole alkaloids	713 × 18.5 (Labomatic)	RP-18 (40 μm)	MeOH-CH_3CN-THF-H_2O	Erdelmeier et al. 1992
Alkaloid glycosides	100 × 10	RP-18 (40–63 μm)	CH_3CN-H_2O 24:76	Pirillo et al. 1995
Alkaloid menthoxyacetates		SiO_2	CH_2Cl_2-MeOH-Et_3N 1990:10:1	Bringmann et al. 1991
Phenolic glycosides	460 × 26 (Büchi)	RP-8	MeOH-H_2O 50:50 → 70:30	Msonthi et al. 1990
Phenylpropanoid glycosides	460 × 50	Cellulose	Hexane-EtOAc-MeOH-H_2O 5:90:15:11	Andary et al. 1989
	460 × 16	Diol (15–25 μm)	Hexane-MeOH-iPrOH 18:5:2	Comte et al. 1996
	713 × 18.5 (Labomatic)	RP-18 (40 μm)	MeOH-H_2O 8:2	Khan et al. 1992
Flavonoids	C.I.G.	RP-18 (20 μm)	MeOH-H_2O 3:1	Shirataki et al. 1991
Flavonol glycosides	460 × 26 (Büchi)	RP-18 (15–25 μm)	MeOH-H_2O 35:65	Ducrey et al. 1995
Dihydroflavonoid glycosides	460 × 26	Polyamide SC-6	Toluene-MeOH	Allais et al. 1995
	460 × 15	RP-18 (20–40 μm)	MeOH-H_2O	
Xanthones	460 × 26 (Büchi)	RP-8 (15–25 μm)	MeOH-H_2O 9:41 → 1:1	Wolfender et al. 1991
	460 × 12 (Büchi)	RP-18 (15–25 μm)	MeOH-H_2O 40:60	Rodriguez et al. 1995a
		Polyamide	$CHCl_3$-MeOH	Tan et al. 1992

Naphthoquinones	460×26 (Büchi)	RP-18 (15–25 μm)	$MeOH-H_2O$ 60:40	Rodriguez et al. 1995b
	460×26 460×16 (Büchi)	RP-18 (15–25 μm)	$MeOH-H_2O$ 13:7, 3:2, 11:9, 1:1, 9:11, 14:11	Gafner et al. 1996b
Pterocarpinoids	920×36 (Büchi)	SiO_2	$CHCl_3$-MeOH 25:2 Petrol ether-EtOAc 3:7	Gunzinger et al. 1988
Furanocoumarins	735×26 (Labomatic)	SiO_2 (5–25 μm)	Hexane-EtOAc-$CHCl_3$-Et_2O 40:0.6:58.5:0.9 (+0.04% H_2O)	Zogg et al. 1989a
Chromones	460×26 (Büchi)	Diol	Hexane-EtOAc 4:1	Ma et al. 1996
Chalcones	800×36	SiO_2	Hexane-tBuOMe-CH_2Cl_2 -EtOH 99:0.4:0.3:0.3	Orjala et al. 1993
Lignan glycosides	100×22 (C.I.G.)	RP-18 (20 μm)	$MeOH-H_2O$ 1:1	Yoshinari et al. 1989
Monoterpenes	460×36 (Büchi)	RP-18 (15–25 μm)	$MeOH-H_2O$ 65:35	Fuzzati et al. 1996
Secoiridoid glycosides	460×26 (Büchi)	RP-18 (25–40 μm)	$MeOH-H_2O$ 1:3	Wolfender et al. 1993
		RP-8 (15–25 μm)	$MeOH-H_2O$ 18:82 → 50:50	
Sesquiterpenes	460×49 460×26 (Büchi)	SiO_2 (63–200 μm)	Petrol ether-EtOAc 65:35	Rücker et al. 1989
		SiO_2 (40–63 μm)	Hexane-Et_2O 1:3 → Et_2O $CHCl_3$-MeOH 20:1 → 5:1	Marco et al. 1992

Table 5.7 (continued)

Substance class	Column size (mm)	Support	Mobile phase	Reference
Diterpenes	460×26 (Büchi)	SiO_2	$CHCl_3$, $CHCl_3$-MeOH 19:1	Zani et al.1993
		SiO_2 (6–35 μm)	CH_2Cl_2-MeOH 99:1 → 97:3	Moulis et al. 1992
	100×22 (C.I.G.)	SiO_2 (10 μm)	Hexane-EtOAc-CH_3CN 7:2:1 C_6H_6-EtOAc 23:2	Itokawa et al. 1989
Quassinoids	360×20	SiO_2	CH_2Cl_2-MeOH 50:1, 20:1	Koike and Ohmoto 1993
Triterpenes	240×20	RP-18 (14–40 μm)	MeOH-H_2O 6:4 → 8:2	Marner et al. 1990
	460×50	SiO_2 (35–70 μm)	CH_2Cl_2-MeOH-HOAc 110:8:3	Taipale et al. 1993
	200×16	SiO_2	CH_2Cl_2-MeOH-HOAc 110:16:3	Taipale et al. 1993
	230×36	RP-8	Acetone-H_2O 1:1 → Acetone	Taipale et al. 1993
Saponins	460×26 (Büchi)	RP-8 (15–25 μm)	MeOH-H_2O 7:3	Maillard et al. 1992
	460×36 (Büchi)	SiO_2 (40–63 μm)	$CHCl_3$-MeOH 9:1 → 8:2	Hamburger et al. 1992
Furanone glucosides	460×70 (Büchi)	RP-18 (40–63 μm)	MeOH-H_2O gradient	Mayerl et al. 1989
Cannabinoids	460×36 (Büchi)	RP-18 (15–25 μm)	MeOH-5% HOAc 9:1	Lehmann and Brenneisen 1992
	160×16 (Büchi)	RP-18 (15–25 μm)	MeOH-5% HOAc 8:2	Lehmann and Brenneisen 1992
Polyacetylenes	500×50 and 100×22 (C.I.G.)	SiO_2 (50 μm)	Hexane-acetone 6:1, $CHCl_3$, C_6H_6-EtOAc 10:1	Hirakura et al. 1992

disulphides were inhibitors of cyclooxygenase and 5-lipoxygenase (Bayer et al. 1989a, b).

For the isolation of sulphur-containing constituents, peeled onion bulbs were squeezed and the juice was extracted with chloroform. Triterpenes were removed from this extract by flash chromatography on RP-8. After centrifugal TLC, the purified active fraction (antiasthmatic) was subjected to MPLC on a 460×49 mm silica gel column eluted with toluene-ethyl acetate 10:2. The thiosulphinate **31** was obtained in a pure state after MPLC on a 460×26 mm column eluted with dichloromethane-acetone 100:1 and the α-sulphinyl-disulphide **32** purified by MPLC on a 460×26 mm column eluted with n-hexane-ethyl acetate 10:3 (Bayer et al. 1989a).

A combination of flash chromatography and MPLC, in which flash is used as a clean-up operation, has been reported for the purification of organic reaction mixtures (Dobler 1983) – a procedure which is also employed for the separation of pure constituents. Following flash chromatography with 40–63 µm Kieselgel 60 (+5% H_2O), MPLC was carried out on a 480×37 mm Labomatic pressure glass column filled with 15–40 µm silica gel at a flow rate of 20 ml/min (CH_2Cl_2) and a pressure of 4.5–5 bar. By this means, a total of 4.6 g biphenyl derivative was obtained after four chromatographic runs, each with ca. 2 g crude product.

5.2.4 High-Pressure LC (HPLC)

The term "high-pressure (or high performance) liquid chromatography" is conventionally applied to columns having plate numbers above 2000, the typical range being 2,000–20,000 (Snyder 1992).

In this section are decribed systems involving microparticulate stationary phases of a narrow size range packed into columns which require high pressure to provide a flow of the mobile phase. Whereas the use of larger particles (50–100 µm), as found in MPLC, leads to easier packing, higher permeability (requiring lower pumping pressures), larger columns and more economic equipment, there is a correspondingly lower separation efficiency. With the smaller (5–30 µm) particles used in preparative HPLC, the complexity and costs of the systems are greater but there is a large gain in separation efficiency. The small k′ values mean small peak volumes and higher concentrations of the compounds in the eluate (Beck and Halasz 1978).

Many isolation problems require the purification of less than 1% of a desired component from a large amount of material. The purification can be so difficult that 10 µm (or smaller) particle size high performance packings are required for final purification and in order to obtain the product, the following sequence may have to be followed:

Preparative separation → semi-preparative separation → analytical separation → PRODUCT

In order to increase throughput, preparative HPLC separations are often run under *overload* conditions.

The applications of HPLC are vast and enter a fair proportion of publications involving the isolation of natural products. Some recent HPLC separations are listed in Table 5.8, providing a summary of the possibilities. As a general rule, HPLC is the final purification step in these examples.

In addition, a tabulation of analytical HPLC separations of natural products (Kingston 1979) and a description of the separation of secondary plant constituents (Hostettmann and Hostettmann 1985) have been published. Other preparative applications of HPLC have been given by Verzele and Dewaele (1985).

"Semi-preparative" is a term coined to include columns of i.d. 8–10 mm, often packed with 10 μm particles, and useful for the separation of 1 mg to 100 mg mixtures (Snyder and Kirkland 1979). Larger amounts can be separated by repeating the injections but since repetitive manual injection and collection is time consuming, automated semi-preparative units have been designed.

Isocratic conditions are most often employed in preparative HPLC because operating problems are thus reduced. However, a fair number of gradient elutions have been reported for those cases in which separations are troublesome e.g. Slimestad et al. 1996, Govindachari et al. 1996.

As described in Sect. 5.1.1, optimization can be performed on analytical HPLC columns before transposition to a semi-preparative scale. This procedure was applied to the separation of ester iridoids (valepotriates) from *Valeriana capense* (Valerianaceae). The interest in these constituents lay in their potent antifungal activity – unusual since *Valeriana* species are more commonly associated with their sedative properties. For the isolation of the active principles, the whole plants were extracted with dichloromethane. This extract was fractionated by silica gel column chromatography and then semi-preparative HPLC, to give four valepotriates (**33–36**) (Fig. 5.19). Valtrate (**35**) was the most active antifungal compound (Fuzzati et al. 1996).

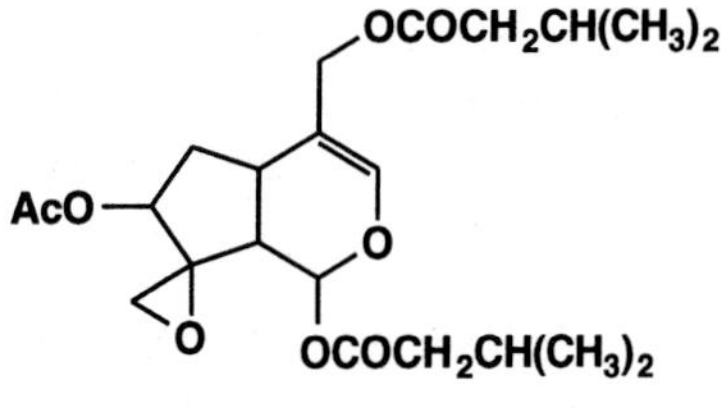

33

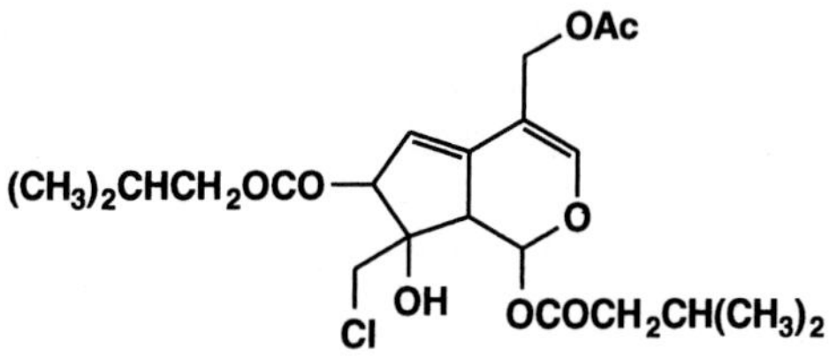

34

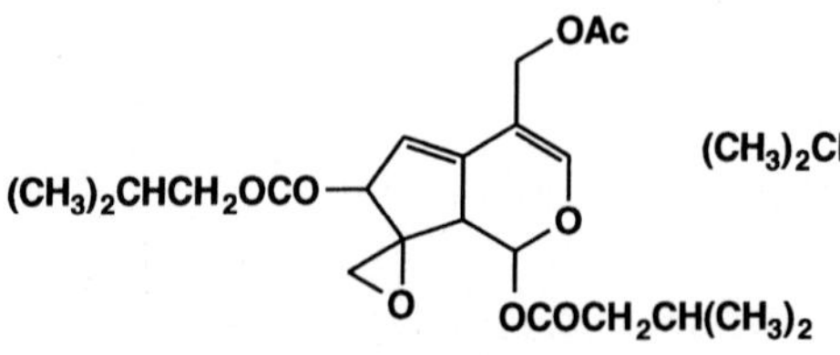

35

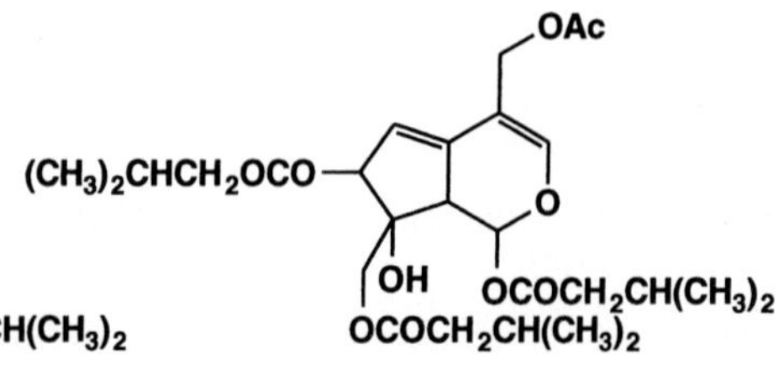

36

Table 5.8. Preparative separations of natural products by HPLC

Substances separated	Column	Dimensions (mm)	Mobile phase	Reference
Alkaloids				
Alkaloids from *Ancistrocladus korupensis* (Ancistrocladaceae)	Dynamax NH_2	250×41.4	CH_2Cl_2-0.1% $(NH_4)_2CO_3$ in MeOH 22:3	Hallock et al. 1994
Guanidine alkaloid from a sponge	CEDI RP-18 (12–40 µm)	300×25	MeOH-H_2O 8:2	Bouaicha et al. 1994
	LiChrosorb Diol (7 µm)	250×10	CH_2Cl_2-MeOH 98:2	
	Waters Radial Pak NH_2	100×8	CH_2Cl_2-MeOH 9:1	
Lignanamide alkaloids from *Jacquemontia paniculata* (Convolvulaceae)	LiChrosorb RP-18 (7 µm)		MeOH-0.5% H_3PO_4 2:3	Henrici et al. 1994
Acetylated alkaloids from *Narcissus pseudonarcissus* (Amaryllidaceae)	LiChrosorb Silica	250×25	nC_6H_{14}-iPrOH-CH_2Cl_2-70:30:5:0.02	Kreh et al. 1995
Homoerythrina alkaloids from *Lagarostrobos colensoi* (Podocarpaceae)	RP-13	250×25	MeOH-H_2O-Et_2NH 35:65:01	Bloor et al. 1996
Aromatics				
Dihydrophenanthrenes from *Juncus effusus* (Juncaceae)	LiChrosorb RP-18		MeOH-H_2O 9:1, 4:1, 3:1	Della Greca et al. 1996
	LiChrosorb RP-8		MeOH-H_2O 1:1, 2:3, 3:7	
	LiChrosorb NH_2		CH_3CN-H_2O 4:1	
Benzofuran from *Eupatorium rugosum* (Asteraceae)	Supelco LC_{18}-DB	250×10	CH_3CN-H_2O 4:6	Beier et al. 1993
	Supelco LC-Si	250×10	$CHCl_3$-EtOAc 99.9:0.1	
General Polyphenols				
Phenolics from *Picea abies* (Pinaceae)	MN Nucleosil 100–7 C_{18}	250×21	MeOH-H_2O 18:82 → 50:50	Slimestad et al. 1996
Phenolics from *Betula* species (Betulaceae)	Diasorb-130-C16T (6 µm)	250×15	EtOH-2.5% HOAc gradient	Ossipov et al. 1996
	Bondapak C_{18} (15 µm)	300×19	CH_3CN-5% HCOOH gradient	

Table 5.8 (continued)

Substances separated	Column	Dimensions (mm)	Mobile phase	Reference
General Polyphenols (continued)				
Alkylphenols from *Plectranthus albidus* (Labiatae)	Spherisorb-ODS (5 µm)	250×20.5	MeOH-5% HCOOH 4:1	Bürgi and Rüedi 1993
Alkenylphenols from cashew nutshell oil	Spherisorb-ODS	250×22.4	A) CH_3CN-H_2O-HOAc 66:33:2 B) THF Gradient 0 → 100% B	Bruce et al. 1990
Phloroglucinols from *Baeckea frutescens* (Myrtaceae)	Shodex SIL-SE		nC_6H_{14}-EtOAc 15:1	Fujimoto et al. 1996
Phloroglucinols from *Hypericum erectum* (Guttiferae)	LiChrosorb RP-18 (7 µm)	250×10	MeOH-H_2O 9:1	Tada et al. 1991
	Asahipak GS310 (gel permeation)	500×7.5	MeOH-H_2O	
Phytoalexins from *Taverniera abyssinica* (Leguminosae)	LiChroprep Diol (7 µm)	250×25	Cyclohexane-TBME 7:3	Stadler et al. 1994
Quinones				
Quinones from *Rumex japonicus* (Polygonaceae)	Develosil 60–10	250×20	$CHCl_3$	Nishina et al. 1993
	Develosil ODS-10	250×20	CH_3CN-H_2O 80:20	
	Develosil ODS-5	250×5	CH_3CN-H_2O 80:20	
Naphthoquinones from *Conospermum incurvum* (Proteaceae)	Rainin Ph		CH_3CN-H_2O 17:3 (+0.1% HOAc)	Dai et al. 1994
	Rainin Si		CH_2Cl_2-MeOH 19:1	
Alkenylbenzoquinones from *Ardisia japonica* (Myrsinaceae)	LiChrosorb RP-2	300×8	MeOH-H_2O-HOAc 75:25:0.6	Fukuyama et al. 1993

Phenylpropanes				
Phenylpropane derivatives from *Cosmos caudatus* (Asteraceae)	LiChrosorb RP-18 (7 µm)	250×16	MeOH-H_2O 13:7 → 4:1	Fuzzati et al. 1995
Phenylethanoid glycosides from *Veronica persica* (Scrophulariaceae)	Develosil Lop-ODS	500×50 (2x)	CH_3CN-H_2O 9:41	Aoshima et al. 1994
Lignans				
Lignan glycosides from *Sesamum indicum* (Pedaliaceae)	Develosil ODS-10 Develosil ODS-5 Develosil SI-60–5	250×20 250×10 250×8	MeOH-H_2O 3:2 MeOH-H_2O 2:3 nC_6H_{14}-EtOAc 1:1	Katsuzaki et al. 1994
Neolignan glycosides from *Codonopsis tangshen* (Campanulaceae)	TSKgel ODS-120T	300×21.5	CH_3CN-H_2O 12:88, 10:90 MeOH-H_2O 3:7, 21:29	Yuda et al. 1990
Chalcones				
Chalcones from *Myrica serrata* (Myricaceae)	LiChrosorb Diol (7 µm) MN Nucleosil 100–7 C_{18}	250×16 250×21	MeOH-H_2O 55:45 MeOH-H_2O 76:24	Gafner et al. 1996a
Chalcones from *Fissistigma lanuginosum* (Annonaceae)	Delta-Pak C_{18} (15 µm)	300×47	MeOH-H_2O 6:4 → 8:2	Alias et al. 1995
Dihydrochalcones from *Piper aduncum* (Piperaceae)	Spherisorb ODS II (5 µm)	250×16	MeOH-H_2O 7:3, 75:25, 65:35	Orjala et al. 1994
Chromones				
Methylchromone from *Aloe ferox* (Liliaceae)	LiChrosorb RP-8	250×10	CH_3CN-H_2O 25:75 → 90:10 CH_3CN-H_2O 35:65 → 100:0	Speranza et al. 1993
Furanochromone glucosides from *Ammi visnaga* (Apiaceae)	Zorbax ODS	250×9.4	EtOH-CH_3CN-H_2O 1:1:8	Tjarks et al. 1989
Chromenes from *Hypericum revolutum* (Guttiferae)	Bondapak C_{18}	300×7.8	MeOH-H_2O 63:37, 82:18	Décosterd et al. 1987a

Table 5.8 (continued)

Substances separated	Column	Dimensions (mm)	Mobile phase	Reference
Coumarins				
Dihydropyranocoumarin from *Peucedanum praeruptorum* (Apiaceae)	Shim-Pack PREP-SIL	250×10	Cyclohexane-EtOAc 4:1	Kong et al. 1996
Coumarin glucoside from *Daphne arisanensis* (Thymelaeaceae)	Develosil C_8 (5 µm)	250×10	MeOH-H_2O 20:80	Niwa et al. 1991
Flavonoids				
Flavone glycosides from *Lysionotus pauciflorus* (Gesneriaceae)	LiChrosorb RP-18	250×10	CH_3CN-H_2O 1:4	Liu et al. 1996
Flavonoid glycosides from *Rosa multiflora* (Rosaceae)	TSKgel ODS-80TM (5 µm)	250×25.4	CH_3CN-H_2O-H_3PO_4 300:700:0.5	Seto et al. 1992
Flavonoid glycosides from *Hebe stricta* (Scrophulariaceae)	Dynamax C-18	250×21.4	MeOH-H_2O-HOAc	Kellam et al. 1993
Flavonoid glucuronides from *Malva sylvestris* (Malvaceae)	Spherisorb ODS II (5 µm)	250×16	CH_3CN-H_2O-THF-HOAc 205:718:62:15	Billeter et al. 1991
Xanthones				
Xanthone from *Senecio mikanioides* (Asteraceae)	LiChrospher 100 Diol	250×10	$CHCl_3$-MeOH-CH_3CN 81:14:5	Catalano et al. 1996
Xanthones from *Hypericum roeperanum* (Guttiferae)	LiChrosorb RP-18 (7 µm)	250×16	MeOH-H_2O 88:12, 73:27	Rath et al. 1996
Xanthones from *Chironia krebsii* (Gentianaceae)	LiChrosorb RP-18 (7 µm)	250×16	MeOH-H_2O 7:3, 7:13	Wolfender et al. 1991

Anthocyanins				
Anthocyanin glycosides	Spherisorb ODS-2 (10 µm)	250×10	MeOH-5% HCOOH	Fiorini 1995
Proanthocyanins and flavans from *Prunus prostrata* (Rosaceae)	Eurospher 80 RP-18 (7 µm)	250×16	CH_3CN-H_2O 1:4, 3:17 (+0.1% TFA)	Bilia et al. 1996
Tannins				
Ellagitannins from *Quercus robur* (Fagaceae)	LiChrospher RP-18 (7 µm)	250×25	MeOH-H_2O-H_3PO_4	Scalbert et al. 1990
Ellagitannins from *Epilobium capense* (Onagraceae)	LiChrospher RP-18 (5 µm)	250×19	CH_3CN-H_2O 16:84 (+0.1% TFA)	Ducrey et al. 1997
Monoterpenes				
Ionol glucosides from *Rubus idaeus* (Rosaceae)	LiChrosorb Diol (7 µm)	250×16	nC_6H_{14}-nBuOH-MeOH-H_2O 65:25:9:1	Pabst et al. 1992
	MN Nucleosil 120-7 C_{18}	250×16	CH_3CN-H_2O 3:17	
Picrocrocin and safranal from *Crocus sativus* (Liliaceae)	Supelcosil PL C-18 (12 µm)	250×21.2	MeOH-H_2O 45:55	Castellar et al. 1993
Iridoids				
Valepotriates from *Valeriana capense* (Valerianaceae)	MN Nucleosil 100–7 C_{18}	250×21	CH_3CN-H_2O 55:45	Fuzzati et al. 1996
Secoiridoids and aromatic acids from *Gentiana algida* (Gentianaceae)	LiChrosorb RP-18 (7 µm)	250×16	CH_3CN-H_2O 47:53 MeOH-H_2O 13:12, 8:17	Tan et al. 1996
Iridoid glycosides from *Phlomis younghusbandii* (Labiatae)	LiChrosorb Si60	250×21.5	$CHCl_3$-MeOH 10:3	Kasai et al. 1994
	TSK-gel ODS 120T	300×21.5	MeOH-H_2O 28:72	
Sesquiterpenes				
Sesquiterpene lactones from *Mikania mendocina* (Asteraceae)	Phenomenex Maxsil 10C_8	500×10	MeOH-H_2O 4:3, 1:1	Bardon et al. 1996
	Phenomenex Ultremex C18 (5 µm)	250×10	MeOH-H_2O 9:1, 1:1	

Table 5.8 (continued)

Substances separated	Column	Dimensions (mm)	Mobile phase	Reference
Sesquiterpenes (continued)				
Sesquiterpenes from *Petasites hybridus* (Asteraceae)	Alltech RSil Silica D	250×10	nC_6H_{14}-TBME 9:1 Cyclohexane-TBME 92:8	Debrunner and Neuenschwander 1994
Sesquiterpenes from *Tussilago farfara* (Asteraceae)	TSK-gel ODS-120T	300×7.8	MeOH-H_2O 2:1	Kikuchi and Suzuki 1992
Diterpenes				
Taxanes from *Taxus × media* (Taxaceae)	Bondapak C_{18} (15 μm)	300×47	CH_3CN-H_2O 1:3	De Bellis et al. 1995
Diterpenes from *Rabdosia trichocarpa* (Labiatae)	Senshu Pak silica-5251-S		CH_2Cl_2-EtOAc-MeOH 470:30:3	Osawa et al. 1994
	Senshu Pak ODS-5251-SH		MeOH-H_2O 11:9 → 1:0	
Diterpenes from *Psiadia arabica* (Asteraceae)	Resolve C_{18}	300×7.8	CH_3CN-H_2O 37:63	El-Domiaty et al. 1993
Diterpene glycosides from *Chrozophora obliqua* (Euphorbiaceae)	ODS-5	250×20	MeOH-H_2O 4:6 → 5:5	Mohamed et al. 1994
Ingol esters from *Euphorbia antiquorum* (Euphorbiaceae)	Chemosorb 5 Si	500×10	nC_6H_{14}-$ClCH_2CH_2Cl$-EtOH 5:2:0.2	Gewali et al. 1989
Diterpenes from *Euphorbia milii* (Euphorbiaceae)	MN Nucleosil (7 μm)	250×8	Cyclohexane-$CHCl_3$ 4:1	Zani et al. 1993
Triterpenes				
Triterpenoids from *Maprounea africana* (Euphorbiaceae)	Dynamax Si Dynamax C-18	250×21.4 250×21.4	Isooctane-EtOH 9:1 MeOH-H_2O 17:3	Chaudhuri et al. 1995
Triterpene caffeates from *Betula pubescens* (Betulaceae)	Nova-Pak C_{18} Radial-Pak cartridge	100×8	CH_3CN-1% HOAc 99:1	Pan et al. 1994
Triterpenes from *Trichocereus pachanoi* (Cactaceae)	MN Nucleosil 60–5	250×10	$CHCl_3$-MeOH 50:1	Takizawa et al. 1993

Azadirachtins from *Azadirachta indica* (Meliaceae)	Shimpack C_{18}	250×50	MeOH-H_2O 60:40 → 70:30	Govindachari et al. 1996
	Shimpack C_{18}	250×20	CH_3CN-H_2O 28:72	
Tetranortriterpenoid from *Azadirachta indica* (Meliaceae)	Phenomenex Silica (5 µm)	250×20	nC_6H_{14}-iPrOH 19:1	Yamasaki et al. 1988
	Phenomenex Ph (5 µm)	250×22.5	CH_3CN-H_2O 2:3	
Limonoids from *Melia azedarach* (Meliaceae)	Bondapak C_{18}		MeOH-H_2O 60:40 → 75:25	Nakatani et al. 1994
Quassinoids from *Quassia indica* (Simaroubaceae)	Capcell Pak ODS SG-120 (Shiseido)	250×10	MeOH-H_2O 2:3	Koike and Ohmoto 1993
Steroids				
Ergostanoids from *Petunia inflata* (Solanaceae)	Dynamax Si	250×21.4	nC_6H_{14}-iPrOH 4:1 nC_6H_{14}-iPrOH-CH_2Cl_2 7:1:2	Elliger et al. 1993
	Dynamax C-18	250×41.4	CH_3CN-H_2O 7:3	
	Dynamax CN	250×10	nC_6H_{14}-iPrOH 4:1	
	Alltech RSil C-18	250×10	CH_3CN-H_2O 7:3	
Brassinosteroids from *Ornithopus sativus* (Leguminosae)	Develosil ODS (5 µm)	250×8	CH_3CN-H_2O 45:55 → 80:20	Schmidt et al. 1993
Sterols from *Dioscorea batatas* (Dioscoreaceae)	Ultrasphere ODS (5 µm)	250×10	MeOH-H_2O 49:1	Akihisa et al. 1991
Saponins				
Saponins from *Phytolacca dodecandra* (Phytolaccaceae)	LiChrosorb RP-8	250×16	CH_3CN-H_2O 38:62	Décosterd et al. 1987a
Saponins from *Chenopodium quinoa* (Chenopodiaceae)	TSK-gel ODS-120T	300×21.5	MeOH-H_2O 68:32	Mizui et al. 1990
Saponins from *Mazus miquelii* (Scrophulariaceae)	Develosil ODS-10/20	250×20	MeOH-H_2O 55:45	Yaguchi et al. 1995
	Develosil PhA-7	250×20	CH_3CN-H_2O 27.5:72.5	

Table 5.8 (continued)

Substances separated	Column	Dimensions (mm)	Mobile phase	Reference
Polyacetylenes				
Polyacetylenes from *Panax ginseng* (Araliaceae)	YMC S-343 RP	300×20	CH_3CN-THF 5:1	Hirakura et al. 1994
Polyacetylene from *Dahlia merckii* (Asteraceae)	LiChrosorb CN (10 μm)	250×16	Heptane	Neuschild et al. 1992
Acetogenins				
Acetogenins from *Annona muricata* (Annonaceae)	μBondapak C_{18} (10 μm)	100×25	MeOH-H_2O 85:15	Gromek et al. 1994
	μPorasil (10 μm)	100×25	CH_2Cl_2-MeOH 97:3	
Acetogenins from *Annona muricata* (Annonaceae)	Dynamax C-18 (60 Å, 8 μm)		nC_6H_{14}-[MeOH-THF (9:1)] 93:7	Rieser et al. 1996
Miscellaneous				
Cyclic peptide from *Psychotria longipes* (Rubiaceae)	Delta-Pak C_{18} (15 μm) (300 Å)	300×47	CH_3CN-H_2O (+TFA) gradient	Witherup et al. 1994
	Spherisorb Phenyl (5 μm) (80 Å)	250×22	CH_3CN-H_2O (+TFA) gradient	
5-Pentyl-2-furaldehyde from an ascomycete	LiChroprep Diol (7 μm)	250×25	Cyclohexane-TBME 4:1	Mayer et al. 1996
Enediyne antibiotic from *Actinomadura verrucosospora*	Dynamax C-18	250×41.4	CH_3CN-MeOH-50 mmol/l NH_4OAc pH 4.4 32.5:32.5:35	Beutler et al. 1994
Castanospermine from *Castanospermum australe* (Leguminosae)	Phenomenex NH_2 (5 μm)	250×22.5	CH_3CN-H_2O gradient	Chen et al. 1990

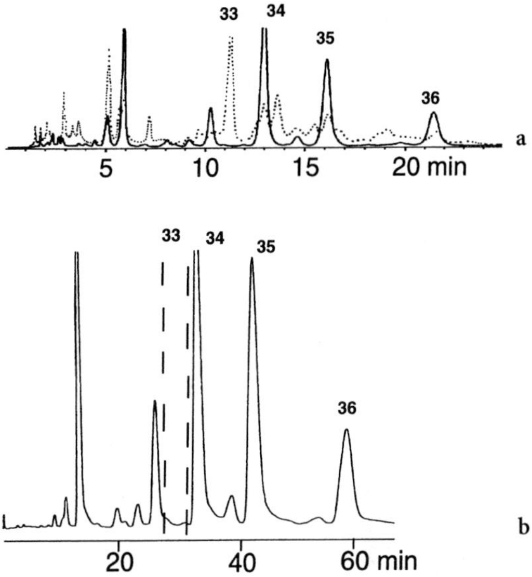

Fig. 5.19 a, b. Transposition of analytical HPLC conditions to semi-preparative HPLC separation of valepotriates from *Valeriana capense* (Valerianaceae). **a** Analytical HPLC. Column: Nucleosil RP-18, 7 μm (250×4 mm). Eluent: CH_3CN-H_2O 55:45, flow rate 1 ml/min. Detection: 254 nm (*solid line*), 210 nm (*dotted line*). **b** Semi-preparative HPLC. Column: Nucleosil RP-18, 7 μm (250×21 mm). Eluent: CH_3CN-H_2O 55:45, flow rate 16 ml/min. Detection: 254 nm (valepotriate **33**, which is not detected at 254 nm, was collected in the indicated region)

Waters has introduced a new line of columns (Symmetry C_8 and C_{18}) which gives identical selectivity for all three granulometries available – 3.5, 5 and 7 μm. This allows direct scale-up from e.g. a 3.5 μm C_{18} 3.9×150 mm column to a SymmetryPrep 7 μm C_{18} 19×150 mm column, as long as the flow rate is adjusted accordingly. The 19 mm i.d. columns are also available in 300 mm lengths.

Highly *polar* and/or water-soluble compounds are most conveniently separated by reversed-phase columns. For marine natural products, some examples are given by Shimizu (1985). In the saponin field, a large proportion of separations involve a final HPLC purification step (Hostettmann and Marston 1995). This is because the saponins have complex, closely-related structures and are very difficult to separate by any other method. Approximately 95% of all reported HPLC applications are on octadecylsilyl phases (Table 5.7). Silica gel columns are not normally considered suitable for polar compounds but Kaizuka and Takahashi (1983) have made use of these columns, with an aqueous solvent, for the reproducible isolation of ginseng saponins. Columns were regenerated by washing with methanol and prolonged use of the columns was claimed to be possible.

Ginsenosides have also been separated on 20 μm octadecylsilyl porous glass. This support gave very good results, due to the optimal pore size of 550 Å. Furthermore, proportionally less organic solvent than with normal silica RP-18 was required in the acetonitrile-water eluent (Kanazawa et al. 1990).

A 31-residue *cyclic peptide* has been isolated from *Psychotria longipes* (Rubiaceae) by an initial HPLC step involving a DeltaPak C_{18}, 300 Å, 15 μm (300×50 mm) column. The eluent was a linear gradient of H_2O (+0.3% TFA) and CH_3CN (+0.3% TFA) at 80 ml/min (0 to 55% CH_3CN over 40 min). Subsequent purification was by means of a Spherisorb phenyl column (80 Å, 5 μm,

250×22 mm) using the same linear gradient as above, with sequential injection of 4×25 mg (Witherup et al. 1994).

Shkarenda and Kuznetsov (1992) have reviewed analytical and preparative liquid chromatography of *coumarins.*

The demand for the cancer chemotherapeutic agent *paclitaxel* (Taxol) (**37**), obtained from the bark of the Pacific yew (*Taxus brevifolia*, Taxaceae) has meant that there has been a tremendous effort to find not only new sources and different analogues but also new ways of producing the drug. This is reflected in the number of separations which have been reported in the literature. There are even special HPLC columns marketed for this purpose, since regular bonded-phase supports give insufficient selectivity and taxanes suffer from low solubility in aqueous solutions. Wu et al. (1995) showed that semi-preparative HPLC (of 12.5 mg crude plant extract injected in 1 ml of mobile phase) with a Zorbx SW-Taxane column (60 Å, 10 μm; methanol-water 60:40) gave better results than on a C-8 column. With heptane-ethanol 75:25 as mobile phase on the SW-Taxane column, better selectivities and recoveries than on the corresponding silica gel column were obtained.

One of the main problems when purifying paclitaxel is the separation of the closely related analogue cephalomannine (**38**). This has been achieved, for example, on a semi-preparative HPLC column (Rainin Dynamax Cyano, 250×21 mm) via a hexane-isopropanol gradient (7:3 → 11:9 → 7:13). Detection was at 270 nm (Cardellina 1991).

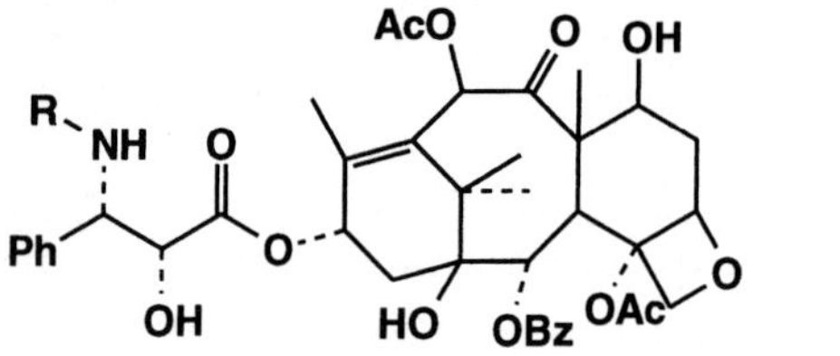

37 R = PhCO (Paclitaxel)

38 R =

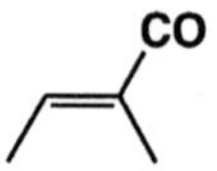

(Cephalomannine)

10-Deacetylbaccatin III (**39**), found in the foliage of several yew species, can be transformed into taxol and another antitumor drug, Taxotere (**40**). A report on the isolation of analogues of 10-deacetylbaccatin III from the European yew (*T. baccata*) has appeared, in which preparative HPLC on a Waters Prep-pak Bondapak C_{18} column (300×47 mm; 15 μm) was used for the separation with acetonitrile-water 25:75 as eluent, at a flow rate of 80 ml/min (Gabetta et al. 1995).

HPLC is well suited to the separation of *essential oils* and sometimes shows significant advantages over currently used open column and GC methods: exposure to air minimal, high temperature degradation products avoided, separation of non-volatile materials, high sample recovery rates. However, essential oils are often composed of very complex mixtures and therefore HPLC is useful either for

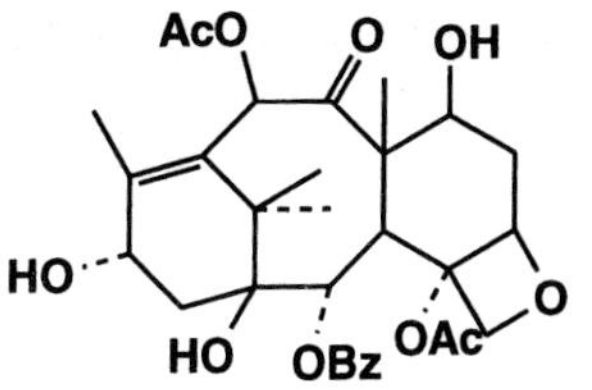

39 (10-Deacetylbaccatin III)

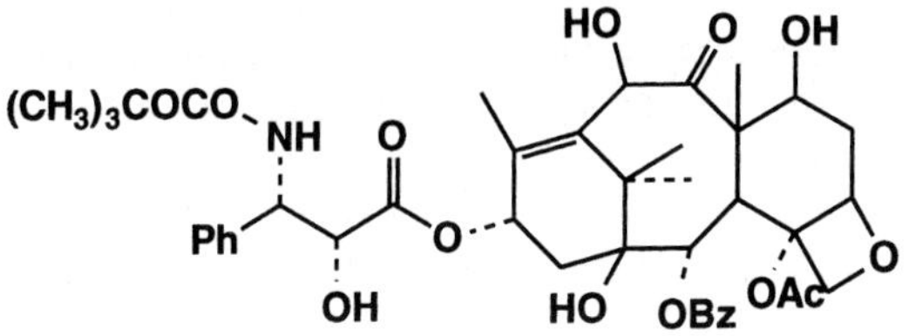

40 (Taxotere)

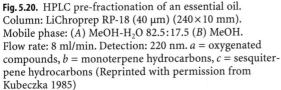

the separation of complex mixtures into classes or the separation of such a class into its components. A preliminary separation of an essential oil is shown in Fig. 5.20. Up to 0.5 ml of essential oil could be separated in one single injection.

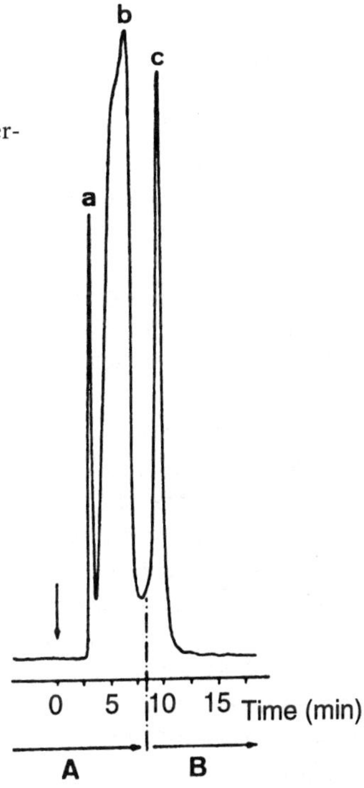

Fig. 5.20. HPLC pre-fractionation of an essential oil. Column: LiChroprep RP-18 (40 μm) (240 × 10 mm). Mobile phase: (*A*) MeOH-H_2O 82.5:17.5 (*B*) MeOH. Flow rate: 8 ml/min. Detection: 220 nm. *a* = oxygenated compounds, *b* = monoterpene hydrocarbons, *c* = sesquiterpene hydrocarbons (Reprinted with permission from Kubeczka 1985)

LiChrosorb Si 60 (7 μm), partially deactivated with 4.8% water, provides good HPLC separations of the individual components of fractions resulting from preliminary open column chromatography of essential oils, with n-pentane as one suitable mobile phase (Kubeczka 1985). Pre-fractionation of lime oils by HPLC before GC-MS analysis has been carried out on three gel columns in tandem (Partisil-10, Partisil-5, Radial Pak 51). The solvent employed was 8% ethyl acetate in dichloromethane-hexane 1:1 (Chamblee et al. 1985).

Preparative HPLC has been employed in the analysis of *plant lipid* fractions, in order to speed up the customary preliminary open-column pre-purification before GC analysis (Anderson et al. 1979). A preparative column, packed with TLC-grade silica gel, permitted the fractionation of a *Teucrium canadense* (Labiatae) petrol ether extract into hydrocarbon, triterpene and sterol, and triglyceride fractions. The triterpenes were then separated from the sterols by semi-preparative HPLC on a Partisil-10 column eluted with n-hexane-ethyl acetate. Triterpene carboxylic acids often, however, suffer from excessive tailing on octadecylsilyl columns. A solution to this problem has been provided by Tischler and Cardellina (1993), who utilized a polybutadiene coated alumina column (Biotage Unisphere-PBD; 250×10 mm) to separate three such triterpenes with the eluent acetonitrile-water 11:9.

Intermediates and final products from organic synthesis lend themselves well to separation by preparative HPLC.

5.2.4.1
Recycling HPLC

In conventional recycle chromatography and its variants, such as external recycle chromatography (Crary et al. 1989), the aim is to enhance the recovery and purity of single-pass separations where the resolutions are not complete.

An instrument dedicated to recycling HPLC is available from Japan Analytical Industry, Ltd., Tokyo: the LC-20 (the successor to the LC-09). Several separations on this instrument have been reported. For example, the sesquiterpene isomers curcumanolides A (**41**) and B (**42**) from the essential oil of *Curcuma heyneana* (Zingiberaceae) rhizomes were separated by recycling 95 times on two Jaigel polystyrene columns with chloroform as solvent (Firman et al. 1988). If a solvent mixture is not used, it is much simpler to recover the eluent.

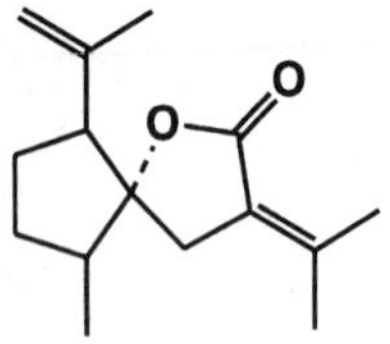

41

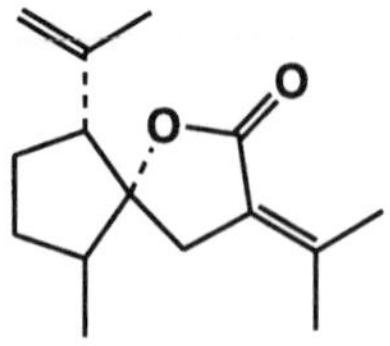

42

Similarly, a guaiane-type sesquiterpene was isolated from the wood of *Viburnum awabuki* (Caprifoliaceae) after recycling HPLC (five times) on a

Jaigel-1H (600×20 mm) column. The eluent was chloroform and flow rate 3.5 ml/min (Fukuyama et al. 1996).

Two Jaigel columns (1H, 500×25 mm and 2H, 500×25 mm) in tandem were insufficient to separate the closely-related *Catha edulis* (Celastraceae) alkaloids cathedulines E3, E4 and E5. However, chromatography on a JAI LC-09 instrument with an Asahipack GS-320 polyvinyl alcohol resin column (500×25 mm) and methanol as eluent was sufficient to provide the required separation (Kubo et al. 1987).

Three anthraquinones, pulmatin (**43**), chrysophanein (**44**) and physcionin (**45**), have been obtained from the root of *Rheum palmatum* (Polygonaceae) by a combination of DCCC and recycling HPLC (JAI LC-09) on an Asahipack GS-320 (500×7.6 mm) column. The mobile phase for this latter step was methanol, at a flow rate of 4.3 ml/min (Kubo et al. 1992).

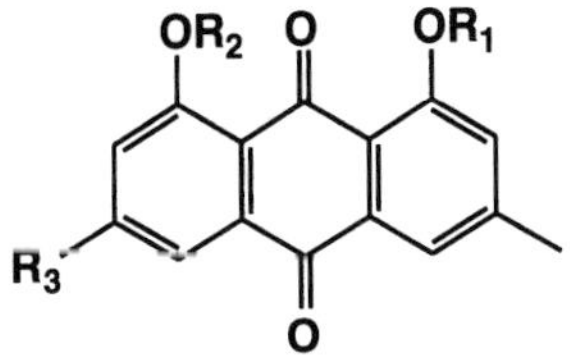

43 R_1 = Glc, R_2 = R_3 = H
44 R_1 = H, R_2 = Glc, R_3 = H
45 R_1 = H, R_2 = Glc, R_3 = OCH_3

Triterpene saponins (Miyase et al. 1996), triterpenes (Kawanishi et al. 1985), limonoids (Ahn et al. 1994) and piperidine alkaloids (Ahn et al. 1992) have also been isolated by procedures which involve a final recycling HPLC step.

5.2.4.2
Preparative Separations with Analytical HPLC Systems

The quantity of purified product required can either be obtained by a single injection onto a column of large dimensions or by repetitive injections onto columns of more modest dimensions. The latter method avoids the need for costly investment in columns, packing materials and accessories, and can make use of existing analytical equipment. Furthermore, scale-up problems are largely avoided and separations are fast.

When only small amounts of pure material (µg to a few mg) are required, separations can be performed on analytical columns without too much trouble. These columns typically have dimensions 250×4.6 mm and often contain reversed-phase material. Samples of ca. 5 µg–100 µg can thus be introduced onto the column until sufficient product is obtained. For example, a glycosidic lupin alkaloid has been isolated from *Lupinus hirsutus* (Leguminosae) using HPLC on a LiChrosorb Si60 5 µm 250×4.6 mm column as the final purification step. The solvent was 25% MeOH in Et_2O-5% NH_4OH (50:1) (Suzuki et al. 1994). A biphenyl and a xanthone were obtained from *Monnina obtusifolia* (Polygalaceae) aerial parts after silica gel CC, Sephadex LH-20 gel filtration and a final separation on a LiChrosorb cyano column (7 μm; 250×4.6 mm). The solvent used for this latter step was n-hexane-i-propanol 17:3 (Pinto et al. 1994).

One field in which separations are often performed on analytical columns is the purification of peptides (see chapter 8). Bioactive peptides often occur in very small quantities and when analytical HPLC is used as the final purification step, there is no danger of overloading the column.

An alternative method is to overload the column intentionally (see Sect. 5.1.11). By this strategy, a 500×6.8 mm column, compatible with conventional equipment, was used to separate 0.1–1 g of mixtures in one single injection at a capacity about ten times greater than generally accepted (Verzele et al. 1982). This column was loaded with 12 g of 10 μm RSiL (Alltech) and since it was reckoned that 30 g of silica gel was required for every gram of mixture to be separated, a sample size of 0.4 g was theoretically possible. However, for close-running spots (TLC), 100 g of silica gel was necessary per gram of sample (Verzele et al. 1982). For reversed-phase operations, a 250×4.6 mm column filled with 5 μm spherical octadecyl silica gel (ROSiL-C_{18}-D) was used. Reasonably good separations could be achieved with up to 30 mg of a binary mixture (Verzele et al. 1982).

To increase efficiency, analytical HPLC columns can be coupled together. Packing material of particle size 20–30 μm should be used to avoid permeability problems, especially with aqueous solvents. With organic solvents such as hexane, the viscosity is low and 10 μm material can be used.

Analytical systems are not capable, however, of supplying the flow rates required for large-scale preparative separations and, in this respect, their use is limited.

In preparative liquid chromatography, two fundamental processes are possible: either *batch introduction* or *continuous introduction*. Most separation problems deal with batchwise introduction of sample but one or two instances of continuous application are known. Methods lying somewhere between these two options have been introduced and in one such case an analytical HPLC unit has been converted into a system which can continuously process a mixture (Hadfield et al. 1983). In this method, small amounts of sample were injected at suitable intervals onto a Whatman Partisil M9 250×9 mm column and the fractions collected – all automatically. A mixture of benzyl-D-glucosides could be separated by this procedure at a rate of 2.5–3 g fractionated material every 24 h. One drawback of the method was that after a few days continuous use, the column needed to be regenerated by washing with water, followed by a mixture of acetic acid and 2,2′-dimethoxypropane.

5.2.4.3
Displacement Chromatography

This, in contrast to elution, is a method whereby sample components are displaced from the stationary phase by a solution of a compound which has a higher affinity for the stationary phase than for all of the sample components (Liao et al. 1987).

First a carrier solvent with a low solvent strength in the separation system has to be chosen. This must dissolve the solute at sufficiently high concentration to allow the introduction of large samples. After equilibrium has been achieved, the sample is introduced into the carrier stream and preconcentrated at the column inlet (the stationary phase must exhibit strong retention of the sample with the carrier solvent as eluent). Then a highly concentrated solution of the displacer is pumped through the column. This causes the sample components to move down the column at speeds determined by the displacer front velocity. Stronger adsorbing components of the sample displace from the surface of the stationary phase those having weaker retention until the separation is achieved. The mixture forms a displacement train which is composed of adjacent square bands of near uniform concentration, all moving at the same velocity.

When applied to analytical columns, displacement chromatography involves the phenomenon of overloading. For example, polymyxin antibiotics have been successfully separated on reversed-phase analytical columns in the displacement mode. Sample sizes of 100 mg were loaded onto the column and then elution was performed by displacement with an aqueous solution of octyldecyldimethylammonium chloride (Kalasz and Horvath 1981).

Although displacement chromatography allows larger throughputs than elution chromatography, the search for a suitable displacer is inconvenient. Regeneration of the column after each separation is also necessary in order to wash out all the displacer.

5.2.4.4
Industrial Separations

Large-scale preparative chromatography (or process scale chromatography, which is generally defined as employing columns of over 10 cm diameter; Jones 1988a) requires specialized equipment such as pumps capable of producing high flow rates and columns structurally designed to withstand the high pressures involved. For this purpose, systems have been designed which can separate up to kilograms of material at pressures of around 150 bar, with flow-rates of 2000 ml/min.

Optimized separations on an analytical scale can be directly transferred to the production scale as long as the following are constant:

- the column height
- the sample loading per unit volume of support
- the steepness of the gradient.

Under these conditions, the flow rate in the preparative system can be calculated using the following equation:

$$\frac{X_a}{r_a^2} = \frac{X_p}{r_p^2} \cdot \frac{1}{C_L}$$

where:
X_a = flow rate in the analytical system
X_p = flow rate in the preparative system
r_a = radius of the analytical column
r_p = radius of the preparative column
C_L = length of the preparative column/length of the analytical column

From this equation, if an analytical system (4.6 mm diameter column) is operated at a flow rate of 1 ml/min, a 20 mm diameter column of the same length would need to be run at a flow rate of about 19 ml/min in order to maintain a comparable linear flow velocity. Scale-up to a 150 mm diameter process column requires a flow rate of over 1000 ml/min to maintain a constant linear flow velocity and produce comparable retention times in the two systems.

Large-scale preparative chromatography of *Bupleurum falcatum* (Umbelliferae) saponins has been performed after preliminary studies on analytical HPLC. Saikosaponins a, c and d were separated on a 100×11 cm axially compressed column packed with 5 kg of YMC 20 µm RP-18 phase. The mobile phase was a step gradient of acetonitrile-water (27:73, 30:70, 35:65) at a flow rate of 210 ml/min. Injection of 10 g crude saponin mixture gave 1210 mg of saikosaponin a, 403 mg of saikosaponin c and 1604 mg of saikosaponin d (Sakuma and Motomura 1987).

Manufacturers of production-scale LC systems include Amicon (K-Prime 3000), YMC (PLC-100 A), Separations Technology (Process 2000; 10–20 cm diameter columns), Varex (PSLC-150, PSLC-300; 15 cm and 30 cm diameter columns, respectively), Prochrom (for example, the Prochrom 150 axially compressed 15 cm i.d. column) and E. Merck (Prepbar 100 and -200 separation systems, accommodating 5–20 cm i.d. cartridges, with flow rates up to 1600 ml/min).

Filling the gap between industrial equipment and smaller laboratory-scale chromatographs are systems like the Lab 800A from Separations Technology, which features a stainless steel column and flow rate capabilities to 800 ml/min at 120 bar. Another small-scale process LC apparatus which is also capable of working with high efficiency microparticulate columns is the PSLC-100 from Varex. The NovaPrep 5000 system from Merck can work at flow rates up to 1000 ml/min at 340 bar; columns ranging from 25 cm×4 mm to 60 cm×20 cm are also included.

The Waters Prep LC 500 chromatograph has been replaced by the Prep LC 2000 system. This offers a 4–300 ml/min flow rate range at a maximum of 140 bar (2000 psi) and is compatible with radial compression cartridges and steel columns. Another system from Waters, the Delta Prep 4000, accommodates both analytical and preparative scale operations and offers a microgram to multigram capability. It is possible to develop gradient HPLC methods, perform isolations and check fraction purity on the same system. A 300 bar (4000 psi) back pressure capacity is available across the entire 0.5–150 ml/min flow rate range. An example of separation on the Delta Prep 4000 system is given by the isolation of 7-*epi*-cephalomannine from *Taxus*×*media* (Taxaceae) needles. Preliminary chromatography of a 80% methanol extract of the needles gave 2 g of a fraction which was dissolved in methanol (40 ml). Preparative HPLC (20 portions, each 2 ml) was performed on a Waters Prep-Pak Bondapak C_{18} column (15 µm; 300×47 mm) with acetonitrile-water (25:75) at a flow rate of 80 ml/min (detection 227 nm). Fractions containing 7-*epi*-cephalomannine were pooled, concentrated and extracted with ethyl acetate, to give 60 mg of the desired product (De Bellis et al. 1995).

Biotage (Charlottesville, VA, USA) also have a range of radial compression cartridges which can be used with the Kiloprep 250 compression module and the Kiloprep 250 pumping module. These cartridges come in two sizes: 300 × 100 mm or 600 × 100 mm (maximum pressure 2000 psi/150 bar).

Pharmacia has fast-flow gels and a monodisperse polymer (Mono-Beads) (see chapter 8) available for production-scale separations, together with customized preparative and process-scale chromatographs. Other companies, such as Whatman and Rainin, are also moving into the process LC market and with the importance of final LC purifications in the areas of protein purification, fermentation products and the isolation of monoclonal antibodies, this particular domain is seeing an enormous expansion.

While academic research is mainly driven by the necessity for publishing results, industrial applications of chromatography are often governed by confidentiality and little of the work is to be found in journals. It is thus difficult to get an overall picture of industrial separations. Certain considerations of applications and costs are given by Jones (1988b).

5.3 Supercritical Fluid Chromatography

The history and principles of supercritical fluid chromatography (SFC) have been the subject of numerous reviews (see, for example: Gere, 1983; White and Houck, 1986; Bevan and Marshall 1994; Perrut 1994).

The technique of preparative SFC has received relatively little attention when compared with analytical SFC. However, the method has several merits which should be further exploited:

- it is faster than HPLC.
- products can be recovered free from solvent easier than by HPLC.
- it has the potential to replace normal-phase preparative HPLC because when coupled with SFE, the extraction, concentration and chromatographic fractionation can be carried out in a single run.
- it has a wider range of applicability than GC (SFC can handle non-volatile, thermally labile compounds).

Thus SFC is a valuable and often complementary tool to GC and HPLC. The complementarity between preparative HPLC and preparative SFC has been shown by Crétier et al. (1994a): in the separation of a minor component from a major component, SFC is preferred if the minor component elutes after the principal peak and HPLC is preferred if the minor component elutes first.

The pressure drop (inlet pressure minus outlet pressure) across a packed chromatographic column is considerably less for a supercritical fluid than for a liquid. For supercritical carbon dioxide, this is one-tenth (or less) than that encountered with typical liquid mobile phases (Gere 1983). In SFC, therefore, a higher linear velocity is achieved more easily with a given pumping system, and the lifetime of the column is extended.

5.3.1
Supercritical Fluids

The most popular and safe supercritical fluid for use in SFC, like SFE, is undoubtedly carbon dioxide. It is chemically inert and allows easy separation of solutes at low temperature in an oxygen-free environment. Carbon dioxide is non-flammable, odour free, available in a state of high purity at low cost and does not present a solvent disposal problem. The critical pressure (73.8 bar) is not difficult to maintain with modern HPLC solvent delivery systems.

The solvent strength of carbon dioxide varies markedly with the density of the fluid. The solvent strength at a density under 0.25 g/cm^3 is less than that of perfluorinated alkanes, while at a density of 0.98 g/cm^3 it exceeds that of hexane (approaching the solvent strength of dichloromethane).

As solvent strength can easily be controlled by changing pressure and/or temperature, mobile phases with a fixed composition can be used for the separations of many types of solutes.

Solute diffusion coefficients are greater in a supercritical fluid than in the liquid phase. This causes narrower chromatographic peaks than in HPLC. Another important feature of supercritical fluids is that they have a much lower viscosity than liquid solvents.

The solvent power and selectivity of supercritical carbon dioxide is low. Cosolvents can be added to avoid these constraints (Brennecke and Eckert 1989).

5.3.2
Load

The loading limit is more likely to be imposed by the solubility of the sample in the supercritical fluid rather than the capacity of the column for the quantity of substance applied.

5.3.3
Special Considerations

Although much of the equipment designed for HPLC can also be used for SFC, there are two problems with SFC which require special attention:

- loading samples
- collection of fractions

5.3.3.1
Loading Samples

In SFC, the volume of sample which can be directly injected without flooding the column and cause band broadening is very limited due to the adverse effects of the sample solvent. The most effective method to get rid of these effects is to eliminate the sample solvent.

One solution to this particular hurdle is to pass the sample solution from the sample injection loop to a packed pre-column. Solvent is then removed by passage of warm helium gas (on-line solvent venting). The carbon dioxide mobile phase is then introduced, dissolving the solutes that have been deposited on the pre-column and transporting them to the preparative column. In order to increase the amount injected without exceeding the volume capacity of the pre-column, successive sample plugs can be loaded onto the pre-column without intermediate removal of the solvent (Crétier et al. 1990; Crétier et al. 1994b).

Hirata et al. (1995) described a variant of the above method with a large volume injection system consisting of a 50×10 mm trap column (pre-column, filled with Wako CQ-3 silica gel 30–50 µm), connected to a 150×10 mm separation column (filled with Jasco 5 µm Superpak Crest SIL), for the separation of fatty acid methyl esters with supercritical carbon dioxide. Up to 100 mg sample in 1 ml hexane was injected. The sample solution was first diluted with liquid carbon dioxide. Then solvent was removed and the sample trapped in the trap column. Finally, the solutes from the trap column were transferred and re-concentrated on the separation column.

5.3.3.2 *Collection of Fractions*

There are several ways to obtain the separated material after SFC:

- depressurization of the supercritical fluid. This can cause blockages with solid carbon dioxide.
- collection in a high-pressure vessel and then slowly reduce the pressure.
- dissolution in a trapping solvent.
- cool the collection vessel to solidify the carbon dioxide.
- adsorption on a solid and subsequent elution with a solvent.

An efficient fractionating and collection system has been described with the modifier as trapping solvent (Heaton et al. 1996).

5.3.4 Large-Scale Systems

A fully automated industrial-scale SFC instrument is available from Prochrom, and Jasco (Japan) manufacture a preparative SFC/SFE system.

Berger and Perrut (1990) have concentrated a lot of effort towards developing industrial scale chromatographs. They have built a pilot unit capable of accommodating 100×6 cm columns and adapted to axial compression. Purification of vitamin intermediates and polyunsaturated fatty acid esters has been performed.

Large diameter SFC columns suffer from the formation of dead volumes and compression techniques have to be used. Perrut has found that dynamic axial compression is the best method of preserving column efficiency (Perrut 1994).

5.3.5 Separations by SFC

Applications of SFC in the isolation of natural products have been reviewed by Bevan and Marshall (1994). Some examples are given below, although it must be emphasised that there will probably be a rapid expansion in the exploitation of SFC over the next few years.

5.3.5.1
Fatty Acids and Lipids

A comparision of the separation of the polyunsaturated fatty acid esters EPA and DHA has shown that preparative SFC has some advantages over HPLC, including higher purities, higher throughput and ease of solvent removal (Perrut 1994).

5.3.5.2
Terpenes

Yamauchi and Saito (1990) described a method for the simple removal of terpenes from lemon-peel oil, using a 50 × 7.2 mm column packed with silica gel (10–20 μm). They were able to separate 0.5 ml lemon-peel oil per run. When the column was at 40 °C, the carbon dioxide density at 2.2 g/min and the column outlet pressure maintained at 100 bar for 9 minutes, the terpenes were eluted in the first fraction.

5.3.5.3
Miscellaneous

Saito and Yamauchi (1990) isolated α- and β-tocopherol from wheat-germ oil by recycle semi-preparative SFC on two 250 × 10 mm columns packed with 5 μm silica gel.

Wünsche et al. (1991) have also purified tocopherols by SFC combined with preparative TLC. They used a Supelco 5 μm C_{18} column (250 × 10 mm) with supercritical carbon dioxide and methanol modifier (2–3%) to separate the tocopherols from a coffee extract. The eluate was collected on a TLC plate which was slowly moved on a conveyor. Migration of the TLC plate gave a second separation capability. The spots on the plate can be detected by spraying with berberine chloride. When the compounds are stripped from the plate and washed from the silica with an organic solvent, the berberine chloride remains on the plate. Alternatively, a soft plastic TLC sheet can be pressed against the thick wet plate to transfer a small quantity of separated compound. By choice of a suitable detection reagent, the mirror image of the compounds of interest is obtained. By this method, several milligrams of pure compounds can be separated but significant losses occur during the transfer from SFC to TLC. A jasmine extract was also resolved into its components by this technique.

5.4 References

Abbott T, Peterson R, McAlpine J, Tjarks L, Bagby M (1989) J Liq Chromatogr 12:2281
Ahn J, Ahn M, Zee O, Kim E, Lee S, Kim HJ, Kubo, I (1992) Phytochemistry 31:3609
Ahn J, Choi S, Lee C (1994) Phytochemistry 36:1493
Akihisa T, Tanaka N, Yokota T, Tanno N, Tamura T (1991) Phytochemistry 30:2369
Alfonso D, Kapetanidis I (1994) Phytochemistry 36:179
Alias Y, Awang K, Hadi AHA, Thoison O, Sevenet T, Païs M (1995) J Nat Prod 58:1160
Allais DP, Chulia AJ, Kaouadji M, Simon A, Delage C (1995) Phytochemistry 39:427
Andary C, Ravn H, Wylde R, Heitz A, Motte-Florac E (1989) Phytochemistry 28:288
Anderson LA, Doggett NS, Ross MSF (1979) J Liq Chromatogr 2:455
Angenot L, Belem-Pinheiro ML, Imbiriba da Rocha AF, Poukens-Renwart P, Quetin-Leclercq J, Warin R (1990) Phytochemistry 29:2746
Aoshima H, Miyase T, Ueno A (1994) Phytochemistry 37:547
Arslanian RL, Harris GH, Stermitz FR (1990) J Org Chem 55:1204
Asari F, Kusumi T, Kakisawa H (1989) J Nat Prod 52:1167
Bardon A, Cardona L, Catalan CAN, Pedro JR (1996) Phytochemistry 41:845
Bashir A, Hamburger M, Msonthi JD, Hostettmann K (1991) Phytochemistry 30:3781
Bashir A, Hamburger M, Gupta MP, Solis PN, Hostettmann K (1993) Phytochemistry 32:741
Bayer T, Breu W, Seligmann O, Wray V, Wagner H (1989a) Phytochemistry 28:2373
Bayer T, Wagner H, Block E, Grisoni S, Zhao SH, Neszmelyi A (1989b) J Am Chem Soc 111:3085
Beck W, Halasz I (1978) Anal Chem 291:340
Beier RC, Norman JO, Reagor JC, Rees MS, Mundy BP (1993) Nat Toxins 1:286
Belofsky GN, Stermitz FR (1988) J Nat Prod 51:614
Benson JR, Woo DJ (1984) J Chromatogr Sci 22:386
Berger C, Perrut M (1990) J Chromatogr 505:37
Beutler JA, Clark P, Alvarado AB (1994) J Nat Prod 57:629
Bevan CD, Marshall PS (1994) Nat Prod Rep 11:451
Bidlingmeyer BA (1987) Preparative liquid chromatography. Elsevier, Amsterdam
Bilia AR, Morelli I, Hamburger M, Hostettmann K (1996) Phytochemistry 43:887
Billeter M, Meier B, Sticher O (1991) Phytochemistry 30:987
Bloor SJ, Molloy BPJ (1991) J Nat Prod 54:1326
Bloor SJ, Benner JP, Irwin D, Boother P (1996) Phytochemistry 41:801
Blunt JW, Calder VL, Fenwick GD, Lake RJ, McCombs JD, Munro MHG, Perry NB (1987) J Nat Prod 50:290
Bouaicha N, Amade P, Puel D, Roussakis C (1994) J Nat Prod 57:1455
Brasseur T, Angenot L (1986) Phytochemistry 25:563
Brennecke JF, Eckert CA (1989) AIChE J 35:1409
Bringmann G, Lisch D, Reuscher H, Aké Assi L, Günther K (1991) Phytochemistry 30:1307
Bruce IE, Long A, Payne PB, Tyman JHP (1990) J Liq Chromatogr 13:2103
Bürgi C, Rüedi P (1993) Helv Chim Acta 76:1890
Buta JG (1984) J Chromatogr 295:506
Cabrera GM, Seldes AM (1995) J Nat Prod 58:1920
Calis I, Rüegger H, Chun Z, Sticher O (1990) Planta Med 56:406
Cardellina JH (1991) J Liq Chromatogr 14:659
Castellar MR, Montijano H, Manjon A, Iborra JL (1993) J Chromatogr 648:187
Catalano S, Luschi S, Flamini G, Goni PL, Nieri EM, Morelli I (1996) Phytochemistry 42:1605
Chamblee TS, Clark BC, Radford T, Iacobucci GA (1985) J Chromatogr 330:141
Charton F, Bailly M, Guiochon G (1994) J Chromatogr 687:13
Chaudhuri SK, Fullas F, Brown DM, Wani MC, Wall ME, Cai L, Mar W, Lee SK, Luo Y, Zaw K, Fong HHS, Pezzuto JM, Kinghorn AD (1995) J Nat Prod 58:1
Chen TM, George RC, Weir JL, Leapheart T (1990) J Nat Prod 53:359
Christensen LP, Lam J (1991) Phytochemistry 30:2663

Clausen TP, Evans TP, Reichardt PB, Bryant JP (1989) J Nat Prod 52:207
Colin H, Hilaireau P, de Tournemire J (1990) LC-GC Intl 3:40
Comte G, Chulia AJ, Vercauteren J, Allais DP (1996) Planta Med 62:88
Coq B, Cretier G, Rocca JL, Vialle J (1981) J Liq Chromatogr 4:237
Corthout J, Pieters L, Claeys M, Geerts S, Vanden Berghe D, Vlietinck A (1994) Planta Med 60:460
Cox GB (1990) LC-GC Intl 3:10
Cox GB, Snyder LR (1988) LC-GC Intl 1:36
Crary JR, Cain-Janicki K, Wijayaratne R (1989) 462:94
Crétier G, Rocca JL (1982) Chromatographia 16:32
Crétier G, Majdalani R, Rocca JL (1990) Chromatographia 30:645
Crétier G, Neffati J, Rocca JL (1994a) J Chromatogr Sci 32:449
Crétier G, Majdalani R, Neffati J, Rocca JL (1994b) Chromatographia 38:330
Curl CL, Price KR, Fenwick GR (1988) J Sci Food Agric 43:101
Dai JR, Decosterd LA, Gustafson KR, Cardellina JH, Gray GN, Boyd MR (1994) J Nat Prod 57:1511
D'Ambrosio M, Mealli C, Guerriero A, Pietra F (1985) Helv Chim Acta 68:1453
D'Ambrosio M, Guerriero A, Pietra F (1987) Helv Chim Acta 70:2019
Damtoft S (1992) Phytochemistry 31:175
De Bellis P, Lovati M, Pace R, Peterlongo F, Zini GF (1995) Fitoterapia 66:521
Debrunner B, Neuenschwander M (1994) Chimia 48:564
Décosterd LA, Dorsaz A, Hostettmann K (1987a) J Chromatogr 406:367
Décosterd LA, Stoeckli-Evans H, Msonthi JD, Hostettmann K (1987b) Helv Chim Acta 70:1694
DeJong AWJ, Potter H, Kraak JC (1978) J Chromatogr 148:127
Della Greca M, Fiorentino A, Monaco P, Pinto G, Pollio A, Previtera L (1996) J Chem Ecol 22:587
Dobler W (1983) GIT Fachz Lab 27:1078
Dodson CD, Stermitz FR, Castro O, Janzen DH (1987) Phytochemistry 26:2037
Dostal J, Taborska E, Slavik J (1992) Fitoterapia 63:67
Du Y, Oshima R, Kumanotani J (1984) J Chromatogr 284:463
Ducrey B, Wolfender JL, Marston A, Hostettmann K (1995) Phytochemistry 38:129
Ducrey B, Marston A, Göhring S, Hartmann RW, Hostettmann K (1997) Planta Med 63:111
Edwards C, Lawton LA, Coyle SM, Ross P (1996) J Chromatogr A 734:163
El-Domiaty MM, El-Feraly FS, Mossa JS, McPhail AT (1993) Phytochemistry 34:467
Elliger CA, Waiss AC, Benson M, Wong RY (1993) Phytochemistry 33:471
Erdelmeier CAJ, Regenass U, Rali T, Sticher O (1992) Planta Med 58:43
Evershed RP, Morgan ED, Thompson LD (1982) J Chromatogr 237:350
Farmakalidis E, Murphy PA (1984) J Chromatogr 295:510
Fiorini M (1995) J Chromatogr A 692:213
Firman K, Kinoshita T, Itai A, Sankawa U (1988) Phytochemistry 27:3887
Franke G, Verillon F (1988) J Chromatogr 450:81
Fu X, Liang W, Tu G (1988) J Nat Prod 51:262
Fujimoto Y, Usui S, Makino M, Sumatra M (1996) Phytochemistry 41:923
Fukuyama Y, Kiriyama Y, Okino J, Kodama M, Iwaki H, Hosozawa S, Matsui K (1993) Chem Pharm Bull 41:561
Fukuyama Y, Minami H, Ichikawa R, Takeuchi K, Kodama M (1996) Phytochemistry 42:741
Fuzzati N, Sutarjadi, Dyatmiko W, Rahman A, Hostettmann K (1995) Phytochemistry 39:409
Fuzzati N, Wolfender JL, Hostettmann K, Msonthi JD, Mavi S, Molleyres LP (1996) Phytochem Anal 7:76
Gabetta B, de Bellis P, Pace R, Appendino G, Barboni L, Torregiani E, Gariboldi P, Viterbo D (1995) J Nat Prod 58:1508
Gaddamidi V, Bjeldanes LF, Shoolery JN (1985) J Agric Food Chem 33:652
Gafner S, Wolfender JL, Mavi S, Hostettmann K (1996a) Planta Med 62:67
Gafner S, Wolfender JL, Nianga M, Stoeckli-Evans H, Hostettman K (1996b) Phytochemistry 42:1315

Ganetsos G, Barker PE (1993) Preparative and production scale chromatography. Marcel Dekker, New York
Gareil P, Rosset R (1982a) Analusis 10:397
Gareil P, Rosset R (1982b) Analusis 10:445
Gere DR (1983) Science 222:253
Gewali MB, Hattori M, Tezuka Y, Kikuchi T, Namba T (1989) Chem Pharm Bull 37:1547
Godbille E, Devaux P (1974) J Chromatogr Sci 12:565
Golshan-Shirazi S, Guiochon G (1994) J Chromatogr A 658:149
Govindachari TR, Gopalakrishnan G, Suresh G (1996) J Liq Chromatogr 19:1729
Gromek D, Hocquemiller R, Cavé A (1994) Phytochem Anal 5:133
Grue MR, Liddell JR (1993) Phytochemistry 33:1517
Guella G, Mancini I, Pietra F (1987) Helv Chim Acta 70:621
Guerriero A, D'Ambrosio M, Cuomo V, Vanzanella F, Pietra F (1988a) Helv Chim Acta 71:57
Guerriero A, D'Ambrosio M, Pietra F (1988b) Helv Chim Acta 71:1094
Guiochon G, Katti A (1987) Chromatographia 24:165
Gunzinger J, Msonthi JD, Hostettmann K (1988) Helv Chim Acta 71:72
Hadfield AF, Dreyer RN, Sartorelli AC (1983) J Chromatogr 257:1
Hallock YF, Dai J, Bokesch HR, Dillah KB, Manfredi KP, Cardellina JH, Boyd MR (1994) J Chromatogr A 688:83
Hamburger M, Slacanin I, Hostettmann K, Dyatmiko W, Sutarjadi (1992) Phytochem Anal 3:231
Härmälä P, Vuorela P, Hiltunen R, Nyiredy S, Sticher O, Törnquist K, Kaltia S (1992) Phytochem Anal 3:42
Hasan A, Waterman P, Iftikhar N (1989) J Chromatogr 466:399
Haywood PA, Munro G (1980) Dev Chromatogr 33
Heath RR, Sonnet PE (1980) J Liq Chromatogr 3:1129
Heaton DM, Bartle KD, Myers P, Clifford AA (1996) J Chromatogr A 753:306
Henke H, Rülke K (1987) Swiss Chem 9:23
Henrici A, Kaloga M, Eich E (1994) Phytochemistry 37:1637
Henschen A, Hupe KP, Lottspeich F, Voelter W (1985) High performance liquid chromatography in biochemistry. VCH, Weinheim
Herbert NR (1991) In:Subramanian G (ed) Preparative and process-scale liquid chromatography. Ellis Horwood, New York, p 10
Heuer C, Hugo P, Mann G, Seidel-Morgenstern A (1996) J Chromatogr A 752, 19
Hirakura K, Morita M, Nakajima K, Ikeya Y, Mitsuhashi H (1992) Phytochemistry 31:899
Hirakura K, Morita M, Niitsu K, Ikeya Y, Maruno M (1994) Phytochemistry 35:963
Hirata Y, Kawaguchi Y, Kitano K (1995) Chromatographia 40:42
Hodge RP, Harris CM, Harris TM (1988) J Nat Prod 51:66
Hostettmann K, Hostettmann M (1985) In:Henschen A, Hupe KP, Lottspeich F, Voelter W (eds) High performance liquid chromatography in biochemistry. VCH, Weinheim, p 599
Hostettmann K, Marston A (1995) Saponins. Cambridge University Press, Cambridge
Hostettmann K, Hostettmann M, Marston A (1986) Preparative chromatography techniques: applications in natural product isolation. Springer, Berlin Heidelberg New York
Hufford CD, Jia Y, Croom EM, Muhammed I, Okunade AL, Clark AM (1993) J Nat Prod 56:1878
Huizing HJ, De Boer F, Malingré TM (1981) J Chromatogr 214:257
Hunt BJ, Rigby W (1967) Chem Ind 1868
Hupe KP, Lauer HH (1981) J Chromatogr 203:41
Hwu JR, Robl JA, Khoudary KP (1987) J Chrom Sci 25:501
Irsch E, Pachaly P, Breitmaier E, Sin K (1993) Liebigs Ann 281
Ishibashi M, Ohizumi Y, Sasaki T, Nakamura H, Hirata Y, Kobayashi J (1987) J Org Chem 52:450
Ishiguro K, Yamaki M, Kashihara M, Takagi S (1987) Planta Med 57:415
Itokawa H, Morita H, Watanabe M, Mihashi S, Iitaka Y (1985) Chem Pharm Bull 33:1148
Itokawa H, Ichihara Y, Kojima H, Watanabe K, Takeya K (1989) Phytochemistry 28:1667
Jansen R, Irschik H, Reichenbach H, Höfle G (1985) Liebigs Ann 822
Jogia MK, Andersen RJ (1989) Can J Chem 67:257

Johnson EL, Stevenson R (1978) Basic liquid chromatography. Varian Associates Inc, Palo Alto, CA, USA
Jones K (1988a) Chromatographia 25:437
Jones K (1988b) Chromatographia 25:547
Kaizuka H, Takahashi K (1983) J Chromatogr 258:135
Kalasz H, Horvath C (1981) J Chromatogr 215:295
Kanazawa H, Nagata Y, Matsushima Y, Tomoda M, Takai N (1990) Chem Pharm Bull 38:1630
Kasai R, Katagiri M, Ohtani K, Yamasaki K, Yang CR, Tanaka O (1994) Phytochemistry 36:967
Kato A, Moriyasu M, Ichimaru M, Nishiyama Y, Juma FD, Nganga JN, Mathenge SG, Ogeto JO (1996) J Nat Prod 59:316
Katsuzaki H, Kawakishi S, Osawa T (1994) Phytochemistry 35:773
Kawanishi K, Hashimoto Y, Qiang W, Zhenwen X (1985) Phytochemistry 24:2051
Kellam SJ, Mitchell KA, Blunt JW, Munro MHG, Walker JRL (1993) Phytochemistry 33:867
Khan IA, Erdelmeier CAJ, Sticher O (1992) J Nat Prod 55:1270
Kikuchi M, Suzuki N (1992) Chem Pharm Bull 40:2753
Kimura Y, Kitamura H, Araki T, Noguchi K, Baba M, Hori M (1981) J Chromatogr 206:563
Kingston DGI (1979) J Nat Prod 42:237
Knox JH, Pyper M (1986) J Chromatogr 363:1
Koike K, Ohmoto T (1993) Phytochemistry 34:505
Kong L, Min Z, Li Y, Li X, Pei Y (1996) Phytochemistry 42:1689
Kouno I, Baba N, Hashimoto M, Kawano N, Takahashi M, Kaneto H, Yang C (1990) Chem Pharm Bull 38:422
Krajewska AM, Powers JJ (1986) J Chromatogr 367:267
Kreh M, Matusch R, Witte L (1995) Phytochemistry 40:1303
Kubeczka KH (1985) In:Vlietinck AJ, Dommisse RA (eds) Advances in medicinal plant research. Wissenschaftliche Verlagsgesellschaft, Stuttgart, p 197
Kubo I, Matsumoto A, Hirotsu K, Naoki H, Wood WF (1984) J Org Chem 49:4644
Kubo I, Kim M, de Boer G (1987) J Chromatogr 402 (1987)
Kubo I, Murai Y, Soediro I, Soetarno S, Sastrodihardjo S (1992) Phytochemistry 31:1063
Kühler TC, Lindsten GR (1983) J Org Chem 48:3589
Lam LKT, Yee C, Chung A, Wattenberg LW (1985) J Chromatogr 328:422
Lehmann T, Brenneisen R (1992) Phytochem Anal 3:88
Lesec J (1985) J Liq Chromatogr 8:875
Leung M, Poon C, Yuen P (1986) Chem Ind 787
Leutert T, von Arx E (1984) J Chromatogr 292:333
Li T, Li J, Li H (1995) J Chromatogr A 715:372
Liao AW, El Rassi Z, LeMaster DM, Cs Horvath J (1987) Chromatographia 24:881
Little CJ, Stahel O (1984) J Chromatogr 316:105
Little JN, Cotter RL, Prendergast JA, McDonald PD (1976) J Chromatogr 126:439
Liu Y, Wagner H, Bauer R (1996) Phytochemistry 42:1203
Loev B, Goodman MM (1967) Chem Ind 2026
Loibner H, Seidl G (1979) Chromatographia 12:600
Lübke M, Le Quéré J, Barron D (1993) J Chromatogr 646:307
Ma WG, Fuzzati N, Xue Y, Yang CR, Hostettmann K (1996) Phytochemistry 41:1287
Maillard M, Recio-Iglesias M, Saadou M, Stoeckli-Evans H, Hostettmann K (1991) Helv Chim Acta 74:791
Maillard M, Adewunmi CO, Hostettmann K (1992) Phytochemistry 31:1321
Makuch B, Gazda K, Cisowski W (1992) J Chromatogr 594:145
Mannila E, Talvitie A, Kolehmainen E (1993) Phytochemistry 33:813
Marco JA, Sanz JF, Carda M (1992) Phytochemistry 31:2163
Marner FJ, Littek A, Arold R, Seferiadis K, Jaenicke L (1990) Liebigs Ann 563
Marston A, Gafner F, Dossaji SF, Hostettmann K (1988) Phytochemistry 27:1325
Marston A, Hamburger M, Sordat-Diserens I, Msonthi JD, Hostettmann K (1993) Phytochemistry 33:809
Matsunaga S, Fusetani N, Konosu S (1985a) J Nat Prod 48:236

Matsunaga S, Fusetani N, Konosu S (1985b) Tetrahedron Lett 26:855
Matsuura H, Kasai R, Tanaka O, Saruwatari Y, Kunihiro K, Fuwa T (1984) Chem Pharm Bull 32:1188
Mayer A, Köpke B, Anke H, Sterner O (1996) Phytochemistry 43:375
Mayerl F, Näf R, Thomas AF (1989) Phytochemistry 28:631
McMaster MC (1994) HPLC, a practical user's guide. VCH, New York
Meyer VR (1984) J Chromatogr 316:113
Meyers AI, Slade J, Smith RK, Mihelich ED, Hershenson FM, Liang CD (1979) J Org Chem 44:2247
Mierzwa R, Truumees I, Patel M, Marquez J, Gullo V (1985) J Liq Chromatogr 8:1697
Miething H, Speicher-Brinker A (1989) Arch Pharm 322:141
Miller L, Bush H, Derrico EM (1989) J Chromatogr 484:259
Mishra NC, Estensen RD, Abdel-Monem MM (1986) J Chromatogr 369:435
Miyase T, Shiokawa K, Zhang DM, Ueno A (1996) Phytochemistry 41:1411
Mizui F, Kasai R, Ohtani K, Tanaka O (1990) Chem Pharm Bull 38:375
Mohamed KM, Ohtani K, Kasai R, Yamasaki K (1994) Phytochemistry 37:495
Mohanraj S, Herz W (1981) J Liq Chromatogr 4:525
Morishita H, Iwahashi H, Osaka N, Kido R (1984) J Chromatogr 315:253
Morita M, Mihashi S, Itokawa H, Hara S (1983) Anal Chem 55:412
Moulis C, Fouraste I, Bon M (1992) J Nat Prod 55:445
Msonthi JD, Toyota M, Marston A, Hostettmann K (1990) Phytochemsitry 29:3977
Musser SM, Gay ML, Mazzola EP, Plattner RD (1996) J Nat Prod 59:970
Mütsch-Eckner M, Meier B, Wright AD, Sticher O (1992) Phytochemistry 31:238
Nakatani M, Huang RC, Okamura H, Naoki H, Iwagwa T (1994) Phytochemistry 36:39
Neuschild K, Christensen LP, Lam J (1992) Phytochemistry 31:4353
Nishina A, Kubota K, Osawa T (1993) J Agric Food Chem 41:1772
Niwa M, Sugino H, Takashima S, Sakai T, Wu YC, Wu TS, Kuoh CS (1991) Chem Pharm Bull 39:2422
Nozawa K, Seyea H, Nakajima S, Udagawa S, Kawai K (1987) J Chem Soc Perkin Trans I 1735
Nyiredy S, Dallenbach-Tölke K, Sticher O (1988) J Planar Chromatogr 1:336
Nyiredy S, Dallenbach-Tölke K, Zogg GC, Sticher O (1990) J Chromatogr 499:453
Okano T, Masuda S, Kusunose S, Komatsu M, Kobayashi T (1984) J Chromatogr 294:460
Onocha PA, Okorie DA, Connolly JD, Rycroft DS (1995) Phytochemistry 40:1183
Orjala J, Wright AD, Erdelmeier CAJ, Sticher O, Rali T (1993) Helv Chim Acta 76:1481
Orjala J, Wright AD, Behrends H, Folkers G, Sticher O (1994) J Nat Prod 57:18
Osawa K, Yasuda H, Maruyama T, Morita H, Takeya K, Itokawa H (1994) Phytochemistry 37:1287
Ossipov V, Nurmi K, Loponen J, Haukioja E, Pihlaja K (1996) J Chromatogr A 721:59
Pabst A, Barron D, Sémon E, Schreier P (1992) Phytochemistry 31:1649
Pan H, Lundgren LN, Andersson R (1994) Phytochemistry 37:795
Pei-Wu G, Fukuyama Y, Yamada T, Rei W, Jinxian B, Nakagawa K (1988) Phytochemistry 27:1161
Perrut M (1994) J Chromatogr A 658:293
Perry NB, Foster LM (1994) Phytomedicine 1:233
Perry NB, Blunt JW, Munro MHG (1987) J Nat Prod 50:307
Perry NB, Albertson GD, Blunt JW, Cole ALJ, Munro MHG, Walker JRL (1991b) Planta Med 57:129
Perry NB, Blunt JW, Munro MHG (1991a) J Nat Prod 54:978
Pietrzyk DJ, Stodola JD (1981) Anal Chem 53:1822
Pietrzyk DJ, Cabill WJ, Stodola JD (1982) J Liq Chromatogr 5:443
Pinto DCG, Fuzzati N, Pazmino XC, Hostettmann K (1994) Phytochemistry 37:875
Pirillo A, Verotta L, Gariboldi P, Torregioni E, Bombardelli E (1995) J Chem Soc Perkin Trans I 583
Pistelli L, Spera K, Flamini G, Mele S, Morelli I (1996) Phytochemistry 42:1455
Poole CF, Poole SK (1991) Chromatography today. Elsevier, Amsterdam
Poppe L, Meisel T, Bela A, Novak L (1988) Chem Abstr 108:58450
Porsch B (1994) J Chromatogr A 658:179

Potterat O, Hostettmann K, Stoeckli-Evans H, Saadou M (1992) Helv Chim Acta 75:833
Prusiewicz K, Kaminski M, Klawiter J (1982) J Chromatogr 238:232
Rahman MMA, Dewick PM, Jackson DE, Lucas JA (1990) Phytochemistry 29:1971
Ramsteiner KA (1988) J Chromatogr 456:3
Rath G, Potterat O, Mavi S, Hostettmann K (1996) Phytochemistry 43:513
Reed JK, Gerrie J, Reed KL (1986) J Chromatogr 356:450
Reichardt PB, Merken HM, Clausen TP, Wu J (1992) J Nat Prod 55:970
Reichenbächer M, Gliesing S, Holpf G (1991) J Prakt Chemie 333:779
Rieser MJ, Gu Z, Fang X, Zeng L, Wood KV, McLaughlin JL (1996) J Nat Prod 59:100
Rodriguez S, Wolfender JL, Odontuya G, Purev O, Hostettmann K (1995a) Phytochemistry 40:1265
Rodriguez S, Wolfender JL, Hakizamungu E, Hostettmann K (1995b) Planta Med 61:362
Rogers CB, Subramony G (1988) Phytochemistry 27:531
Rücker G, Mayer R, Lee KR (1989) Arch Pharm 322:821
Rüedi P (1985) J High Res Chrom 8:256
Russell GB, Sirat HM, Sutherland ORW (1990) Phytochemistry 29:1287
Sachdev-Gupta K, Radke CD, Renwick JAA, Dimock MB (1993) J Chem Ecol 19:1355
Saito M, Yamauchi Y (1990) J Chromatogr 505:257
Sakuma S, Motomura H (1987) J Chromatogr 400:293
Sarker M, Yun T, Guiochon G (1996) J Chromatogr A 728:3
Scalbert A, Duval L, Peng S, Monties B, Du Penhoat C (1990) J Chromatogr 502:107
Schaufelberger D, Hostettmann K (1985) J Chromatogr 346:396
Schaufelberger D, Hostettmann K (1988) Planta Med 54:219
Schaufelberger D, Gupta MP, Hostettmann K (1987) Phytochemistry 26:2377
Schaufelberger DE, Koleck MP, Beutler JA, Vatakis AM, Alvarado AB, Andrews P, Marzo LV, Muschik GM, Roach J, Ross JT, Lebherz WB, Reeves MP, Eberwein RM, Rodgers LL, Testerman RP, Snader KM, Forenza, S (1991) J Nat Prod 54:1265
Schmidt J, Spengler B, Yokota T, Adam G (1993) Phytochemistry 32:1614
Schwarzenbach R (1985) J Chromatogr 334:35
Scott RPW (1995) Techniques and practice of chromatography. Marcel Dekker, New York
Scott RPW, Kucera P (1976) J Chromatogr 119:467
Seto T, Yasuda I, Akiyama K (1992) Chem Pharm Bull 40:2080
Shimizu Y (1985) J Nat Prod 48:223
Shirataki Y, Naguchi M, Yakoe I, Tominori T, Komatsu M (1991) Chem Pharm Bull 39:1568
Shkarenda VV, Kuznetsov PV (1992) Khim Prir Soedin 155
Simpson CF (1982) Techniques in liquid chromatography. John Wiley, Chichester
Slimestad R, Marston A, Mavi S, Hostettmann K (1995) Planta Med 61:562
Slimestad R, Marston A, Hostettmann K (1996) J Chromatogr A 719:438
Smith RM (1984) J Chromatogr 291:372
Smith AB, Barbosa J, Wong W, Wood JL (1996) J Am Chem Soc 118:8316
Snyder LR (1992) In:Heftmann E (ed) Chromatography, 5th ed (Journal of Chromatography Library Series, Vol 51A). Elsevier, Amsterdam, p 1
Snyder LR, Kirkland JJ (1979) Introduction to modern liquid chromatography, 2nd ed. John Wiley, New York
Sordat Diserens I, Rogers C, Sordat B, Hostettmann K (1992a) Phytochemistry 31:313
Sordat-Diserens I, Hamburger M, Rogers C, Hostettmann K (1992b) Phytochemistry 31:3589
Speranza G, Manitto P, Cassara P, Monti D, de Castri D, Chialva F (1993) J Nat Prod 56:1089
Stadler M, Dagne E, Anke H (1994) Planta Med 60:550
Still WC, Kahn M, Mitra A (1978) J Org Chem 43:2923
Stuppner H, Müller EP (1993) Phytochemistry 33:1139
Stuppner H, Müller EP, Mathuram V, Kundu AB (1993) Phytochemistry 32:375
Suzuki H, Koike Y, Takamatsu S, Sekine T, Saito K, Murakoshi I (1994) Phytochemistry 37:591
Swager TM, Cardellina JH (1985) Phytochemistry 24:805
Taber DF (1982) J Org Chem 47:1351
Tada M, Chiba K, Yamada H, Maruyama H (1991) Phytochemistry 30:2559

Taipale HAT, Vepsalainen J, Laatikainen R, Reichard PB, Lapinjoki SP (1993) Phytochemistry 34:755
Takizawa T, Kinoshita K, Koyama K, Takahashi K, Kondo N, Yuasa H, Kawai KI (1993) J Nat Prod 56:2183
Tan P, Hou CY, Liu YL, Lin LJ, Cordell GA (1992) Phytochemistry 31:4313
Tan RX, Wolfender JL, Ma WG, Zhong LX, Hostettmann K (1996) Phytochemistry 41:111
Tischler M, Cardellina JH (1993) J Liq Chromatogr 16:15
Tjarks LW, Spencer GF, Seest EP (1989) J Nat Prod 52:655
Tokuyama T, Nishimori N, Shimada A, Edwards MW, Daly JW (1987) Tetrahedron 43:643
Tomas-Barberan FA, Msonthi JD, Hostettmann K (1988) Phytochemistry 27:753
Unger KK (1994) Handbuch der HPLC. Teil 2:Präparative Säulenflüssig-Chromatographie. GIT Verlag, Darmstadt
Unger KK, Janzen R (1986) J Chromatogr 373:227
van Beek TA, Subrtova D (1995) Phytochem Anal 6:1
Verzele M, Dewaele C (1985) LC Magazine 3:22
Verzele M, Dewaele C (1986) Preparative high performance liquid chromatography. A practical guideline. RSL Europe, Eke, Belgium
Verzele M, Geeraert E (1980) J Chromatogr Sci 18:559
Verzele M, Dewaele C, van Dijck J, van Haver D (1982) J Chromatogr 249:231
Verzele M, De Conick M, Vindevogel J, Dewaele C (1988) J Chromatogr 450:47
Villegas M, Vargas D, Msonthi JD, Marston A, Hostettmann K (1988) Planta Med 54:36
Von Arx E, Richert P, Stoll R, Wagner K, Wuest KH (1982) J Chromatogr 238:419
Wainer IW, Crabos M, Cloix JF (1985) J Chromatogr 338:417
Wang Y, Hamburger M, Gueho J, Hostettmann K (1989) Phytochemistry 28:2323
Wang T, Hartwick RA, Miller NT, Shelly DC (1990a) J Chromatogr 523:23
Wang Y, Toyota M, Krause F, Hamburger M, Hostettmann K (1990b) Phytochemistry 29:3101
Wang Y, Hamburger M, Gueho J, Hostettmann K (1992) Helv Chim Acta 75:2699
Ward TJ, Armstrong DW (1986) J Liq Chromatogr 9:407
Wawrzynowicz T, Waksmundzak-Hajnos M (1990) J Liq Chromatogr 13:3925
West RR, Cardellina JH (1991) J Chromatogr 539:15
Weyerstahl P, Christiansen C, Marschall H (1995) Liebigs Ann 1039
Weyerstahl P, Marschall H, Collin G (1996) Liebigs Ann 99
White CM, Houck RK (1986) J High Res Chromatogr 9:4
Williams PJ, Strauss CR, Wilson B, Massy-Westropp RA (1982) J Chromatogr 235:471
Witherup KM, Bogosky MJ, Anderson PS, Ramjit H, Ransom RW, Wood T, Sardana M (1994) J Nat Prod 57:1619
Witt MF, Hart LP, Pestka JJ (1985) J Agric Food Chem 33:745
Wolfender JL, Hamburger M, Msonthi JD, Hostettmann K (1991) Phytochemistry 30:3625
Wolfender JL, Hamburger M, Hostettmann K, Msonthi JD, Mavi S (1993) J Nat Prod 56:682
Wu D, Lohse K, Greenblatt HC (1995) J Chromatogr A 702:233
Wünsche L, Keller U, Flament I (1991) J Chromatogr 552:539
Yaguchi E, Miyase T, Ueno A (1995) Phytochemistry 39:185
Yamasaki RB, Klocke JA, Lee SM, Stone GA, Darlington MV (1986) J Chromatogr 356:220
Yamasaki RB, Ritland TG, Barnby MA, Klocke JA (1988) J Chromatogr 447:277
Yamauchi Y, Saito M (1990) J Chromatogr 505:237
Yoshinari K, Sashida Y, Shimomura H (1989) Chem Pharm Bull 37:3301
Yuda M, Ohtani K, Mizutani K, Kasai R, Tanaka O, Jia M, Ling Y, Pu X, Saruwatari, Y (1990) Phytochemistry 29:1989
Zani CL, Marston A, Hamburger M, Hostettmann K (1993) Phytochemistry 34:89
Zhao W, Xu J, Qin G, Xu R, Wu H, Wenig G (1994) J Nat Prod 57:1613
Zief M, Crane LJ, Horvath J (1982a) International Lab 72
Zief M, Crane LJ, Horvath J (1982b) J Liq Chromatogr 5:2271
Zogg GC, Nyiredy Sz, Sticher O. (1989a) Chromatographia 27:591
Zogg GC, Nyiredy Sz, Sticher O. (1989b) J Liq Chromatogr 12:2031
Zogg GC, Nyiredy Sz, Sticher O. (1989c) J Liq Chromatogr 12:2049

CHAPTER 6

Preparative Gas Chromatography

The technique of gas chromatography (GC) is well known as an analytical tool but preparative applications are less common. Small-scale preparative GC is essentially analytical GC with the addition of a sample collection system at the column outlet; milligram or sub-milligram quantities can be conveniently handled. Isolation by repetitive injection is normally required. However, larger quantities of sample can also be handled on larger diameter columns (Henly and Royer 1969).

6.1 Columns

Most preparative-scale gas chromatography is run on *packed* columns (Zlatkis and Pretorius 1971). As in preparative liquid chromatography, a high sample throughput is the primary objective. To achieve this, two possibilities are:

- increasing the column diameter at constant length,
- increasing the column length at constant diameter.

For simple separations, a short and wide packed column is best, e.g. a column 1–3 m long and 6–10 cm i.d. For more complex separations, higher column efficiencies are required and long, narrow columns are used, e.g. 10–30 m long and 0.5–1.5 cm i.d. Alternatively, recycle chromatography can be applied in this latter case.

The support material is coarse with a narrow particle size distribution (35–40 mesh). High carrier gas flow rates (100–1000 ml/min) are used and a coarse column packing is needed to permit operation at a reasonable inlet pressure. As the stationary phase loading is greater and the column length may be longer than those used in analytical separations, operation at higher temperatures is also common.

The chromatographic performance of preparative columns is inferior to that of analytical columns, due to the radial uneveness of the packing structure. During the packing of large diameter columns the solid support particles segregate preferentially according to their size, with the larger particles being closer to the wall. The packing in this region is less dense due to the physical constraint of the wall. This leads to uneven flow profiles for the carrier gas and sample through the packed bed, inducing band broadening. For these reasons, various solutions have been proposed for a better column packing (e.g. Albrecht and Verzele 1970).

Separations on packed columns are usually carried out *isothermally*. Temperature programming is not possible with wide bore columns.

Open tubular columns have a lower sample capacity than packed columns but produce much higher resolutions. They are useful for isolating small quantities of components from complex mixtures, as shown in several examples below. Because of the narrow peak widths of the eluting compounds, a rapid switching between traps and waste is a necessity. Automation of repetitive injection and fraction collection is suggested in this case. Temperature gradients are often applied during separations on open tubular columns.

6.2 Injection

Injection volumes of 0.1 – 10.0 ml are common with packed columns and normal analytical GC injection techniques are impractical. The injection process is controlled by time (due to the limited thermal capacity of injection block heaters and their inability to flash vaporize large sample volumes) and an automated injector is often employed. An evaporation device between the injector and the column is introduced, maintained at a temperature 50 °C above the boiling point of the least volatile sample component. The sample should be transferred to the column as a rectangular plug diluted in the minimum volume of carrier gas. If it is not completely vaporized, a homogeneous distribution of sample over the cross-sectional area of the column is not obtained and excessive band broadening may result.

Vaporization problems may be avoided by direct and slow on-column injection without carrier gas flow. This is often the most practical approach for injecting large sample volumes on analytical instruments.

6.3 Sample Collection

Detection of the separated components is usually performed with a thermal conductivity detector or a flame-ionization detector. As FID is destructive, the exit of the column is fitted with a splitter so that a small quantity of the effluent is diverted to the detector and the remainder can be collected.

Collection of the vapours in the carrier gas is performed manually or automatically (Roeraade et al. 1986). In commercial, preparative-scale gas chromatographs, injection and sample collection are automated, while both processes are normally manual with an analytical instrument.

There are several methods (Poole and Poole 1991) for trapping the sample from the transfer line:

- packed and unpacked cold traps,
- solution and entrainment traps,
- total effluent and adsorption traps,
- Volman traps,
- electrostatic precipitators.

The equilibrium vapour pressure of the solute can be reduced by lowering the temperature, by dissolving the solute in a solvent or by adsorbing it onto a material of high surface area. The most common method is to reduce the temperature of the effluent, either in a trap filled with a material such as glass beads or column packing, or in a simple open tube trap (this can take the form of a U-tube cooled in dry ice-acetone or liquid nitrogen).

6.4 Applications

The main area of application of preparative GC is obviously the isolation of volatile compounds, in particular those which are not easily recovered in high yield from common solvents. It is also suitable for the isolation of milligram amounts of pure substances from mixtures too complex to be separated by liquid chromatography (Nitz et al. 1988).

Pheromones and substances important for flavour and fragrances are those classes most frequently encountered in preparative GC applications. Table 6.1 lists some recent separations of natural products by isolation procedures which involve the technique.

To illustrate the possibilities, one or two examples will be described in detail.

1) Volatile β-chromene from *Wisteria sinensis* flowers (Joulain and Tabacchi 1994).

Flowers of *Wisteria sinensis* (Leguminosae) give off a rosy, peppery and spicy fragrance. Vacuum cryogenic trapping (the "vacuum headspace" method) was used to obtain the volatile components; GC-MS and GC-FTIR allowed the location of the component with the characteristic odour and showed it to have a phenol group. For structure elucidation, freshly harvested flowers (4800 g) were extracted with hexane and the phenol fraction separated by standard procedures, to give 260 mg of a dark yellow liquid. This was subjected to preparative GC (30 µl injections) with thermal conductivity detection on a 3 m×4 mm glass column coated with 5% polymethylsiloxane on Chromosorb G, AW, DMCS (carrier gas helium at 32 ml/min). The temperature was programmed from 120 to 250 °C at 5 °C/min. A β-chromene was isolated and its structure determined as **1**.

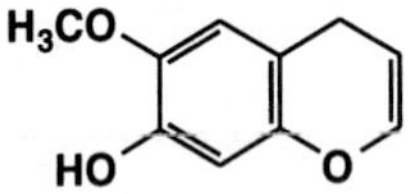

1

2) Sesquiterpenes from *Cupressus bakeri* (Cupressaceae) (Cool and Jiang 1995).

The foliage of the tree was extracted with n-hexane and the resulting oil was hydrodistilled over saturated NaCl solution with the addition of sodium bicarbonate. Final purification of thujopsenol α-epoxide was by preparative GC over a 4 m×4 mm column packed with 15% Carbowax 20 M/Chromosorb W under isothermal conditions at 190 °C. (+)-2-Ethylmenthone was obtained

Table 6.1. Separations of natural products by preparative gas chromatography.

Substances separated	Column[a]	Column dimensions	Reference
β-Chromenes from *Wisteria sinensis* (Leguminosae)	5% Polymethylsiloxane on Chromosorb G	3 m×4 mm	Joulain and Tabacchi 1994
Sesquiterpenes from *Aquilaria agallocha* (Thymelaeaceae)	PEG 20 M	2 m×3 mm	Ishihara et al. 1993
Sesquiterpenes from *Cupressus bakeri* (Cupressaceae)	15% Carbowax 20M on Chromosorb W; 10% SE-30 on Chromosorb W	4 m×4 mm	Cool and Jiang 1995
Sesquiterpenes from *Meum athamanticum* (Apiaceae)	Phase A on Chromosorb W-HP; Phase B on Chromosorb G-HP	1.8 m× 4.3 mm 2.0 m× 5.3 mm	König et al. 1996a
Sesquiterpene from liverworts	10% SE-30	1.8 m× 4.3 mm	Hardt et al. 1995
Sesquiterpene from liverwort *Lophocolea heterophylla* (Jungermanniales)	Phase C on Chromosorb W-HP	2.05 m× 5.1 mm	Rieck and König 1996
Sesquiterpenes from liverwort *Preissia quadrata*	Phase C on Chromosorb W-HP	2.05 m× 5.1 mm	König et al. 1996b
Sesquiterpenes from liverwort *Lophocolea bidentata* (Jungermanniales)	Phase A on Chromosorb W-HP	1.8 m× 4.3 mm	Rieck et al. 1995
Sesquiterpene alcohol from *Streptomyces citreus*	OV-1	25 m× 0.53 mm	Gansser et al. 1995
Unsaturated hydrocarbon from beetle *Psacothea hilaris*	HP-1	5 m× 0.53 mm	Fukaya et al. 1996
Phenolic compounds from beaver *Castor canadensis*	OV-101 Carbowax-20 M	3 m×10 mm 6 m×6 mm 3 m×6 mm 2 m×6 mm 1 m×6 mm	Tang et al. 1993

[a] Phase A = 5% heptakis-(2,6-di-*O*-methyl-3-*O*-pentyl)-β-cyclodextrin-OV-1701 (1:1 w/w); Phase B = 2.5% heptakis-(6-*O*-dimethylthexylsilyl-2,3-di-*O*-methyl)-β-cyclodextrin-SE-52 (20:80); Phase C = 6% octakis-(6-*O*-methyl-2,3-di-*O*-pentyl)-γ-cyclodextrin-PS-086 (1:1 w/w).

using columns of the same dimensions, first over 10% SE-30/Chromosorb W (isothermal 185 °C) and then over the Carbowax column (isothermal 165 °C). Finally, a new dimethylhexahydroindenyl isopropyl ketone (**2**) was purified over the SE-30 column under isothermal conditions at 170 °C. In all cases, the helium carrier gas flow rate was 50 ml/min, the injector temperature 190 °C and the detector temperature 200 °C.

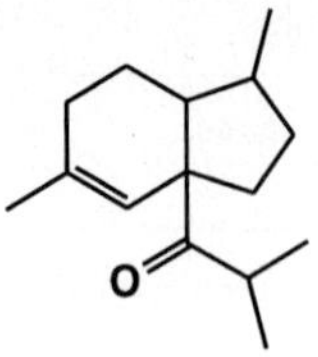

2

3) Sesquiterpene alcohol from *Streptomyces citreus* CBS 109.60 (Gansser et al. 1995).

Compounds from the fermentation broth of *S. citreus* were adsorbed onto activated Lewatit 1064. After separation of adsorbent, these compounds were desorbed with pentane-dichloromethane (2:1). Silica gel open-column chromatography gave ca. 16 mg of a pentane-dichloromethane (1:3) fraction containing seven major compounds. Preparative GC of this fraction was performed on a MCS Gerstel Series II instrument with cold injection system and autosampler. A Carbowax 20 M pre-column (1.5 m × 0.53 mm, film thickness 2 μm) was connected to an OV-1 fused silica capillary column (25 m × 0.53 mm, film thickness 2 μm). Two ports (PC1 and PC2), allowing the removal of undesired components from the sample, were located between the cold injection system and the pre-column, and between the pre-column and the main column. The temperature of the cold injection system (45 °C) after injection was rapidly raised (10 °C /sec) to 250 °C and then held for 1 min. The oven temperature was programmed as follows: 130 °C isothermal for 1 min, linear gradient to 220 °C at 3 °C /min, isothermal at 220 °C for 10 min. Other temperatures were: FID 200 °C, supply pipe and distributor 240 °C, trap cooling 0 °C. The carrier gas was hydrogen at 5 ml/min. PC1 was switched to blow-out status 1 min after injection and PC2 after 18 min. The sesquiterpene alcohol **3** (4*S*,7*R*-germacra-1(10)*E*,5*E*-diene-11-ol) was eluted after 16.5 min and collected in cooled glass traps (yield 5.9 mg).

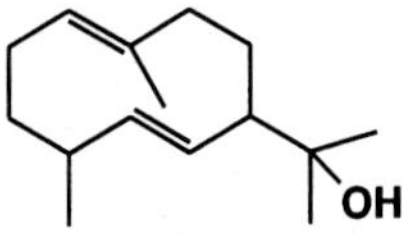

3

4) Sesquiterpenes from liverworts (Rieck and König 1996).

Extensive use of preparative GC has been made by König and co-workers for the isolation of sesquiterpenes from liverworts. These structurally diverse sesquiterpenes are the major constituents of the essential oils in this plant family. Packed GC columns with coatings of modified cyclodextrins, as developed for the preparative separation of enantiomers (Hardt and König 1994; see also Chap. 9), have been used. For example, the liverwort *Lophocolea heterophylla* (Jungermanniales) has been investigated. It is widespread in Europe and known for its mossy fragrance. The essential oil was prepared by steam distillation of aqueous homogenates of the fresh plant using n-hexane as collection solvent.

Isolation of a furano-eudesmane alcohol (**4**) was performed by preparative GC on a Varian 1400 instrument equipped with a stainless steel column (2.05 m×5.1 mm) coated with 6% octakis-(6-*O*-methyl-2,3-di-*O*-pentyl)-*γ*-cyclodextrin-PS-086 (1:1 w/w) on Chromosorb W-HP. Helium was used as carrier gas at a flow rate of 240 ml/min.

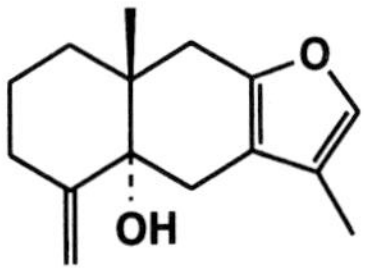

4

5) Female contact sex pheromone of the yellow-spotted longicorn beetle (*Psacothea hilaris*) (Fukaya et al. 1996).

The female yellow-spotted longicorn beetle releases a sex pheromone with at least two components with different functions. In order to identify these components, the elytra from 180 females were dipped in diethyl ether for 24 h. The extract was chromatographed on silica gel, eluting successively with hexane, a step gradient of diethyl ether in hexane, diethyl ether. The active fractions were fractionated on silica gel impregnated with 10% silver nitrate; elution was with hexane, followed by 0.5%, 1%, 2% and 5% diethyl ether in hexane. The active 0.5% ether fraction was purified by preparative GC on a large-diameter capillary column (HP-1, 5 m×0.53 mm, 2.65 μm film thickness) connected to the cool on-column injector. The GC column effluent (30 ml/min) was split at the outlet to the transfer tube and the FID in a ratio of 5:1. The compounds from the transfer tube were collected in a glass capillary (12.7 cm×1 mm) at room temperature. Helium was used as carrier gas. The active compound, inducing typical mating behaviour in males (holding, mounting and abdominal bending) was identified as (*Z*)-21-methyl-8-pentatriacontene.

6) Phenolic compounds from male castoreum of the North American beaver (*Castor canadensis*) (Tang et al. 1993).

Castoreum is a strong smelling brown paste contained in the castor sacs of beaver. It is used in beaver for scent communication and also for medicinal purposes in humans. Male castor sacs were extracted with dichloromethane and the phenols were separated by standard procedures. Two different packed columns were used for isolations of the phenolics: a) 5% OV-101 on Chromosorb W (3 m×10 mm); temperature programme for low-boiling fractions 75 °C (0 min), 2 °C /min, 210 °C (10 min); temperature programme for high-boiling fractions 100 °C (0 min), 4 °C /min, 230 °C (5 min); b) 4% Carbowax-20 M on Chromosorb G (6, 3, 2 and 1 m×6 mm; for less volatile components, a shorter column was used); temperature programme 100 °C (0 min), 8 °C /min, 220 °C (10 min). The GC fractions were collected in glass capillary tubes (30 cm×3 mm) using a Brownlee-Silverstein thermal gradient collector.

A total of nine different phenolics were isolated by the two types of preparative GC column: phenol, *o*-cresol, *m*-cresol, *p*-ethylphenol, *p*-methylguaiacol, *p*-propylphenol, *p*-ethylguaiacol, *p*-n-propylguaiacol, 2,6-dimethoxy-4-methylphenol.

6.5 References

Albrecht J, Verzele M (1970) J Chromatogr Sci 8:586
Cool LG, Jiang K (1995) Phytochemistry 40:177
Fukaya M, Yasuda T, Wakamura S, Honda H (1996) J Chem Ecol 22:259
Gansser D, Pollak FC, Berger RG (1995) J Nat Prod 58:1790
Hardt IH, Rieck A, König WA, Muhle H (1995) Phytochemistry 40:605
Henly RS, Royer DJ (1969) Methods Enzymology 14:450
Ishihara M, Tsuneya T, Uneyama K (1993) Phytochemistry 33:1147
Joulain D, Tabacchi R (1994) Phytochemistry 37:1769
König WA, Rieck A, Saritas Y, Hardt IH, Kubeczka KH (1996a) Phytochemistry 42:461
König WA, Bülow N, Fricke C, Melching S, Rieck A, Muhle H (1996b) Phytochemistry 43:629
Nitz S, Drawert F, Albrecht M, Gellert U (1988) J High Res Chromatogr 11:322
Poole CF, Poole SK (1991) Chromatography today. Elsevier, Amsterdam
Rieck A, König WA (1996) Phytochemistry 43:1055
Rieck A, Bülow N, König WA (1995) Phytochemistry 40:847
Roeraade J, Blomberg S, Pietersma HDJ (1986) J Chromatogr 356:271
Tang R, Webster FX, Müller-Schwarze D (1993) J Chem Ecol 19:1491
Zlatkis A, Pretorius V (1971) Preparative gas chromatography. Wiley, New York

CHAPTER 7

Countercurrent Chromatography

Countercurrent chromatography (CCC) is a separation technique which relies on the partition of a sample between two immiscible solvents, the relative proportions of solute passing into each of the two phases being determined by the respective partition coefficients. It is an all-liquid method which is characterized by the absence of a solid support, and thus has the following advantages over other chromatographic techniques:

- no irreversible adsorption of the sample (solutes are readily retained by a solid support; the adsorption effects are often manifested by elution curves which show tailing of the substances under consideration),
- quantitative recovery of the introduced sample,
- risk of sample denaturation greatly reduced,
- low solvent consumption,
- favourable economics.

However, elimination of a solid support and maintenance of solvent flow (e.g. countercurrent flow) create a number of problems (Ito and Bowman 1971; Mandava and Ito 1988):

a) how to retain the stationary phase against a steady flow of the mobile phase,
b) how to divide the chromatography column into numerous partition units,
c) how to minimise laminar flow spreading of sample bands,
d) how to increase interfacial contact,
e) how to reduce mass transfer resistance, i.e. how to mix the phases.

This chapter will show how these problems have been overcome, mainly due to the pioneering work of Ito, who introduced the term CCC (Ito 1981) to describe this liquid chromatography technique, which has a liquid stationary phase and an immiscible liquid mobile phase (Berthod 1995).

Countercurrent chromatography is basically an outgrowth of countercurrent distribution (CCD), a method developed in the 1940s and 1950s for the batchwise (Craig Distribution) (Craig 1944) or continuous (O'Keefe Distribution) fractionation of mixtures (Hecker 1955). In practice, the CCD apparatus consists of several hundred elements, in each of which a partition of the solute between two liquid layers is performed before transferring one of the layers to the next element. In CCD, equilibration is complete before phase transfer. Countercurrent chromatography, on the other hand, is a continuous non-equilibrium process comparable to LC, since the flow is continuous. Furthermore, the considerable dilution of the sample which occurs during the repetition of the partition process during CCD is avoided to a large measure in CCC.

Since the first edition of this book, there has been an explosion of work on the new CCC techniques and five monographs have been published (Conway 1990; Conway and Petroski 1995; Foucault 1995; Ito and Conway 1996; Mandava and Ito 1988). Some of the more widely-used instruments and their applications will be described in this chapter.

7.1 Droplet Countercurrent Chromatography

Droplet countercurrent chromatography (DCCC), as described by Tanimura et al. (1970), exploits the observation that a light liquid phase with low wall surface affinity forms discrete droplets that rise through a heavy non-miscible phase with visible evidence of very active interfacial motion. Under ideal conditions, each droplet could become a "plate" if kept more or less discrete through the system (Ito and Bowman 1970a). By partitioning a solute between the stationary phase and the droplets, a separation is achieved. Many applications of this chromatographic technique in the fields of natural product chemistry and biochemistry, for the separation of very diverse molecules, have been reported over the last few years (Hostettmann 1980; Hostettmann et al. 1984b; Marston and Hostettmann 1994).

When the partition coefficients of two compounds dispersed between two immiscible solvents differ greatly, a one-step extraction process with a separatory funnel may be sufficient to separate the compounds. However, as the structures of the compounds approach one another, the differences in their partition coefficients become smaller and mixtures are more difficult to separate, ultimately requiring multistep extractions. To achieve such separations, countercurrent distribution (CCD) instruments, so-called because both solvents move in opposite directions, have been designed (Craig and Craig 1956). These machines involve the rather complicated mechanical process of agitating two immiscible liquids in a series of tubes, allowing the layers to separate and then transferring the layers to neighbouring tubes, also containing solvent.

Droplet countercurrent chromatography, however, allows the continuous passage of droplets of a mobile phase through a stationary liquid phase, effectively giving a constantly changing interface. Consequently the solute is repeatedly partitioned between two phases. This method is much less cumbersome and complex than conventional countercurrent distribution machines and avoids the problems of emulsion or foam formation. In addition, separation times are shorter and solvent consumption is considerably reduced (Hostettmann 1980).

Mixtures ranging from milligram to gram quantities can be separated by DCCC and there is no loss of material when performing a DCCC run.

7.1.1 Apparatus

In a typical DCCC apparatus, consisting of 200 to 600 vertical columns (20–60 cm in length) of narrow-bore silanized glass tubing, interconnected in series by capillary Teflon tubes, droplets of mobile phase are passed through the columns filled with stationary phase. Collection of mobile phase is effected at the end of the battery of columns (Fig. 7.1).

In practice, a suitable two-phase solvent system for the separation of a mixture is chosen and the whole instrument is then filled with the selected stationary phase. The sample, dissolved in the lighter phase or the heavier phase or in a mixture of both phases, is injected into the sample chamber. Mobile phase is subsequently pumped via the sample chamber into the first of the columns, displacing the sample and forming a stream of droplets in the immiscible stationary phase. Depending on the choice of solvents for the mobile and stationary phases, these droplets are made either to ascend or descend through the columns. The capillary tubing connecting the columns does not allow entry of the stationary phase, except at the very beginning of the separation.

As the mobile phase moves through the column in the form of droplets, turbulence within the droplets promotes efficient partitioning of the solute between the two phases. Separation occurs according to the difference in the partition coefficients of the components of the sample. The size and mobility of the droplets are functions of the column bore, the flow rate of the mobile phase, the diameter of the inlet tip, the difference in specific gravities of the two liquid phases, the viscosity of the solvents and the surface tension. In general, columns of less than 1 mm internal diameter produce flow plugs, in which the entire contents of the tube are displaced.

Columns with an internal diameter of 2 mm are most often used for separations but internal diameters of 2.7, 3.0 and 3.4 mm are now commercially avail-

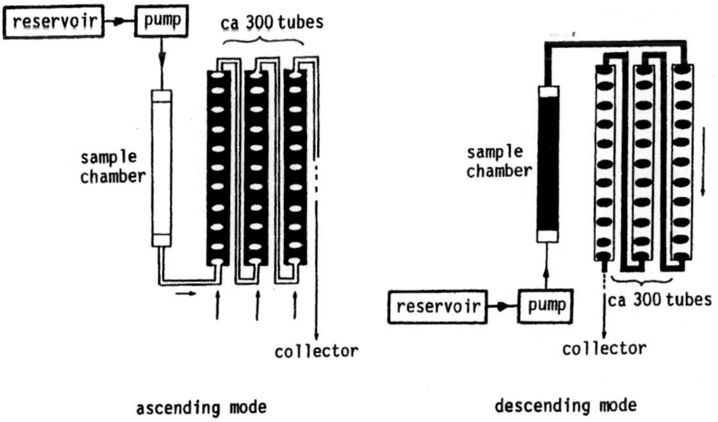

Fig. 7.1. Schematic illustration of the principle of DCCC. (Reprinted with permission from Hostettmann 1980)

able (Hostettmann et al. 1984b). Columns with a wider bore have the advantages that higher rates of flow are possible and that there is a higher sample loading capacity. Increasing the flow rate actually leads to an *increase* in resolution (Hostettmann et al. 1984a). Resolution can also be improved by increasing the number of columns but separation times are lengthened accordingly. Care must be taken, however, to avoid a very high flow rate, as this can lead to elution of the stationary phase.

Sample loadings of up to 6.4 g of crude extract on columns of i.d. 3.4 mm have been performed (Hostettmann et al. 1984b) and even sample sizes of 16 g have been possible with 10 mm i.d. columns (Komori et al. 1983).

7.1.2 Choice of Solvents

Binary solvent systems are impractical for the formation of suitable droplets because of the large difference in polarity between the two components. Ternary (or quaternary) systems are required for the preparation of the two phases, such that the addition of a third (or fourth) component, miscible with the other components, diminishes the difference in polarity between the two phases (Fig. 7.2) (Hostettmann et al. 1984b). The selectivity of the system is thus increased, allowing the separation of closely-related substances. Furthermore, the interfacial tension and the viscosity of the system should be diminished by addition of the third component. The most frequently employed components of DCCC two-phase systems are indicated in Table 7.1.

Methanol is often used as auxiliary solvent because a small change in the amount of methanol leads to a large change in the polarity of the system.

The frequently-used Craig countercurrent distribution system methanol-hexane-water is unsuitable because of a tendency to form plugs in the narrow-bore columns.

The use of aqueous solvent systems introduces limitations concerning the range of compounds which may be separated by DCCC. Weakly polar or non-polar substances cannot generally be chromatographed under these conditions because they tend to elute with the solvent front, even when the organic phase is the mobile phase. In addition, precipitation of weakly polar compounds may occur in the DCCC apparatus. However, *non-aqueous* systems capable of droplet formation do exist (Hostettmann et al. 1984b). Basically the same criteria hold as for aqueous systems, i.e. a third or fourth component of the

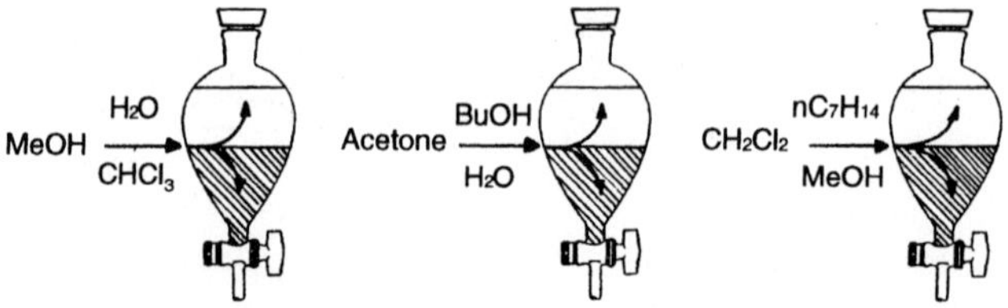

Fig. 7.2. The preparation of ternary solvent systems for DCCC

Table 7.1. Solvent systems for droplet countercurrent chromatography

Substance class	Basic binary system	Auxiliary solvent
Very polar	*n*-Butanol-water	Methanol, ethanol, acetone, acetic acid, (pyridine)
Polar	Chloroform-water	Methanol, *n*-propanol, *i*-propanol
Lipophilic	*n*-Heptane-methanol	Chlorinated hydrocarbons, acetone
	n-Heptane-acetonitrile	Chlorinated hydrocarbons, acetone

solvent is required to decrease the interfacial tension, leading to better droplets (Table 7.1). This variant is important for the separation of weakly polar compounds which are unstable in the presence of water or which decompose during chromatography on silica gel.

Numerous solvent systems have been used for the separation of natural products by DCCC (Hostettmann 1980; Hostettmann et al. 1984b; Marston and Hostettmann 1994) and some representative examples are listed in Table 7.2. Since the glass tubes and interconnecting Teflon capillaries are chemically inert, the DCCC apparatus is compatible with solvent systems containing acids, alkalis and all organic solvents.

A rapid TLC method exists for the choice of two-phase aqueous solvent system that is most applicable to the separation of a given mixture or extract. A thin-layer chromatogram on silica gel is run, with the water-saturated organic layer (i.e. the organic phase of the two-phase solvent system) as eluent. If the R_f values of the compounds to be separated are higher than about 0.5 (less polar solutes), given that there is a suitable separation of the components of the mixture, the less polar layer is suitable for use as the mobile phase. For more polar solutes ($R_f < 0.5$), the more polar layer should be used as the mobile phase (Fig. 7.3).

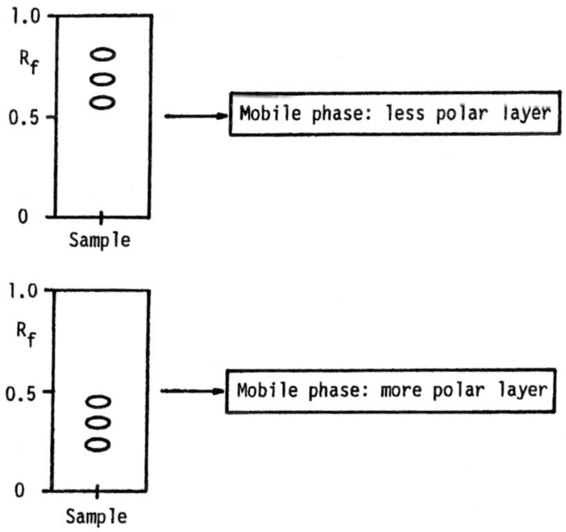

Fig. 7.3. Choice of DCCC modes by TLC. Support:silica gel. Eluent: less polar layer of 2-phase solvent system

Table 7.2. Aqueous and non-aqueous DCCC solvent systems

Solvent system	Substances separated
Aqueous systems	
$CHCl_3$-MeOH-H_2O 7:13:8 and other proportions	Glycosides, tannins, alkaloids, antibiotics etc.
$CHCl_3$-MeOH-acetate buffer pH 3.6 9:12:8	Alkaloids
$CHCl_3$-MeOH-1% aq. NH_3 7:12:8	Basic steroid saponins
$CHCl_3$-MeOH-5% HCl 5:5:3	Alkaloids
$CHCl_3$-MeOH-*n*PrOH –0.5% aq. $CaCl_2$ 50:60:6:40	Glycolipids
$CHCl_3$-MeOH-*i*PrOH-H_2O 5:6:1:4	Saponins, flavonoid glycosides
$CHCl_3$-MeOH-*n*PrOH-H_2O 9:12:1:8	Saponins, iridoid glycosides, xanthone glycosides, flavonoid glycosides, anthraquinone glycosides
$CHCl_3$-MeOH-*n*BuOH-H_2O 10:10:1:6	Flavonoid glycosides
$CHCl_3$-MeOH-*n*PrOH-EtOH-H_2O 9:6:1:8:8	Saponins
$CHCl_3$-MeOH-C_6H_5Me-H_2O 5:7:5:2	Alkaloids
CH_2Cl_2-MeOH-H_2O 8:13:7	Saponins
*n*BuOH-HOAc-H_2O 4:1:5	Anthocyanins, flavonoid glycosides, peptides
*n*BuOH-MeOH-H_2O 5:1:5	Naphthalide glycosides
*n*BuOH-*n*PrOH-H_2O 2:1:3	Tannins
*n*BuOH-EtOH-H_2O 4:1:5	Iridoid and secoirdoid glycosides
*n*BuOH-Me_2CO-H_2O 33:10:50	Sennosides
*n*BuOH-TFA-H_2O 120:1:160	Peptides
Hexane-EtOAc-EtOH-H_2O 5:2:4:1	Gibberellins, anthraquinones
Hexane-Et_2O-*n*PrOH-95% EtOH-H_2O 4:8:3:5:4	Diterpenoids, coumarins
Non-aqueous systems	
Hexane-EtOAc-$MeNO_2$-MeOH 8:2:2:3	Essential oils
Heptane-$C_2H_4Cl_2$-MeOH 47:6:72	Triterpenes, steroids
Heptane-Me_2CO-MeOH 5:1:4	Triterpenes, steroids, depsides
Heptane-CH_2Cl_2-CH_3CN 10:3:7	Triterpenes, steroids

An example of the elution pattern with different modes, using the same two-phase solvent system, is shown in Fig. 7.4, for the separation of xanthones.

DCCC of a fraction from the crude extract of *Gentiana strictiflora* (Gentianaceae) with $CHCl_3$-MeOH-*n*PrOH-H_2O 9:12:1:8 as solvent system gave the upper elution curve (Fig. 7.4a) when the mobile phase was the more polar upper layer (ascending mode) and the lower elution curve (Fig. 7.4b) when the mobile phase was the less polar lower layer (descending mode) (Hostettmann et al. 1979). In the ascending mode, the more polar compound (**1**) elutes first, whereas the less polar compound (**2**) elutes first in the descending mode. When the R_f values lie in the range 0.40–0.60, the separation can be achieved by using either the more polar layer or the less polar layer as the mobile phase. If the R_f values of the compounds to be separated are either too low (<0.20–0.30) or too high (>0.70–0.80), no elution will take place in a

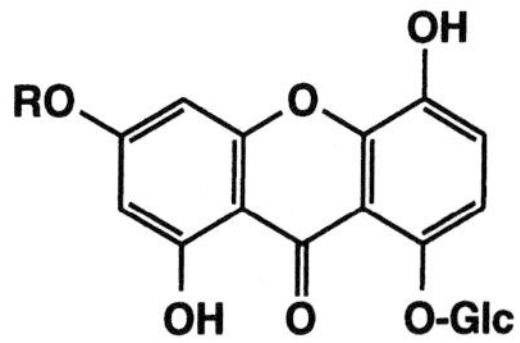

1 R = H (R_f = 0.35 on TLC with $CHCl_3$-MeOH-nPrOH-H_2O 9:12:1:8 lower phase)
2 R = CH_3 (R_f = 0.41 on TLC with $CHCl_3$-MeOH-nPrOH-H_2O 9:12:1:8 lower phase)

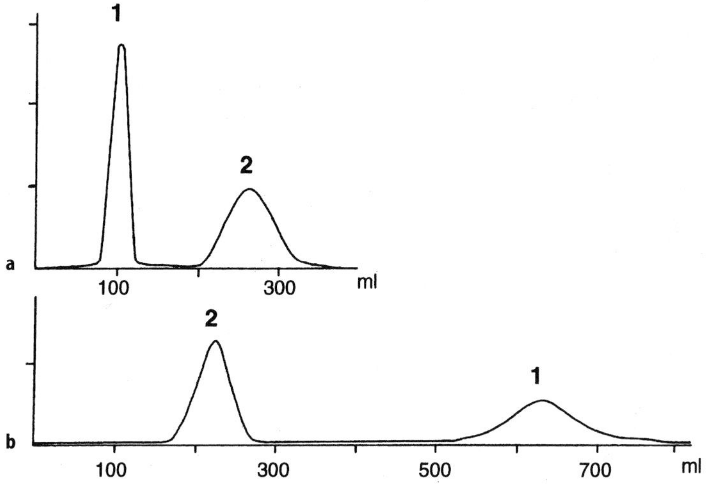

Fig. 7.4 a, b. DCCC separation of xanthone *O*-glycosides with $CHCl_3$-MeOH-*n*PrOH-H_2O 9:12:1:8: **a** ascending mode; **b** descending mode. (Reprinted with permission from Hostettmann et al. 1979)

reasonable time, although the compounds can be recovered quantitatively from the remaining stationary phase.

The separation of the components of a mixture having a wide range of polarities is possible by DCCC if first the organic phase is used as mobile phase, to elute less polar compounds. The polar constituents can then be recovered from the apparatus using the aqueous phase as the mobile phase for their separation ("reversed-phase" operation).

For non-aqueous solvent systems, TLC on silica gel is not suitable as the separation is based on adsorption. Thus chemically-bonded phases, e.g. RP-8, may be substituted for silica gel plates (Domon et al. 1982). For example, *n*-heptane-containing systems (Table 7.2) can be investigated by HPTLC with the lower layer as solvent (Domon et al. 1982).

An alternative method of selecting suitable solvents is to monitor the distribution of 5–10 mg of sample between 5–10 ml of each of the two phases constituting the solvent system (Snyder et al. 1984). A system in which 15–25% of the sample is distributed in one of the phases is chosen. The stationary phase is chosen as the phase that contains the most sample.

7.1.3 Applications

The first documented use of DCCC was in the base-line separation of an artificial mixture of seven DNP-amino-acids, with $CHCl_3$-HOAc-0.1 mol/l HCl 2:2:1 as solvent system, in the ascending mode. Resolution was moderate (900 theoretical plates for dinitrophenyl-alanine) (Tanimura et al. 1970). Subsequent applications have, in the main, involved the particular suitability of the method for the preparative separation of polar compounds (solid support not required, direct introduction of crude extracts possible, wide range of solvents etc.).

The separation of polar natural products by DCCC is now a routine laboratory technique and has met with considerable success, especially in the fields of polyphenols, monoterpene glycosides and saponins (Hostettmann 1980; Hostettmann et al. 1984b; Marston and Hostettmann 1991, 1994). However, with the continuing improvements made to centrifugal partition chromatography, the importance of DCCC is now declining.

Chloroform-methanol-water systems of varying compositions are the most widely used, as a result of their good droplet-forming characteristics and convenient viscosities (see Table 7.2). In addition, as previously mentioned, small differences in the methanol contents lead to large polarity differences of the systems. Chlorinated solvents, which have high densities and low viscosities, are particularly suited for use as descending mobile phases.

With non-aqueous solvent systems, more separations of less polar compounds are possible, e.g. mono-, di- and triterpenes, iridoids and steroids (Hostettmann et al. 1984a).

7.1.3.1 *Polyphenols*

The acidic hydroxyl groups in polyphenols often cause irreversible adsorption on solid stationary phases during column chromatographic procedures and it would appear that DCCC is ideally suited to the separation of this class of compounds. Furthermore, polyphenols often occur as polar glycosides and these are readily soluble in the solvent systems employed in DCCC. Indeed, many polyphenol glycosides have been successfully purified by DCCC, the flavone glycosides providing a large number of applications of the method. In Table 7.3, a number of solvent systems for polyphenols are listed and it is immediately obvious that chloroform-methanol-water mixtures of different compositions are the most widely-used. The viscosities of chloroform-methanol-water mixtures are ideal and the speeds of separation are relatively high.

Droplet countercurrent chromatography of *flavonoids* has been reviewed (Hostettmann and Hostettmann 1982) and here, too, chloroform-methanol-water systems predominate. For example, $CHCl_3$-MeOH-H_2O 13:7:8 (lower organic phase as mobile phase) was used in the isolation of 11 kaempferol, quercetin and isorhamnetin glycosides from the leaves of *Quercus pubescens*, *Q. cerris* and *Q. ilex* (Cupuliferae) (Romussi et al. 1991).

Table 7.3. Selected DCCC solvent systems for polyphenols

Class of polyphenol	Solvent
Flavonol aglycones	$CHCl_3$-MeOH-*n*BuOH-H_2O 10:10:1:6
Flavonol glycosides	$CHCl_3$-MeOH-H_2O 7:13:8
	$CHCl_3$-MeOH-*i*PrOH-H_2O 5:6:1:4
Flavone glycosides	$CHCl_3$-MeOH-*n*PrOH-H_2O 5:6:1:4
C-Glycosylflavonoids	$CHCl_3$-MeOH-*n*BuOH-H_2O 10:10:1:6
Isoflavone glycosides	$CHCl_3$-MeOH-H_2O 7:13:8
Anthocyanins	*n*BuOH-HOAc-H_2O 4:1:5
Tannins	*n*BuOH-*n*PrOH-H_2O 2:1:3
Anthraquinones	Hexane-EtOAc-EtOH-H_2O 5:2:4:1
Anthraquinone glycosides	$CHCl_3$-MeOH-H_2O 5:5:3
Phenyl glycosides	$CHCl_3$-MeOH-H_2O 7:13:8
Biphenyl glycosides	$CHCl_3$-MeOH-H_2O 5:5:3

Dihydroflavonol glucosides have been isolated from the needles of Norway spruce (*Picea abies*, Pinaceae) with DCCC as an intermediate purification step. The 80% methanol extract of the needles was applied to Amberlite XAD-7 resin and eluted first with water and then with methanol. The methanol fraction was chromatographed by DCCC (250 columns; 40 cm × 3.4 mm; flow rate 16 ml/h) with the upper phase of $CHCl_3$-MeOH-*n*BuOH-H_2O 10:10:1:8. Aromadendrin 7-*O*-*β*-D-glucopyranoside (**3**) was obtained directly from DCCC fractions 14–20 after gel filtration over Sephadex LH-20 (40 cm×10 mm; MeOH-H_2O 1:1; 18 ml/h) (Slimestad et al. 1994).

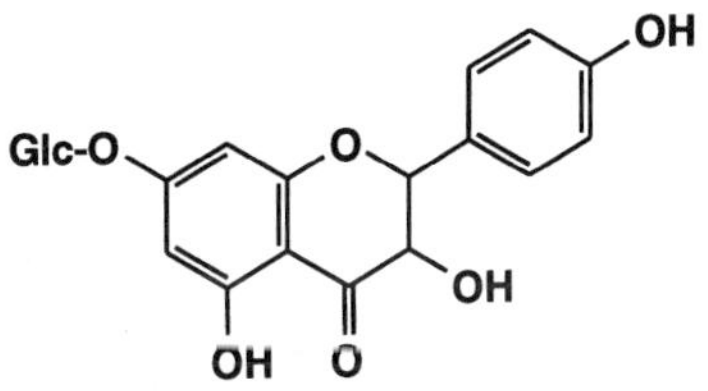

3

Other separations of flavonoids include:

- lipophilic flavonoids from *Orthosiphon spicatus* (Lamiaceae) ($CHCl_3$-MeOH-H_2O 13:7:4; descending) (Malterud et al. 1989),
- flavonol glycosides from the leaves of *Alangium premnifolium* (Alangiaceae) ($CHCl_3$-MeOH-*i*PrOH-H_2O 9:12:2:8) (Kijima et al. 1995),
- quercetin 3-*O*-glucuronide from *Vaccinium uliginosum* (Ericaceae) ($CHCl_3$-MeOH-PrOH-5% HOAc 31.2:37.5:6.2:25; descending) (Gerhardt et al. 1989),
- *O*-glycosyl-*C*-glycosylflavones from *Galipea trifoliata* (Rutaceae) ($CHCl_3$-MeOH-*n*BuOH-H_2O 10:10:2:6; descending) (Bakhtiar et al. 1994),
- biflavonoids from the moss *Dicranoloma robustum* (Dicranaceae) ($CHCl_3$-MeOH-*n*BuOH-H_2O 10:10:1:6; descending) (Markham et al. 1988).

Tannins often exist as complicated mixtures in plants and are very difficult to purify. The operation of DCCC with solvents of high polarity has meant that this technique lends itself particularly well to the separation of tannins. Typical for this class of natural products are the separations of the ellagitannins geraniin (nBuOH-Me_2CO-H_2O 7:2:11; ascending) and galloylgeraniin (nBuOH-PrOH-H_2O 2:1:3; descending) from *Spondias mombin* (Anacardiaceae) (Corthout et al. 1991).

Anthocyanins form another class of polyphenolics which gives separation problems. This is a result of their instability under any but acidic conditions and their irreversible adsorption to conventional column packing materials. DCCC with acid-containing solvents has proved ideal for the separation of anthocyanins, as shown in the isolation of cyanidin-3-sambubioside (**4**) from the fruits of the common elder *Sambucus nigra* (Caprifoliaceae). The fruits were extracted with acidified methanol and then partitioned against ethyl acetate. The polar fraction was chromatographed on an Amberlite XAD-7 column and then injected onto an Eyela Tokyo Rikakikai Model A DCCC instrument (300 glass columns, 40 cm × 2 mm). Elution was with the lower phase of nBuOH-HOAc-H_2O (4:1:5). The pure anthocyanin was obtained by a final Amberlite XAD-7 step, followed by Sephadex LH-20 gel filtration (Andersen et al. 1991).

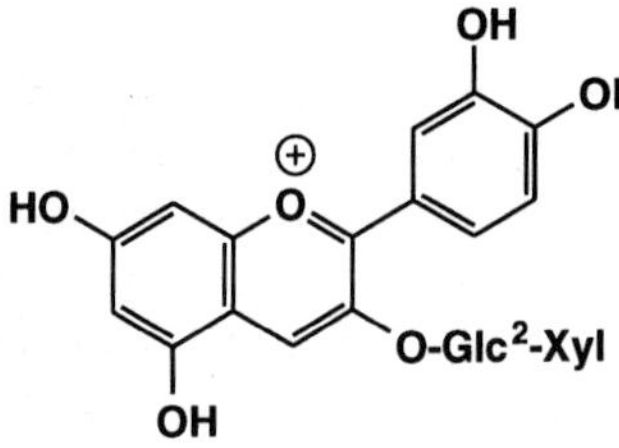

4

Proanthocyanidins have been isolated from the stem bark of *Pavetta owariensis* (Rubiaceae) using DCCC as an intermediate fractionation step, with the solvent system BuOH-PrOH-H_2O 2:1:3 (ascending mode) (Baldé et al. 1991).

In the separation of *anthraquinones*, it was possible to resolve the in vitro oxidation products of aloin by DCCC. Thus, the diastereoisomers 10-hydroxyaloin A (**5**) and 10-hydroxyaloin B (**6**) were separated on 300 DCCC columns

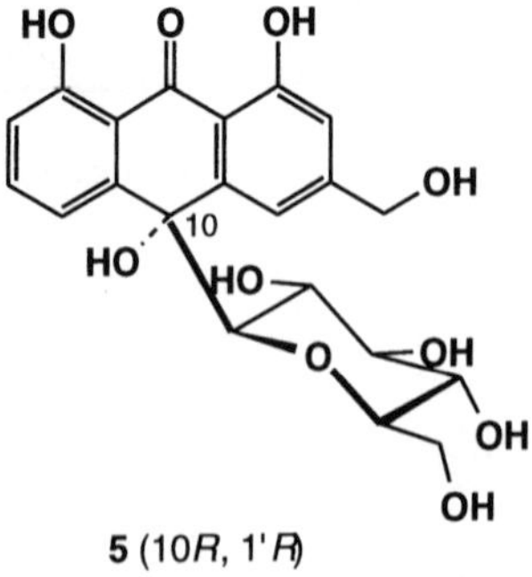

5 (10*R*, 1'*R*)

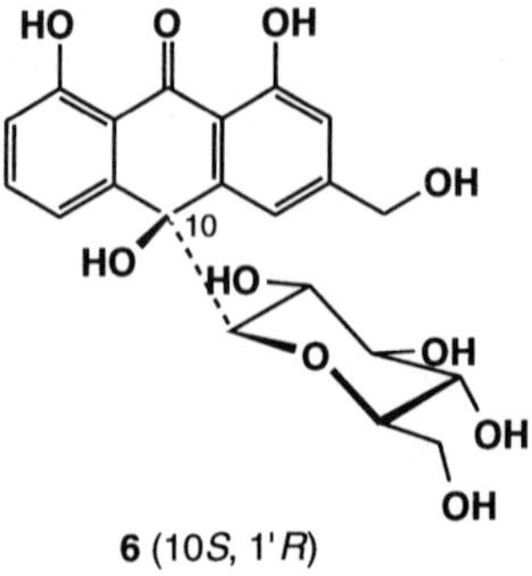

6 (10*S*, 1'*R*)

(40 cm × 2 mm) with the solvent $CHCl_3$-MeOH-H_2O 7:13:8 (descending mode; 638 mg sample; flow rate 20 ml/h) (Rauwald and Lohse 1992).

A bianthrone *C*-glycoside has been directly isolated from a diethyl ether extract of the tubers of *Asphodelus ramosus* (Liliaceae) by DCCC. The extract was first eluted with $CHCl_3$-MeOH-H_2O 4:4:3 in the ascending mode and then the contents of the tubes were discharged as six fractions. The *C*-glycoside was found in the last fraction discharged (Adinolfi et al. 1989).

Two DCCC steps were employed in the purification of the *chromone* furoaloesone (**7**) from Cape aloe (*Aloe ferox* and hybrids, Liliaceae). The separations were performed on a Büchi 670 apparatus (300 columns; 40 cm × 2.7 mm), first with $CHCl_3$-MeOH-H_2O 4:4:3 (ascending mode, 60 ml/h) and then with the non-aqueous system hexane-$MeNO_2$-EtOAc-MeOH 8:2:2.5:3 (descending mode, 30 ml/h). Final purification was by semi-preparative HPLC (Speranza et al. 1993).

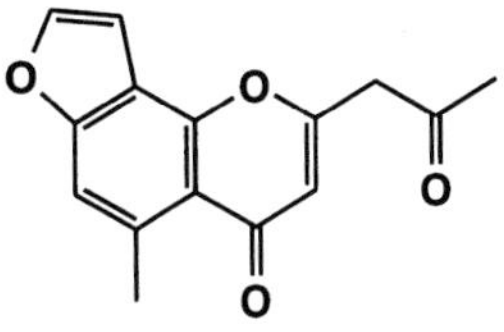

7

During the separation of *chalcone* glycosides from the flowers of *Bidens pilosa* (Asteraceae), a *step-gradient* DCCC operation was involved. After initial Sephadex LH-20 gel filtration, DCCC was performed with the solvent system $CHCl_3$-MeOH-H_2O 13:7:4 (descending mode). During the run, the solvent composition was changed to $CHCl_3$-MeOH-*i*PrOH-H_2O 26:13:1:8 and then finally to $CHCl_3$-MeOH-*i*PrOH-H_2O 13:6:1:4. Final HPLC purification gave three okanin glucosides from the different DCCC fractions (Hoffmann and Hölzl 1989).

A new *phloroglucinol* derivative (**8**) with antifungal and antimalarial activity has been obtained from the European ornamental plant *Hypericum calycinum* (Guttiferae) by a strategy involving a combination of DCCC, LPLC and gel filtration. A petroleum ether extract of the aerial parts was subjected to fractionation by DCCC on a modified apparatus with 870 columns (290 columns 3 mm i.d., 580 columns 2.7 mm i.d.), using the solvent system petrol ether-96% EtOH-EtOAc-H_2O 83:67:33:17 (ascending mode). Two portions (4 g and 6 g) of extract were chromatographed, giving a total of 255 mg of a fraction containing the phloroglucinol. This was further purified by Lobar LPLC on diol and Sephadex LH-20 (Decosterd et al. 1991).

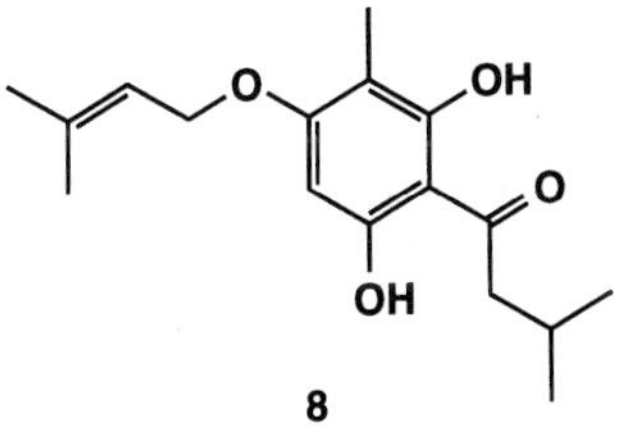

8

Other phenolics which have been separated by DCCC include:

- phenylpropanoid glycosides from *Eurya tigang* (Theaceae) ($CHCl_3$-MeOH-H_2O 7:13:8; ascending) (Khan et al. 1992),
- phenylpropanes and lignans from *Urtica dioica* (Urticaceae) ($CHCl_3$-MeOH-H_2O 5:9:7; descending) ($CHCl_3$-MeOH-$C_6H_5CH_3$-H_2O 8:8:5:2; descending) ($CHCl_3$-MeOH-*i*PrOH-H_2O 5:6:1:4; descending) (Chaurasia and Wichtl 1986),
- stilbene glycosides from *Rheum palmatum* (Polygonaceae) ($CHCl_3$-MeOH-H_2O 7:13:8; descending) (Kubo et al. 1991),
- benzoic acid derivatives and phenolic glycosides from *Alangium platanifolium* (Alangiaceae) ($CHCl_3$-MeOH-H_2O 5:6:4) ($CHCl_3$-MeOH-*n*PrOH-H_2O 9:12:2:8) (Otsuka et al. 1989),
- naphthoxirene glucosides from *Sesamum angolense* (Pedaliaceae) ($CHCl_3$-MeOH-*i*PrOH-H_2O 5:6:1:4; ascending) (Potterat et al. 1987),
- dihydrophenanthrene glucosides from *Juncus effusus* (Juncaceae) ($CHCl_3$-MeOH-H_2O 13:7:4, 26:14:5; ascending) (Della Greca et al. 1995).

7.1.3.2
Monoterpenes

The main application of DCCC in this field has been the separation of iridoid and secoiridoid glycosides. Otsuka and coworkers have reported the isolation of numerous iridoid glycosides from *Premna* species (Verbenaceae), *Alangium* species (Alangiaceae) and *Linaria japonica* (Scrophulariaceae), using mainly the solvent system $CHCl_3$-MeOH-*n*PrOH-H_2O. For example, a series of acylglycosylcatapols, including **9**, have been isolated from *Premna japonica*. The stems were extracted with methanol and this extract was partitioned between water, ethyl acetate and butanol. The butanol fraction was passed over Diaion HP-20 with increasing proportions of methanol in water. The 60 % MeOH eluate was chromatographed over silica gel ($CHCl_3$-MeOH 9:1), followed by DCCC with $CHCl_3$-MeOH-*n*PrOH-H_2O 9:12:1:8, to give 65 mg of **9** (Otsuka et al. 1991).

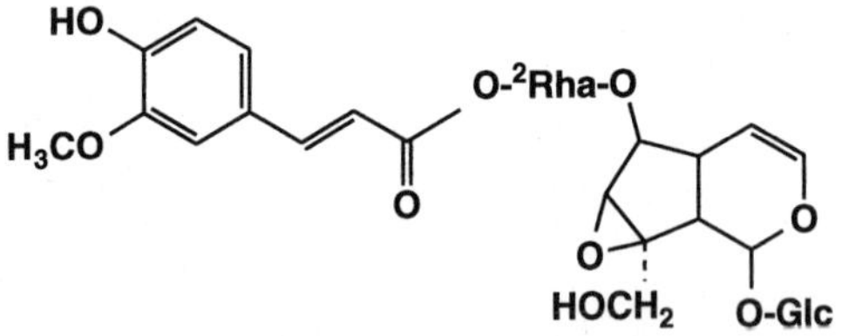

9

A similar procedure was employed for the isolation of iridoid glycosides from *Linaria japonica*. In one report, four iridoid esters of glucose were obtained from a methanol extract of the whole plants. After Diaion HP-20 chromatography, the iridoid-containing fractions were fractionated by DCCC (Tokyo Rikakikai, 500 columns; 40 cm × 2 mm) with $CHCl_3$-MeOH-*n*PrOH-H_2O 9:12:2:8 (ascending mode). Final purification was achieved by semi-preparative HPLC (Otsuka 1995).

Iridoid glucosides have also been obtained from *Sesamum angolense* (Pedaliaceae) with the DCCC solvent system $CHCl_3$-MeOH-iPrOH-H_2O 5:6:1:4 (ascending mode) (Potterat et al. 1988). Geniposidic acid was isolated from the leaves of *Canthium gilfillanii* (Rubiaceae) in one step by DCCC with the same solvent (Nahrstedt et al. 1995).

Other solvent systems which have been used for the separation of iridoid and secoiridoid glycosides are as follows:

- $CHCl_3$-MeOH-H_2O 4:6:5 (ascending mode) (Abe et al. 1988),
- $CHCl_3$-MeOH-H_2O 43:37:20 (ascending mode) (Do et al. 1987),
- $CHCl_3$-MeOH-H_2O 5:6:4 (descending mode) (Gering-Ward and Junior 1989),
- *n*BuOH-EtOH-H_2O 4:1:1 (ascending mode) (Tanahashi et al. 1989),
- *n*BuOH-*n*PrOH-H_2O 2:1:3 (ascending mode) (Franke and Rimpler 1987).

As well as a neolignan glycoside and iridoid glycosides from *Alangium platanifolium* (Alangiaceae) (Otsuka et al. 1996a), DCCC has also been used to purify *ionol* (C_{13}-norisoprenoid) glycosides from *A. premnifolium* (Otsuka et al. 1994, 1995). The leaves of the plant were extracted with methanol and this extract was submitted to a preliminary purification (solvent partition, Diaion HP-20, silica gel CC, reversed-phase open-column CC) before droplet countercurrent chromatography under the same conditions as for the separation of iridoid glycosides i.e. $CHCl_3$-MeOH-*n*PrOH-H_2O 9:12:2:8 (ascending mode). Alangionoside A (**10**) was obtained after a final HPLC step but alangionoside B (**11**) was obtained directly after DCCC (Otsuka et al. 1994).

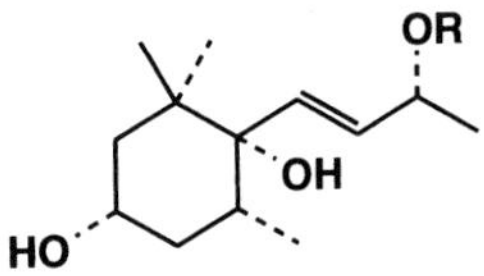

10 R = Glc
11 R = Glc[6]-Api

Ionone glycosides have been fractionated by DCCC with *n*BuOH-Me_2CO-H_2O 60:18:22 (De Tommasi et al. 1992).

Monoterpenes and norisoprenoids are the major volatile aroma compounds in wine. The parent grapes often contain these secondary metabolites in the form of glycoconjugates and DCCC has proved to be ideal for the separation of these glycosides from grape juice and wine. This was the case for the glycosides of, e.g. vitispirane (**12**) and β-damascenone (**13**) (both aglycones and their derivatives are important contributors to the aroma of bottle-aged Riesling wine), which were obtained after DCCC with the solvent system $CHCl_3$-MeOH-H_2O

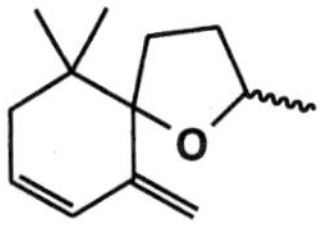

12

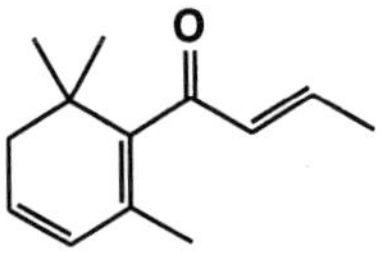

13

7:13:8 (ascending mode) (Winterhalter et al. 1990; Waldmann and Winterhalter 1992).

7.1.3.3
Triterpenes and Steroids

Because of the relatively low polarity of these classes of compounds, the application of aqueous DCCC systems to their isolation has been limited. However, a certain number of separations involving the more polar representatives, e.g. ecdysteroids, have been reported.

A one-step DCCC purification procedure was sufficient to purify calmodulin-inhibitory fasciculols from the poisonous mushroom *Naematoloma fasciculare* (Strophariaceae). Fresh fruiting bodies of the mushroom (400 g) were extracted with methanol and this extract was separated into ethyl acetate-soluble and water-soluble portions. The ethyl acetate fraction (2.4 g) was dissolved in 10 ml of a 1:1 mixture of the two solvent phases and introduced into a Tokyo Rikakikai DCCC instrument (300 columns; 40 cm×2 mm) filled with the stationary phase of C_6H_6-$CHCl_3$-MeOH-H_2O 5:5:7:2. Elution with the mobile phase in the ascending mode gave the three pure fasciculols (**14–16**) (Fig. 7.5) (Kubo et al. 1985).

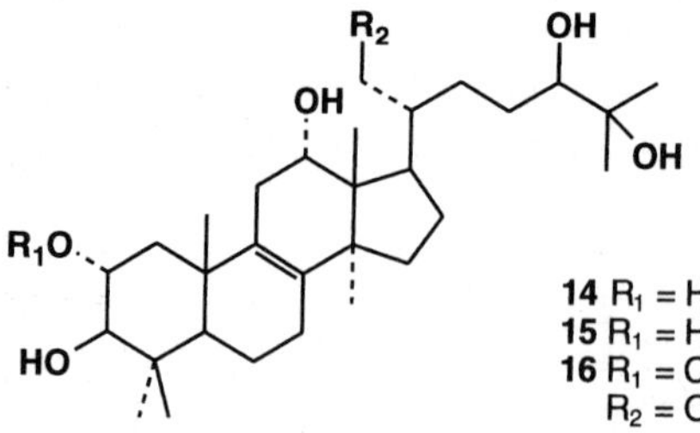

14 R_1 = H, R_2 = OH (Fasciculol C)
15 R_1 = H, R_2 = H (Fasciculol B)
16 R_1 = CO-CH_2-C(OH)(CH_3)-CH_2-CONH-CH_2-$COOCH_3$, R_2 = OH (Fasciculol F)

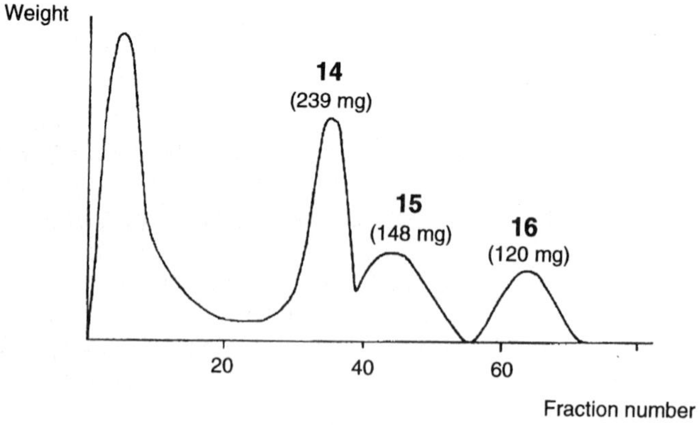

Fig. 7.5. DCCC separation of fasciculols from the mushroom *Naematoloma fasciculare*. Solvent system: C_6H_6-$CHCl_3$-MeOH-H_2O 5:5:7:2 (ascending mode); sample 2.4 g. (Reprinted with permission from Kubo et al. 1985)

Szendrei et al. (1988) report the isolation of 5-deoxykaladasterone from *Leuzea carthamoides* (Asteraceae) as an 11-dehydro artefact of ajugasterone C produced during initial Al_2O_3 chromatography of the ecdysteroid mixture. The pure compound was obtained by DCCC with C_6H_6-$CHCl_3$-MeOH-H_2O 15:15:23:7 in the descending mode.

Another ecdysteroid has been isolated from both *Silene dioica* and *S. otites* (Caryophyllaceae) using DCCC as the preliminary fractionation step ($CHCl_3$-MeOH-H_2O 13:4:4; descending) (Girault et al. 1996).

Polyoxypregnane ester derivatives of general structure **17** were characterized from *Leptadenia hastata* (Asclepiadaceae). A total of 34 new derivatives were isolated, possessing 3–5 sugar residues at the C-3 position. Droplet countercurrent chromatography played a major rôle in the separation strategy, using the following solvent systems: BuOH-EtOH-HOAc-H_2O 8:4:2:1 (ascending); $CHCl_3$-MeOH-H_2O 7:13:8 (ascending); $CHCl_3$-MeOH-EtOH-*n*PrOH-H_2O 9:6:8:1:8 (ascending) (Aquino et al. 1996).

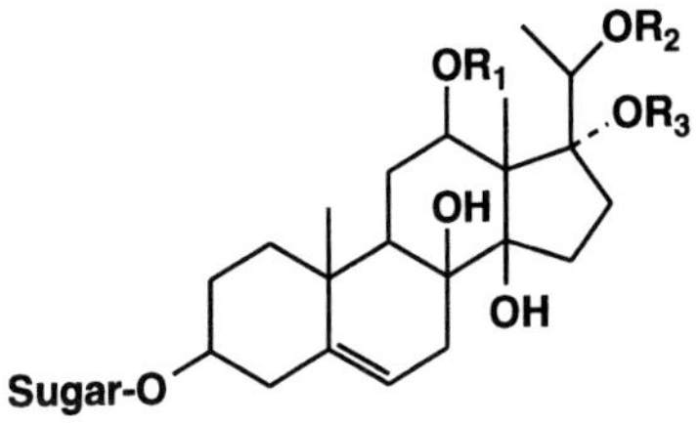

17

In the separation of triterpenes, a C_6H_6-$CHCl_3$-MeOH-H_2O system, as used for steroids, has been employed. DCCC (C_6H_6-$CHCl_3$-MeOH-H_2O 1:3:5:2) of a prepurified extract of *Simaba multiflora* (Simaroubaceae) afforded tricin, 8-hydroxycanthin-6-one, simalikalactone D, vanillic acid and the two tirucall-7-ene triterpenes hispidol and 3*S*,23*R*,25-trihydroxytirucall-7-en-24-one (Arisawa et al. 1987).

7.1.3.4
Saponins

The saponins represent a very large group of complex natural products, often with important biological activities (Hostettmann and Marston 1995). Interest in these glycosidic compounds is growing rapidly, especially as the introduction of modern separation techniques has greatly simplified the problems associated with their isolation. DCCC, in conjunction with other separation techniques, provides a useful tool for obtaining pure saponins. Many applications in the separation of triterpene glycosides, steroid glycosides and basic steroid saponins are known but, of these, triterpene glycosides provide the most examples (Hostettmann and Marston 1995).

The root bark of *Diospyros zombensis* (Ebenaceae) from Malawi, for example, has furnished two oleanane glycosides (**18, 19**). These were isolated after a combination of silica gel open-column chromatography, LPLC, gel filtration and

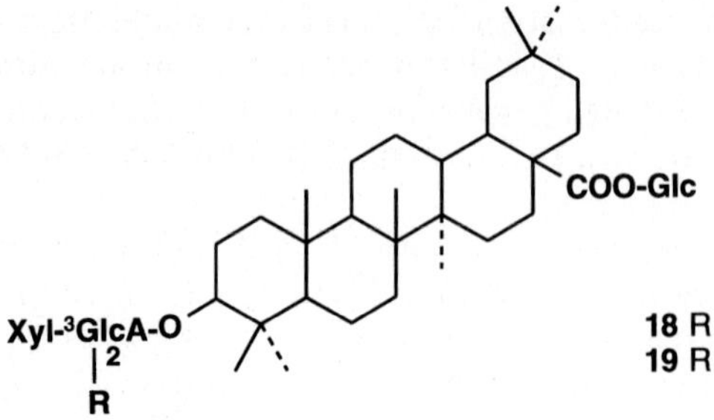

18 R = H
19 R = Xyl

DCCC with the solvent system *n*BuOH-Me_2CO-H_2O 33:10:50 (descending mode) (Gafner et al. 1987).

Other recent applications of DCCC to the separation of saponins include the characterization of four jujubogenin glycosides from the leaves of *Zizyphus spina-christi* (Rhamnaceae), in which the DCCC solvent was $CHCl_3$-MeOH-*n*PrOH-H_2O 9:12:1:8 (descending mode) (Mahran et al. 1996) and the separation of six oleanane glycosides from *Hydrocotyle ranunculoides* (Apiaceae) ($CHCl_3$-MeOH-H_2O 26:14:5; ascending mode) (Della Greca et al. 1994). In addition, four oleanane-type saponins, including the new glycoside **20**, were isolated from the leaves of *Ardisia japonica* (Myrsinaceae) ($CHCl_3$-MeOH-*n*PrOH-EtOH-H_2O 9:6:1:8:8; descending mode) (Piacente et al. 1996).

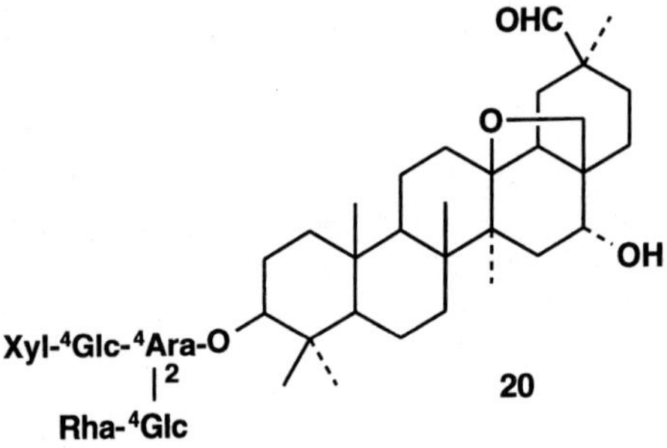

20

In the last few years a variety of hydroxylated steroids, steroid sulphates and sulphated steroids has been isolated from marine organisms (Minale et al. 1993). Many of the separations, especially from the group of Minale, have involved DCCC (D'Auria et al. 1993). Most frequently used have been the solvent systems *n*BuOH-Me_2CO-H_2O 3:1:5 (descending) and $CHCl_3$-MeOH-H_2O 7:13:8 (descending) (Riccio et al. 1985, 1986; Iorizzi et al. 1993).

7.1.3.5
Alkaloids

In view of their polar nature and frequent existence as salts, applications of liquid-liquid chromatography to the separation of alkaloids are numerous.

Pyrrolizidine alkaloids, which are important for their toxicity to humans, have a tendency to adsorb irreversibly to solid chromatographic supports such as alumina, silica gel or reversed-phase silica gel and it is advantageous to use DCCC for their separation. The most frequently employed solvents are $CHCl_3$-MeOH-H_2O systems and $C_6H_5CH_3$-$CHCl_3$-MeOH-H_2O 5:5:7:2 . For example, the alkaloids integerrimine (**21**) and trichodesmine (**22**) have been isolated from *Crotalaria sessiliflora* (Leguminosae) by this approach. The seeds were defatted and extracted with 65% ethanol (containing 0.5% HOAc). The extract was acidified to pH 1–2 and extracted with chloroform. The aqueous layer was brought to pH 9–10 with NH_4OH and extracted with chloroform. After crystallization of monocrotaline, the other two alkaloids were obtained by DCCC with $CHCl_3$-MeOH-H_2O 5:5:3 (ascending) (Röder et al. 1992).

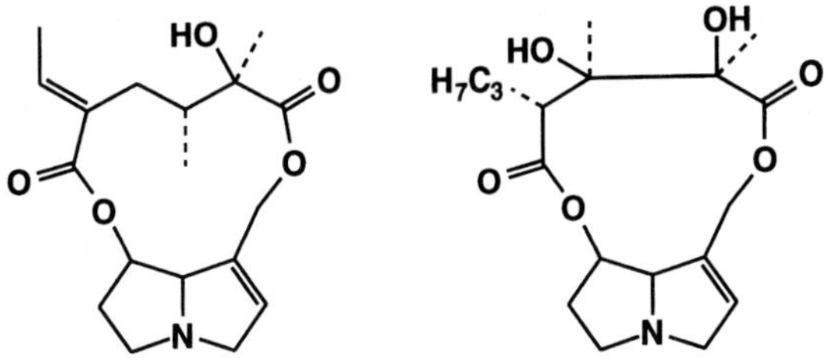

21 **22**

Similarly, tussilagine and isotussilagine (Passreiter et al. 1992), tussilaginic acid and isotussilaginic acid (Passreiter 1992) have been isolated from the flower heads of *Arnica* species (Asteraceae) with DCCC ($CHCl_3$-MeOH-H_2O 5:6:4; descending) as an intermediate separation step.

Ten 12-membered macrocyclic pyrrolizidine alkaloids (esters of retronecine or otonecine) have been isolated from *Senecio anonymus* (Asteraceae) by DCCC. During this work, it was possible to separate *cis* and *trans* isomers, e.g. senkirkine (**23**) and neosenkirkine (**24**) were separated with the DCCC solvent system C_6H_6-$CHCl_3$-MeOH-H_2O 5:5:7:2 (descending) (Zalkow et al. 1988).

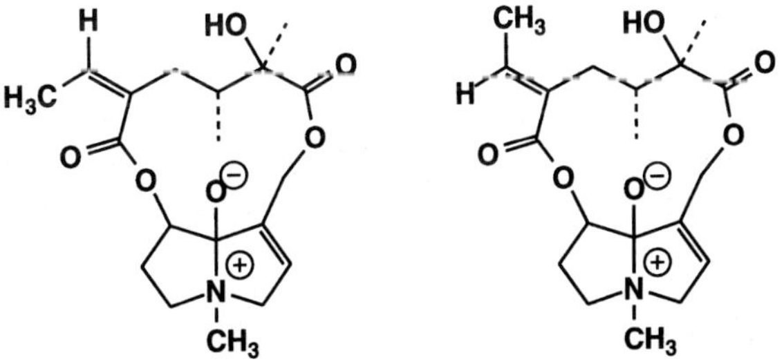

23 **24**

Further separations of alkaloids by DCCC include:

- aporphine alkaloids from *Ocotea caesia* (Lauraceae) ($CHCl_3$-MeOH-H_2O 5:5:3; descending) (Vilegas et al. 1989),
- benzophenanthridine alkaloid from *Chelidonium majus* (Papaveraceae) ($CHCl_3$-MeOH-H_2O 5:6:4; descending) (De Rosa and Di Vincenzo 1992),

- tetrahydroisoquinoline alkaloids from *Cephaelis ipecacuanha* (Rubiaceae) ($CHCl_3$-MeOH-*n*PrOH-H_2O 45:60:2:40; descending) (Itoh et al. 1991),
- chromone alkaloids from *Schumanniophyton magnificum* (Rubiaceae) (*n*BuOH-MeOH-H_2O 5:1:5; descending) (*n*BuOH-MeOH-dil.NH_4OH 5:1:5; ascending) (Houghton and Hairong 1987),
- brominated indole alkaloids from marine sponge *Orina* sp. (Bifulco et al. 1995) and peptide alkaloids from marine sponge *Anchinoe tenacior* (Demospongiae) (Casapullo et al. 1994) (*n*BuOH-Me_2CO-H_2O 3:1:5; descending).

An interesting development is *ion-pair* droplet countercurrent chromatography of alkaloids (Tani et al. 1975; Hermans-Lokkerbol and Verpoorte 1986; Van der Heijden et al. 1987). By running a gradient in which there is a decreasing concentration of counter ion, the retention of the alkaloids can be selectively reduced. This method has been applied, for example, in the isolation of eight alkaloids from suspension cultures of *Tabernaemontana elegans* (Apocynaceae), although a final preparative TLC step was necessary to purify some of the compounds (Van der Heijden et al. 1987). The solvent system comprised McIlvaine buffer (0.025 mol/l citrate and 0.05 mol/l phosphate, adjusted to pH 4.2 with phosphoric acid) containing sodium perchlorate-methanol-chloroform (3:5:5), with the aqueous phase as the mobile phase (ascending mode). Elution of the alkaloids was started with 1 mol/l sodium perchlorate in the buffer and then changed successively to 0.75 mol/l, 0.50 mol/l, 0.25 mol/l and finally 0 mol/l (aqueous phase with no perchlorate). The more acidic the mobile phase becomes, the more the alkaloids are protonated and the more they pass into the aqueous phase.

7.1.3.6
Other Applications

Sesquiterpenes

- Parthenolide (sesquiterpene lactone) from *Chrysanthemum parthenium* (Asteraceae): C_6H_6-$CHCl_3$-MeOH-H_2O 5:5:7:2 (descending); trichloroethylene-CH_3CN-$CHCl_3$-MeOH-H_2O 4:2:2:5:2 (descending) (Kery et al. 1988).
- Sesquiterpene glycosides from *Eriobotrya japonica* (Rosaceae): $CHCl_3$-MeOH-H_2O 7:13:8 (ascending) (De Tommasi et al. 1990).
- Sesquiterpene glycosides from *Alangium premnifolium* (Alangiaceae): $CHCl_3$-MeOH-*n*PrOH-H_2O 9:12:2:8 (ascending) (Otsuka et al. 1996b).

Diterpenes

- Labdane glycosides from *Mitraria coccinea* (Gesneriaceae): $CHCl_3$-MeOH-H_2O 7:13:8 (ascending) (Cardenas et al. 1992).
- Furanoid diterpene glycosides from *Jateorhiza palmata* (Menispermaceae): $CHCl_3$-MeOH-H_2O 7:13:8 (ascending) (Yonemitsu et al. 1987).
- Furanoid diterpene glycosides from *Tinospora tuberculata* (Menispermaceae): $CHCl_3$-MeOH-H_2O 7:13:8 (ascending); $CHCl_3$-MeOH-H_2O 13:7:4 (ascending) (Fukuda et al. 1993).
- Clerodane diterpenoids from *Portulaca pilosa* (Portulacaceae): $CHCl_3$-MeOH-H_2O 13:7:4 (ascending) (Ohsaki et al. 1991).

- Limonoids from *Toona ciliata* (Meliaceae): hexane-EtOH-EtOAc-H_2O 5:4:2:1 (ascending) (Neto et al. 1995).
- Casbane diterpenoid from *Agrostistachys hookeri* (Euphorbiaceae) cyclohexane-Et_2O-*i*PrOH-EtOH-H_2O 7:16:6:10:8 (ascending) (Choi et al. 1988).

Cardenolides
- Cardenolides from *Ornithogalum boucheanum* (Liliaceae): $CHCl_3$-MeOH-H_2O 5:10:6 (ascending and descending) (Ghannamy et al. 1987).
- Cardenolides from *Digitalis subalpina* (Scrophulariaceae): $CHCl_3$-MeOH-H_2O 5:6:4 (descending) (Lichius et al. 1991).

Cyanogenic glycosides
- Cyanogenic glycosides from *Perilla frutescens* (Lamiaceae): $CHCl_3$-MeOH-*n*BuOH-H_2O 9:12:2:8 (ascending) (Aritomi et al. 1985).
- Cyanogenic glycosides from *Olinia* species (Oliniaceae): $CHCl_3$-MeOH-PrOH-H_2O 5:6:1:4 (ascending) (Nahrstedt and Rockenbach 1993).

7.1.4 Non-Aqueous Solvent Systems

Weakly polar or apolar compounds, especially those which are unstable in the presence of water or decompose during chromatography on silica gel, can be conveniently separated by DCCC with non-aqueous solvent systems. This type of solvent system is also important for natural products which are insoluble in water.

For example, the triterpenes oleanolic acid and hederagenin were separated from an acid-hydrolysed extract of *Hedera helix* (Araliaceae) with *n*-heptane-methanol-1,2-dichloroethane 37.6:4.8:57.6 in the descending mode (Domon et al. 1982) and a mixture of betulinic acid, betulin, β-amyrin and cholesterol could be separated with *n*-heptane-acetonitrile-dichloromethane 10:7:3 (descending) (Domon et al. 1982).

The very simple system *n*-hexane-methanol 1:1 has also proved useful. For example, DCCC elution in the descending mode effectively separates anisaldehyde, pseudoisoeugenyl-(2-methylbutyrate), anethole and sesquiterpenes from a petroleum ether extract of Spanish anise fruits, *Pimpinella anisum* (Apiaceae). Because the methanol phase is the mobile phase, the most polar component (anisaldehyde) elutes first and the least polar components (the sesquiterpenes) elute last (Miething and Seger 1989). With the same system in the descending mode, it was also possible to separate foeniculine and safrole, which have R_f values of 0.79 and 0.78 on TLC plates (developing solvent: toluene) (Miething and Seger 1989).

Furoacridone alkaloids, including the new compound rutagravin (**25**), have been obtained from tissue cultures of *Ruta graveolens* (Rutaceae) by a procedure using DCCC as the key separation step. A crude dichloromethane extract of the callus tissue was chromatographed with the solvent system *n*-heptane-cyclohexane-dichloromethane-toluene-acetonitrile before purification of the

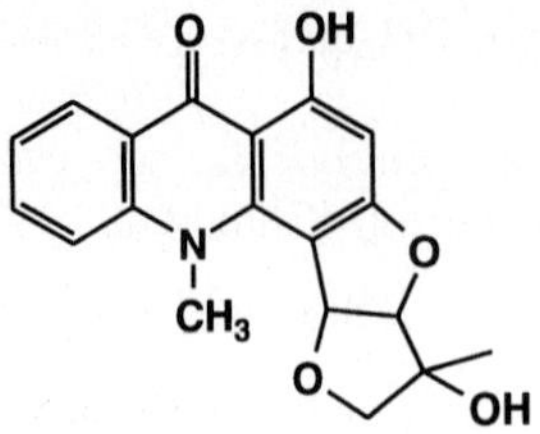

25

individual components by silica gel column and thin-layer chromatography (Nahrstedt et al. 1985).

7.1.5 Perspectives

Droplet countercurrent chromatography has found widespread use in the separation of natural products. The first described applications were with polar compounds and, since these are frequently difficult to purify, DCCC will continue to be used for the isolation of, e.g. terpenoid glycosides and polyphenolic glycosides, even though CPC appears now to be dominating the field.

The most widely employed solvent systems in DCCC (approximately 80% of all separations) involve $CHCl_3$-MeOH-H_2O mixtures. These have been applied to compounds with a wide range of polarities. However, the introduction of new systems, such as *n*-hexane-$CHCl_3$-MeOH-H_2O and *n*-hexane-Et_2O-*n*PrOH-EtOH-H_2O mixtures, means that separations involving less polar constituents is possible. Non-aqueous systems also increase the applicability of the technique.

DCCC provides reproducible and efficient separations. Milligram to gram quantities and crude extracts can be handled. Acidic and basic separation conditions can be applied without problem. As no solid separation matrix is present, irreversible adsorption and band spreading is avoided. Solvent consumption is generally lower than in preparative HPLC but separation times are very long and resolution is low. An attempt to speed up the washing of the DCCC instrument after each separation has been reported (Kubo et al. 1988) but the technique will remain inherently slow because it relies on droplet formation for its effectiveness.

Apparatus for performing DCCC separations is commercially available and not technically difficult to use. Maintenance is kept to a minimum but it is advisable to keep the glass separation columns clean. When the surfaces of the columns are dirty and droplet formation is impeded, washing can be performed first with methanol and then with a 1 mol/l solution of NaOH (containing 0.5% sodium hypochlorite) (Kubo et al. 1988).

7.2 Rotation Locular Countercurrent Chromatography

This technique, like droplet countercurrent chromatography, relies on liquid-liquid partition, and thus on a two-phase solvent system, for the separation of a mixture. Following the original paper on the method (Signer et al. 1956), various developments were made (Ito and Bowman 1970b; Snyder et al. 1984) and a commercial apparatus was produced. Many different classes of natural products have successfully been separated by rotation locular countercurrent chromatography (RLCC) but the method has now largely been superseded by the newer centrifugal partition chromatography techniques.

7.2.1 Description of the Method

The RLCC apparatus consists of 16 glass columns (50 cm × 1.1 cm i.d.) arranged around a rotational axis (Fig. 7.6b) and connected in series with Teflon tubing (1 mm i.d.). Both the speed of rotation of the columns and their angle of inclination (Ø) can be varied (Fig. 7.6a). Solvent enters and leaves the bundle of columns by means of rotating seals.

Each column is divided into 37 compartments (loculi) by Teflon discs and, as each disc is perforated by a small hole (1 mm diameter) in the centre, solvent can flow between the compartments (Fig. 7.6a). The columns are moved to the vertical position and stationary phase is pumped into each column in turn from the bottom, using a constant flow pump connected to the inlet tube. At the same time, the set of 16 columns is rotated about the central axis and care must be taken to remove all air bubbles from the loculi. The columns are then usually tilted to an angle of 20–40° from the horizontal and mobile phase is pumped into the system. If separations are to take place in the *ascending mode*, the mobile phase is the lighter of the two phases and is pumped in through the bottom of the columns, as shown in Fig. 7.6a. While the mobile phase enters the columns at a rate of 15–50 ml/h, the columns rotate about the central axis at 60–80 rpm. The liquid entering the first loculus displaces stationary phase until it reaches the level of the hole leading to the next loculus. The lighter, mobile phase then passes into the second loculus and so the process continues until the mobile phase emerges from the uppermost loculus. It is then directed to the bottom of the next column and the displacement of stationary phase is repeated until all the columns are charged with mobile phase.

For separations which are to proceed in the *descending mode*, the columns are filled with the less dense layer of a two-phase solvent system and the more

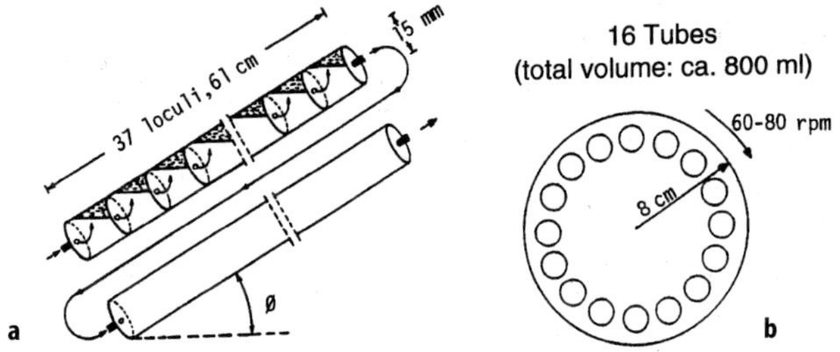

Fig. 7.6a, b. RLCC apparatus. (Reprinted with permission from Snyder et al. 1984)

dense mobile phase is pumped into the loculi in the opposite direction to that shown in Fig. 7.6a – first into the uppermost loculus.

The sample is dissolved in ca. 4 ml mobile or stationary phase or a mixture of mobile and stationary phases and is introduced into the first column. Then further mobile phase is pumped after the sample. As the mobile phase percolates through the compartments, there is a partitioning of the solute between the mobile and stationary phases. This partitioning is aided by the rotatory movement of the columns, which serves to increase the contact between the two phases. However, no shaking is involved and, as long as the rotation is not too fast, there are no problems of emulsion formation. Mobile phase which elutes from the apparatus is collected and analysed, either by a UV detector and/or by TLC. Any solute remaining in the stationary phase can be eluted by pumping additional stationary phase through the columns.

Separating power is dependent on three variables: a) the speed of rotation; b) the inclination of the columns; c) the flow rate (Ito and Bowman 1970b). Optimal resolution is observed for rotation speeds of ca. 70 rpm. At faster speeds, the interface is disturbed and separating power decreases. The angle of inclination of the columns affects both the resolution and retention volume of the products insofar as a difference in the proportions of mobile and stationary phases is produced in each loculus.

RLCC is a very simple technique to use and separations with resolutions of the order of 250–300 theoretical plates can be obtained in a matter of hours.

7.2.2
Solvent Selection

In contrast to DCCC, the formation of droplets in a two-phase solvent system is not a necessary condition of RLCC. Consequently, a wide range of solvent systems is available. Moreover, ethyl acetate-water systems, which are often incompatible with DCCC, present no problems in RLCC. Binary systems may be employed but, for reasons of selectivity, ternary or quaternary systems are preferred.

The selection of appropriate biphasic solvent systems can be performed by TLC (Snyder et al. 1984), in which each of the two phases is used in turn. If the sample has R_f values greater than 0.8 in one of the phases and between 0.2 and 0.4 in the other phase, the solvent system is suitable for RLCC. A small amount, e.g. 5 mg, of sample is subsequently mixed with 5–10 ml of each of the two phases and a solvent system is chosen such that 15–25% of the sample remains in one phase. The phase containing the majority of sample is chosen as the stationary phase. This distribution can be quantified by HPLC. Separations can be completed within 36 h if these guidelines are followed; if equilibration gives a more equal distribution of sample, separations are quicker but resolution consequently suffers. A distribution giving less than 15% in the mobile phase results in longer separation times and a subsequent increase in solvent consumption.

7.2.3
Applications

Rotation locular countercurrent chromatography is little used nowadays but has still provided (and provides) quite a variety of applications in the natural product field. Some of these are listed in Table 7.4.

For the separation of *flavonoid glycosides*, a solvent system containing ethyl acetate (more polar than chloroform) has proved useful: EtOAc-nPrOH-H_2O 4:2:7 (Hostettmann and Hostettmann 1981). DCCC systems containing large proportions of ethyl acetate cause problems associated with poor droplet formation and RLCC is, in this respect, more flexible. Kubo and co-workers have employed a *gradient* with ethyl acetate in order to separate rutin (**26**) from

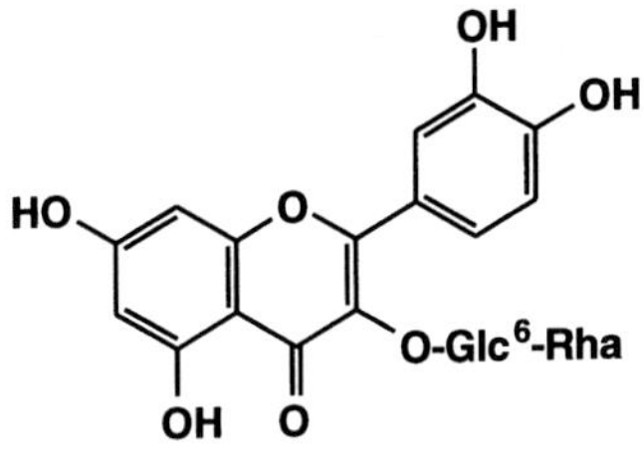

26

leaves of *Esenbeckia pumila* (Rutaceae) (Kubo 1991). The RLCC apparatus was filled with water as the stationary phase and a methanol extract of the plant was dissolved in water and injected. Elution with mobile phase was performed in the following sequence: a) water-saturated diethyl ether; b) water-saturated ethyl acetate; c) upper layer of ethyl acetate-propanol-water (10:1:2); d) upper layer of ethyl acetate-propanol-water (4:1:1).

An ethyl acetate-containing solvent was also employed for the purification of *saponins* from *Phytolacca dodecandra* (Phytolaccaceae). RLCC proved to be a very efficient step in the isolation of the bidesmosidic triterpene glycosides **27**–**29**. Dried berries of the plant were extracted successively with petroleum ether, chloroform and methanol. A portion (4 g) of the methanol extract was

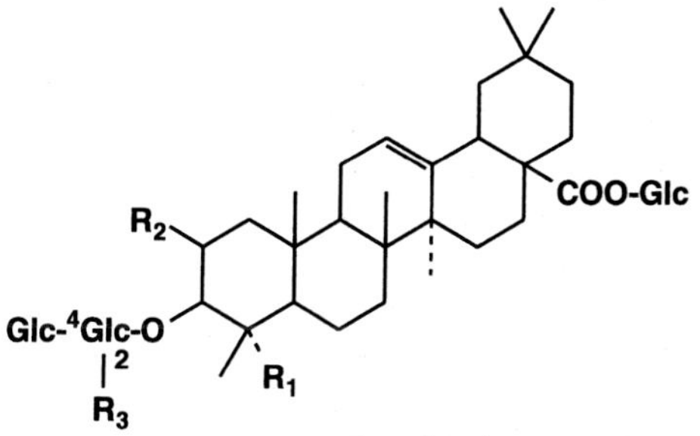

27 R_1 = CH_2OH, R_2 = OH, R_3 = H
28 R_1 = CH_3, R_2 = H, R_3 = Glc
29 R_1 = CH_2OH, R_2 = H, R_3 = Glc

Table 7.4. Separations of natural products by RLCC

Substances separated	Solvent system	Mobile phase	Reference
Flavonoids			
Flavonoids from *Baccharis trimera* (Asteraceae)	$CHCl_3$-MeOH-H_2O 13:7:4	Upper	Soicke and Leng-Peschlow 1987
Flavonoid glycosides from *Arnica* species (Asteraceae)	$CHCl_3$-MeOH-*n*BuOH-H_2O 10:10:1:6	Lower	Merfort and Wendisch 1987
	CH_2Cl_2-MeOH-H_2O 7:13:8	Lower	
	EtOAc-*n*PrOH-H_2O 4:2:7	Upper	
Flavonol glucuronides from *Arnica montana* (Asteraceae)	EtOAc-*n*PrOH-H_2O 4:2:7	Upper	Merfort and Wendisch 1988
Other polyphenols			
Caffeoylquinic acids from *Arnica* species (Asteraceae)	EtOAc-*n*PrOH-H_2O 4:2:7	Upper	Merfort 1992
Phenylpropanoid glycosides from *Mussatia* species (Bignoniaceae)	$CHCl_3$-MeOH-H_2O 7:13:8	Upper	Jimenez et al. 1987
Chalcone glucosides from *Bidens pilosa* (Asteraceae)	$CHCl_3$-MeOH-*i*PrOH-H_2O 8:7:1:6	Lower	Hoffmann and Hölzl 1988
Alkaloids			
Benzyl-aporphine dimers from *Thalictrum fauriei* (Ranunculaceae)	$CHCl_3$-MeOH-pH 3.0 acetate buffer 5:5:3	Upper	Lee and Doskotch 1996
Pyrrolizidine alkaloids from *Heliotropium spathulatum* (Boraginaceae)	$CHCl_3$-MeOH-H_2O 5:5.3	Upper, lower	Röder et al. 1991
Bishordeninyl terpene alkaloids from *Zanthoxylum coriaceum* (Rutaceae)	Et_2O-Me_2CO-H_2O 2:2:1	Lower	Marcos et al. 1990
Monoterpenes			
Secologanin from *Lonicera tatarica* (Caprifoliaceae)	EtOAc-*n*PrOH-H_2O 4:2:7	Upper	Hermans-Lokkerbol and Verpoorte 1987
	*n*BuOH-*n*PrOH-H_2O 2:1:3	Upper	

Norisoprenes			
Vomifoliol glycoside from *Malus sylvestris* (Rosaceae)	$CHCl_3$-MeOH-H_2O 9:12:8	Lower	Schwab and Schreier 1990 Herderich et al. 1992
Ionol glucoside from *Ribes uva crispa* (Grossulariaceae)	$CHCl_3$-MeOH-H_2O 7:13:8	Upper	Humpf et al. 1992
Ionol gentiobioside from *Cydonia oblonga* (Rosaceae)	$CHCl_3$-MeOH-H_2O 7:13:8	Upper	Winterhalter et al. 1991
Diterpenes			
Phorbol esters from *Croton tiglium* (Euphorbiaceae)	Petrol ether-Et_2O-MeOH-H_2O 18:2:15:0.75, 18:1:15:0.3	Lower	Pieters and Vlietinck 1986
Steroids			
Ecdysteroids from *Vitex strickeri* (Verbenaceae)	EtOAc-*n*PrOH-H_2O 6:6:1	Upper	Zhang et al. 1992
Ecdysones from *Helleborus odorus* (Ranunculaceae)	$C_6H_5CH_3$-$CHCl_3$-MeOH-H_2O 1:13:7:4	Lower	Kissmer and Wichtl 1987
Steryl glycosides from *Urtica dioica* (Urticaceae)	$C_6H_5CH_3$-$CHCl_3$-MeOH-H_2O 1:13:7:4	Lower	Chaurasia and Wichtl 1987
Saponins			
Saponins from *Phytolacca dodecandra* (Phytolaccaceae)	EtOAc-EtOH-H_2O 2:1:2	Upper	Domon and Hostettmann 1984
Glycoalkaloids from *Solanum incanum* (Solanaceae)	EtOAc-*n*BuOH-H_2O gradient	Upper	Kubo 1991
Miscellaneous			
Anthranilic acid glucoside from *Bromelia plumieri* (Bromeliaceae)	$CHCl_3$-MeOH-H_2O 7:13:8	Upper	Parada et al. 1996
2-Ethyl-3-methylmaleimide *N*-glucoside from *Garcinia mangostana* (Guttiferae)	$CHCl_3$-MeOH-H_2O 7:13:8	Upper	Krajewski et al. 1996
Essential oils	*n*-Hexane-MeOH-H_2O 150:135:18	Upper, lower	Hefendehl and Kuhne 1984

chromatographed by RLCC with EtOAc-EtOH-H_2O 2:1:2, using the upper phase as the mobile phase (45 ml/h; 60–70 rpm; 20° inclination). Further purification of the fractions from RLCC by Lobar reversed-phase liquid chromatography yielded glycosides of oleanolic acid (**28**), hederagenin (**29**) and bayogenin (**27**) (Domon and Hostettmann 1984).

Several reports on the isolation of *alkaloids* by RLCC have appeared. In one instance, an indole glycoalkaloid, raucaffricine (**30**), was isolated directly by

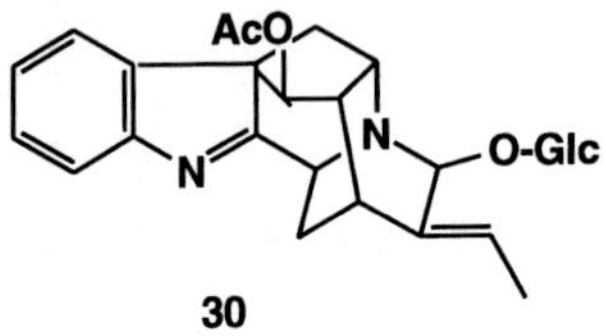

30

RLCC of an extract of freeze-dried cells from a suspension culture of *Rauwolfia serpentina* (Apocynaceae). A portion (24 g) of the crude extract was introduced into the instrument and eluted with the lower phase of $CHCl_3$-MeOH-H_2O 43:37:20 to give 2.5 g of pure raucaffricine. The flow rate was 70 ml/h, the rotation speed 80 rpm and the angle of inclination of the columns 35°. By repeating the separation 16 times, a total of 40 g raucaffricine was obtained (Schübel et al. 1984).

In the isolation of the precursors of aroma compounds from fruits, extensive use of RLCC has been made by Schreier, Winterhalter and co-workers. For example, a norisoprenoid has been obtained from apple fruit pulp (*Malus sylvestris*, Rosaceae). The methanol extract was passed through Amberlite XAD-2 and washed with water. The required compound, a vomifoliol trisaccharide (**31**), was eluted with methanol. This fraction was separated in 5 batches by RLCC (16 columns; 30° inclination; 1 ml/min; 80 rpm) with $CHCl_3$-MeOH-H_2O 9:12:8 with first the lower layer as mobile phase. The polar glycosides, including **31**, were then collected by using the upper layer as mobile phase. Final separation was by semi-preparative HPLC on C-18 (Herderich et al. 1992).

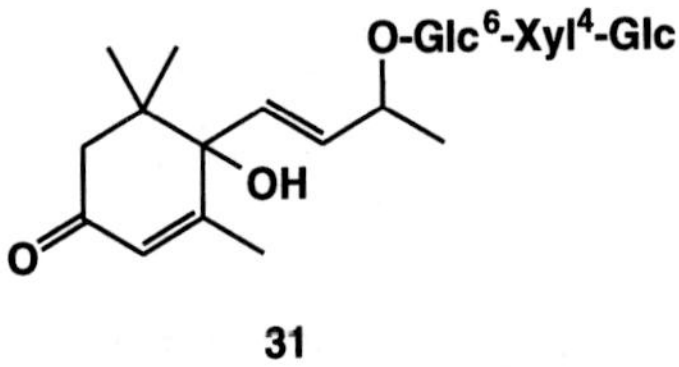

31

7.2.4 Perspectives

Rotation locular countercurrent chromatography has the advantage over DCCC that droplet formation is not a condition for a good separation. It is also compatible with two-phase solvent systems which are not suitable for DCCC, such as those containing large proportions of ethyl acetate. RLCC has a larger sample capacity than DCCC (Schübel et al. 1984; Zhang et al. 1992) and is suitable for handling crude natural product extracts. For this reason, a number of applications have been described.

However, the RLCC instrument suffers from mechanical shortcomings and the rotating seals which connect the spinning columns and the body of the apparatus have been especially troublesome. Consequently, widespread acceptance of the technique has not been forthcoming and the future of RLCC is not promising.

7.3 Centrifugal Partition Chromatography

There are two basic types of countercurrent chromatography (Ito 1981):

- the hydrostatic equilibrium system (HSES),
- the hydrodynamic equilibrium system (HDES).

In the static column system (Fig. 7.7), the coil is filled with the stationary phase of a biphasic solvent system and the other phase is pumped through the coil at a suitable speed. A point is reached where no further displacement of stationary phase occurs, the apparatus contains approximately 50% of each of the two phases and steady pumping-in of mobile phase results in elution of mobile phase alone. Both droplet countercurrent chromatography (DCCC) and rotation locular countercurrent chromatography (RLCC) are extensions of this gravity method (Fig. 7.8). Their disadvantage is that they are slow techniques, with separations taking two days or more.

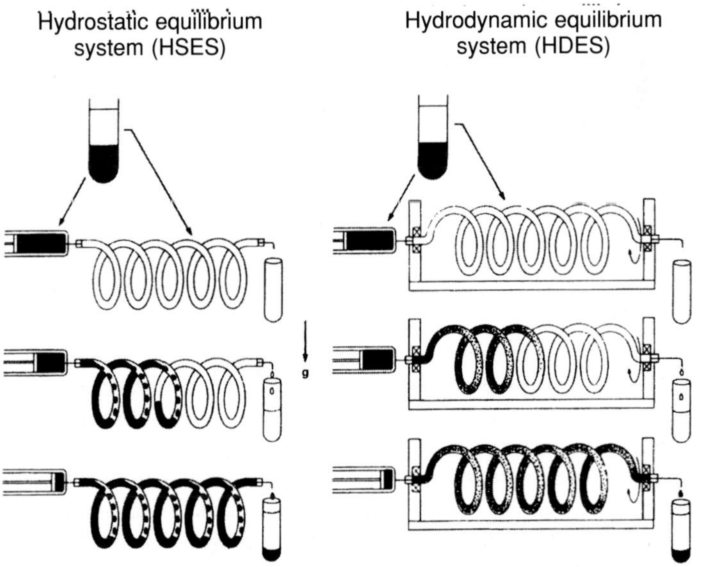

Fig. 7.7. Basic countercurrent chromatography systems. (Reprinted with permission from Ito 1981)

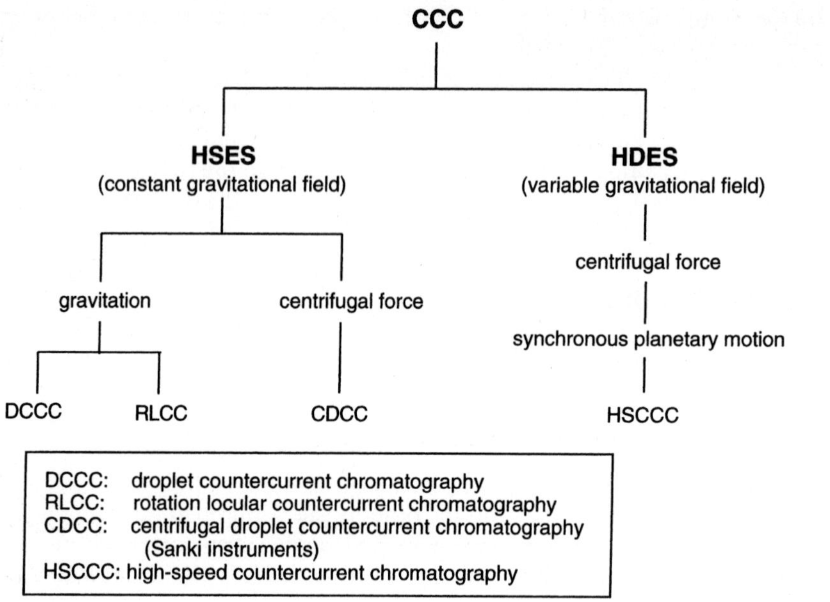

Fig. 7.8. Classification of different CCC instruments

This basic CCC system uses only 50% of the efficient column space for actual mixing of the two phases. A more effective way of using the column space is to *rotate* the coil while eluting the mobile phase. A hydrodynamic equilibrium (Fig. 7.7) is rapidly established between the two phases and almost 100% of the column space can be used for their mixing. Thus the interfacial area of the phases is dramatically increased (Ito 1981). Solutes, which may be injected as solutions in the mobile phase, stationary phase or a mixture of both, are partitioned between the two phases and are separated according to their partition coefficients. In fact, as there is no solid support, solute retention depends *only* on the partition coefficient.

The number of partition units in an HDES column system can be augmented by increasing the number of coils, by reducing the inernal diameter of the coiled column and by reducing the helical diameter. The rotating coil then requires a *centrifugal* force to prevent excessive loss of stationary phase due to displacement by the mobile phase. Another modification is to twist the column instead of coiling it (Ito 1981). Once the flow pressure has been established, care must be taken to maintain the centrifugal field constant within fairly narrow limits to prevent sudden changes in flow.

The original instruments, called flow-through coil-planet centrifuges, were equipped with rotating seals at the inlet and outlet of the coil (Ito and Bowman 1970a) but due to leakage and contamination problems, *seal-free* systems have been developed. By clever positioning of the coils, a continuous tubing arrangement is possible (Ito 1981; Ito and Conway 1984). Some of the different positions for seal-free operation of separation coils are shown in Fig. 7.9. The rota-

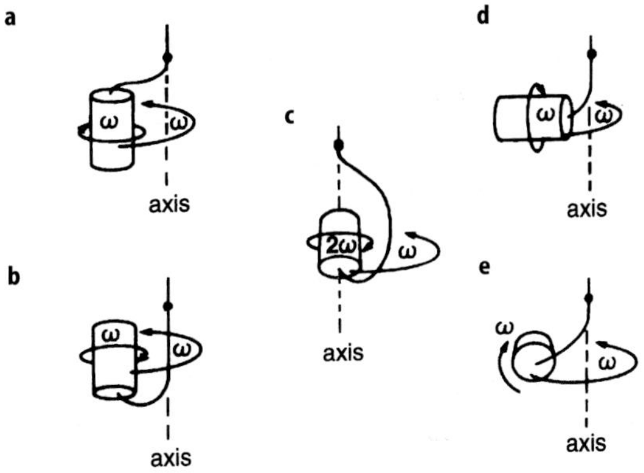

Fig. 7.9a–e. Arrangements of seal-free countercurrent chromatography separation columns

Table 7.5. Centrifugal countercurrent chromatography apparatus

Apparatus	Method (see Fig. 7.9)	Reference
Flow-through coil-planet centrifuge	a	Ito and Bowman 1971
Horizontal flow-through coil-planet centrifuge	b	Ito and Bowman 1977
Combined horizontal flow-through coil-planet centrifuge	a + b	Ito 1980a
Toroidal coil-planet centrifuge	b	Ito 1980b
High-speed CCC	b	Ito and Conway 1996
Toroidal coil centrifuge	c	Ito and Bowman 1978
Rotating coil assembly	c	Ito and Bhatnagar 1981
Cross-axis CCC	d, e	Ito 1987

tion of the coil alone and its revolution about a central centrifugal axis are described by ω.

A selection of the different instruments which have been constructed around the various seal-free arrangements is listed in Table 7.5.

All the methods discussed here involve centrifugal forces acting on the biphasic solvent system: hence the terms *centrifugal partition chromatography* (CPC) or *centrifugal countercurrent chromatography* (CCCC). Since most instruments have a stationary phase and a mobile phase, true countercurrent work is not performed and the term centrifugal partition chromatography (CPC) is a more accurate description of the technique.

7.3.1 Instruments

A whole range of CPC instruments has been introduced over the last few years (Conway 1990; Foucault 1995) but only a few have received any widespread commercial interest. A short description of the main representatives will be given here.

7.3.1.1
Hydrodynamic Equilibrium Systems (HDES)

The most frequently encountered HDES arrangement is that shown in Fig. 7.9b. The column (or bobbin) is made by winding PTFE tubing around a cylindrical holder hub to make multiple layers of coils. The holder revolves around the central axis of the centrifuge and simultaneously rotates about its own axis at the same angular velocity (Fig. 7.10). The motion of the coil causes vigorous agitation of the two solvent phases and a repetitive mixing and settling process, ideal for solute partitioning, occurs at over 13 times per second (Ito 1984). This rapid interchange explains why efficient separations are possible with small volumes of solvent. However, the mechanism of hydrodynamic distribution of the solvent phases in the coil is not known (Ito 1984). The technique is also known as high-speed countercurrent chromatography (HSCCC). One of the most important experimental parameters (β) is defined as the ratio of the coil radius (r) to the orbital radius of revolution (R) (Fig. 7.10). This determines the hydrodynamic distribution of the two immiscible solvent phases in the rotating coil.

Retention of the stationary phase is strongly dependent on interfacial tension, density difference and viscosity of the two phases. Introduction of larger samples alters these parameters and leads to decreased retention of stationary phase. However, the available instruments are able to separate multigram quantities of sample within a few hours.

Multilayer coils can be characterized as having "head" and "tail" ends. The head is defined as the end to which a bead will travel when the coil is rotated. When the direction of rotation is reversed, the head and tail positions are interchanged (Fig. 7.11).

Several machines are presently available; these are listed in Table 7.6. Depending on the manufacturer, the number of coils (or columns) in the in-

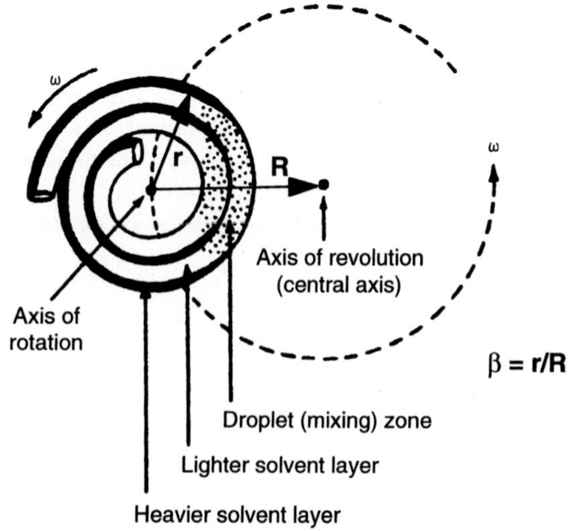

Fig. 7.10. High-speed countercurrent chromatography (ω = rate of rotation)

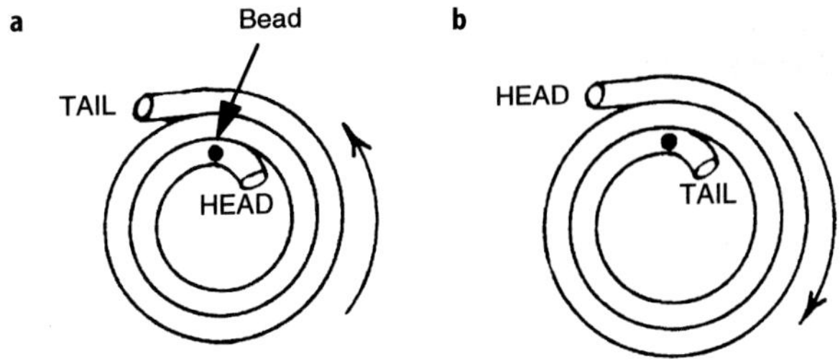

Fig. 7.11 a, b. Definition of: **a** "head"; **b** "tail" in a multilayer coil

Table 7.6. Rotating coil CPC instruments

Name	Columns	Capacity	Rotation speed
MKII Multilayer Countercurrent Chromatograph (PC Inc.)	1 (with counter-weight)	300 ml (1.68 mm tube) 400 ml (2.60 mm tube)	up to 1200 rpm
Pharma-Tech CCC-800	3	1500 ml	800 rpm
Pharma-Tech CCC-1000	3	850 ml	up to 1500 rpm
Pharma-Tech CCC-2000	3	260 ml	1400 rpm
Pharma-Tech CCC-3000	3	40 ml	up to 3000 rpm
Quattro (AECS)	2	630 ml (1.6 mm tube)	up to 1000 rpm
Kromaton (SEAB)	2	1200 ml	600 rpm
Kromaton II (SEAB)	2	1000 ml + 100 ml	800 rpm

strument can vary from 1 to 3. By placing multiple columns symmetrically around the rotary frame, the counterweight can be eliminated.

P.C. Inc. Instrument

One instrument is available from P.C. Inc. Coiled column assemblies can readily be interchanged on this chromatograph: a 300 ml coil, a 400 ml coil and a combination multilayer coil (15 ml + 75 ml + 215 ml coils on one assembly). The 15 ml coil is used for preliminary tests of solvent systems or for separating small (mg) amounts of sample. A combined sample injector/pressure monitor is available.

Pharma-Tech Instruments

These are constructed with three columns (Ito and Chou 1988; Ito et al. 1989, 1991) which can be serially connected without the risk of twisting the flow tubes (Fig. 7.12 shows the arrangement for two of the columns). Model CCC-3000 has a total capacity of 40 ml and is used for investigating solvent systems or separating very small amounts of sample. The other three, CCC-2000, CCC-1000 and CCC-800, are suitable for semi-preparative and preparative applications.

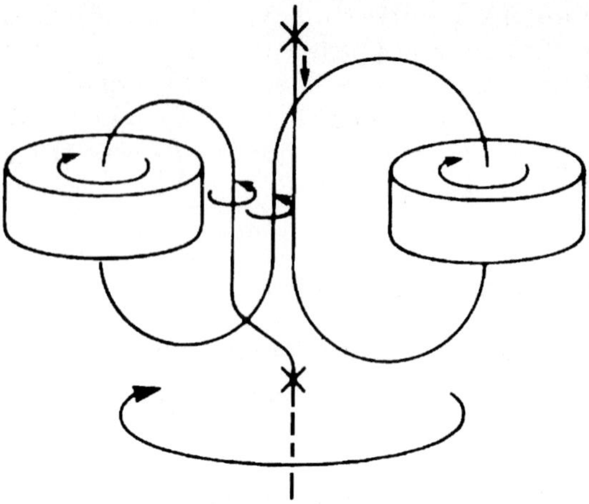

Fig. 7.12. Design principle of the multi-column chromatograph without rotary seals

Quattro
This instrument has two bobbins but no central shaft, which allows for higher β values. This means that mixing efficiency is better and that there are less problems of stationary phase loss. The name "Quattro" arises from the fact that four coils are incorporated: 50 ml and 250 ml on one bobbin, 100 ml and 200 ml on the other bobbin (although custom-made versions with different volume combinations are possible). Consequently, the coils can be interconnected in various combinations or, alternatively, used to run 4 independent samples. Another feature of the instrument is that temperature control is possible.

Kromaton
This machine is manufactured in two variants: the Kromaton has a single preparative coil of 1200 ml on two bobbins; the Kromaton II has a low-volume coil of 100 ml for investigation of separation conditions and a high-volume coil (1000 ml) which can be connected to the small coil for preparative work.

Cross-Axis CCC
In this method, the axis of column rotation is perpendicular to the axis of revolution (arrangements d and e in Fig. 7.9) (Ito and Zhang 1988; Menet et al. 1994). High flow rates and biphasic polar solvent systems can be employed. Cross-axis instruments are suitable for the separation of proteins with aqueous-aqueous polymer phase systems.

Other Instruments
Most of the different CPC coil arrangements conceived by Ito (Mandava and Ito 1988; Conway 1990) have remained at the prototype phase. Certain of these were discussed in the first edition of this book, together with their applications. For this reason, they will not be described here. Mention should be made, however, of a horizontal flow-through coil planet centrifuge and an HSCCC produced by Zhang at the Beijing Institute of New Technology Application (China) (Ito and Conway 1996).

In the new generation of all-liquid chromatography instruments, only one solvent phase is mobile. However, there is now available a particular combination of coil orientation and planetary motion which produces a unique hydrodynamic effect, permitting a true countercurrent flow. This has been named *dual countercurrent chromatography* and certain applications have been reported (Lee et al. 1988, 1989).

7.3.1.2
Hydrostatic Equilibrium Systems (HSES)

This variant of centrifugal CCC differs from HDES systems in that the column units are *fixed* in a centrifuge and do not themselves rotate, i.e. they only have a *single* axis.

Cartridge and Disk Instruments

Murayama and collaborators at Sanki Engineering, Kyoto, Japan, described the operation of their chromatograph in a paper dating from 1982 (Murayama et al. 1982) and Sanki is still the only manufacturer of this technology.

In the original instruments, the separation columns, in the form of *cartridges*, are arranged around the rotor of a centrifuge with their longitudinal axes parallel to the direction of the centrifugal force and connected by narrow-bore tubes (Fig. 7.13). The columns are filled with stationary phase and, when the rotor is spinning, mobile phase is pumped into the separation columns, forming a stream of droplets which crosses the stationary phase. Rotary seals at the upper and lower axes of the centrifuge are necessary to allow passage of solvent into the apparatus under pressure. When the heavier phase of a two-phase system is used as stationary phase, the mobile phase is pumped in a clockwise direction round the rotor. The phases can be switched without interrupting the separation – pumping is simply changed to the opposite direction.

The apparatus is a modular system, with 12 cartridges. The number of cartridges can be reduced to increase the speed of separation when high separation efficiencies are unnecessary.

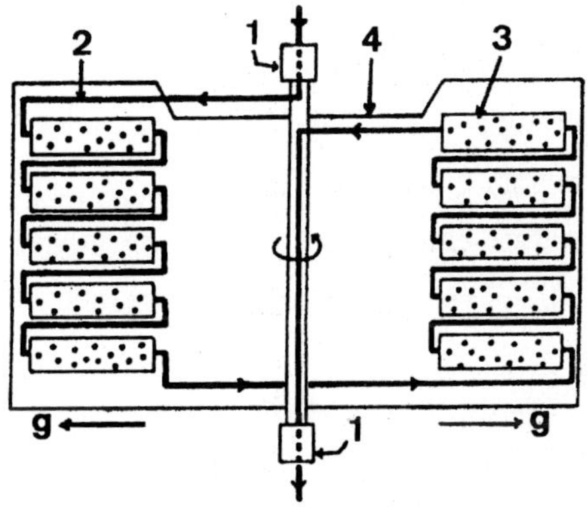

Fig. 7.13. The Sanki cartridge centrifugal chromatograph. 1: rotatory seal joint. 2: connecting tube. 3: separation column. 4: rotor. g: centrifugal force. (Reprinted with permission from Murayama et al. 1982)

The different parameters of the cartridge instrument have been investigated (Berthod and Armstrong 1988). A high rotational speed gives better peak resolution. Suitable separations are only achieved when not more than 20–30% of the stationary phase is eluted with the mobile phase. Judicial choice of rotor speed and pumping pressure is necessary to fulfil this condition. Surprisingly, efficiency increases with increasing flow rates.

There are two sizes of instrument:

- Type 250 W with 400 channels per cartridge (total volume of cartridge 18 ml and total volume of instrument ca. 220 ml).
- Type 1000E with 40 channels per cartridge (total volume of cartridge 75 ml and total volume of instrument ca. 900 ml).

The *disk* (high performance CPC, HPCPC or Series 1000) system was introduced in 1992 and consists of two packs, each of six disks, in a centrifuge. The disk has 89 channels and the total volume of the 2 disks is ca. 220 ml (Foucault 1995).

Sanki Engineering now offers the following options:

- LLB-M (up to 2000 rpm) with 2136 partition channels and a total volume of 230 ml.
- LLL-1.5 (up 1500 rpm) with 1296 partition channels and a total volume of 1360 ml.
- LLI-7 (up to 1300 rpm) with 1040 partition channels and a total volume of 5400 ml.

Operation of Sanki CPC instruments has been extensively described in a book edited by Foucault (1995).

The disadvantages of the Sanki instruments are the presence of rotary seals and the tendency towards high back pressures, which can attain 60 kg/cm^2 with viscous solvents. This increases the risk of connecting tube breakages. However, a wider range of solvent systems is possible with these instruments than in the case of rotating coil apparatus.

7.3.2 Choice of Solvent

The correct choice of solvent system is crucial to the success of a CPC separation. The most efficient combinations are those that yield partition coefficients of 0.2–5; values approaching 1 are best (Conway 1990). It is also vital that the settling time for the two phases is short. For HSCCC, this should be shorter than 30 s (Oka et al. 1991). It is also advantageous if the chosen solvent system provides roughly equal volumes of upper and lower phases. Selection of the nature and composition of the solvents can be guided by several means, described below.

7.3.2.1 *Reference to Existing Systems*

A survey of the specialist literature reveals numerous examples of solvent systems used in different countercurrent chromatographic separations (Mandava and Ito 1988; Conway 1990; Marston and Hostettmann 1994). Consultation of these and related references is the most rapid method of finding a system which may solve the separation problem in question. To start with, a search should be made for literature separations of the required substance class. Once these have been found, the associated chromatography solvent can be tested.

A solvent selection guide in which 13 systems are classified by the polarity of their upper and lower phases has been described (Abbott and Kleiman 1991). Chloroform-methanol-water or less polar n-hexane-ethyl acetate-methanol-water mixtures can be chosen as starting points and, by modifying the relative proportions of each individual solvent, it is possible finally to obtain the required distribution of sample between the two phases.

Oka et al. (1991) describe systematic solvent searches starting with the combinations *n*-hexane-ethyl acetate-*n*-butanol-methanol-water and chloroform-methanol-water, while Conway et al. (1988) list common chloroform-methanol-water mixtures.

Ternary diagrams for chloroform-methanol-water have been published (Conway 1990; Foucault 1995). From these, suitable solvent ratios can be calculated. In the book by Foucault (1995), a large number of the ternary diagrams for other solvent systems from the compilation of Sorensen and Arlt (1979) are reproduced. Ternary diagrams also give the compositions of the individual phases for a defined mixture. This allows preparation of a single phase: mobile phase can thus be prepared independently, for example (Miething and Rauwald 1990).

As chloroform-based systems provide large density differences and relatively high interfacial tensions between the two solvent phases, they are frequently employed in CPC. Furthermore, their short settling times allow a reduction in the amount of displaced stationary phase and normally produce satisfactory retention. One drawback, however, is that these systems can lead to overpressure problems with cartridge instruments. Moreover, halogenated solvents pose a health risk. There is a move, therefore, to employ non-hazardous solvents alone: heptane, ethyl acetate, methyl *t*-butyl ether, acetone, methyl isobutyl ketone, methanol, butanol and water.

7.3.2.2
Thin-Layer Chromatography

As in DCCC, the selection of appropriate biphasic solvent systems can be made by TLC. When using the organic layer of a two-phase aqueous system as TLC solvent, the R_f values should lie between 0.2 and 0.5. However, this method only gives an approximate indication of the utility of a particular solvent because TLC involves both partition and adsorption mechanisms, whereas countercurrent chromatography is based on purely liquid-liquid partition phenomena. All the same, TLC is a valuable initial tool. When a small amount of the sample is partitioned between the two phases of the solvent system, TLC is also useful for checking the distribution between the phases: if the sample is found almost exclusively in one phase, the solvent system is of no use.

Cellulose TLC plates have been proposed for solvent selection in the separation of tannins (Okuda et al. 1986).

7.3.2.3
High-Performance Liquid Chromatography

A knowledge of the partition coefficients of the solutes in question is of great benefit for the choice of suitable solvent systems. In fact, it is theoretically possible in CPC to predict the locations of eluted solute peaks once their partition coefficients are known (Ito and Bowman 1971). Measurement is possible by UV spectrophotometry (Conway et al. 1988) and an analytical HPLC method has been described for the determination of partition coefficients of the components of a mixture (Conway and Ito 1984). The solutes are partitioned between two immiscible liquid phases and their respective concentrations are established by reversed-phase HPLC. The partition coefficient of each component is calculated from the detector response following injection of a solution of the compounds before and after extraction with the lower phase of the solvent (Fig. 7.14). This procedure is of value for natural product mixtures, in which the identities of the individual components may not be known.

7.3.2.4
Partition Ratio of Biological Activity

This method, limited to bioactive compounds, is based on the distribution of the biological activity of the mixture to be separated. The sample is first shaken with the two-phase solvent system under test and then the upper and lower phases are screened for the activity in question. A solvent system which gives a fairly good distribution of activity between the two phases can be considered a candidate for the separation. One disadvantage is the length of time required for collection of the results from the biological test. The method has found certain applications, mainly in the isolation of antibiotics (Brill et al. 1985).

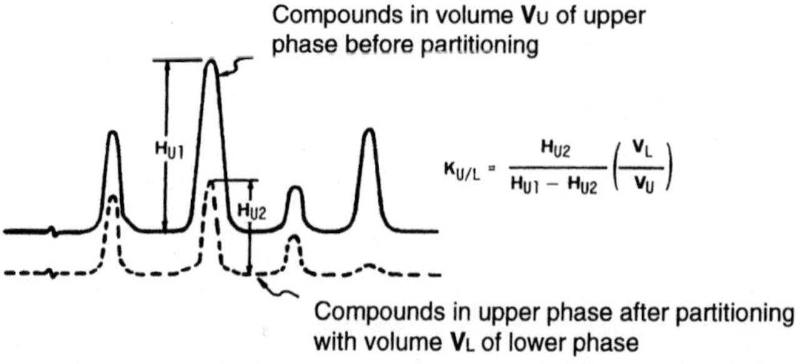

Fig. 7.14. Estimation of partition coefficients by HPLC, using peak heights (H)

7.3.2.5
Analytical CPC

A valid method for the choice of solvent systems to be used in preparative-scale separations is the application of analytical CPC. For example, rapid information on the suitability of a solvent system can be obtained on a Pharma-Tech Model CCC-3000 analytical instrument (coil capacity ca. 40 ml) or on the analytical coil (50 ml) of a Quattro instrument.

Care is necessary, however, since differences between the instruments (coil i.d. etc.) can lead to variations when passing from one machine to another. Although carry-over of a viscous stationary phase may not affect the quality of a preparative separation, leakage of only a very minor amount of this phase in the small coil of an analytical apparatus results in a dramatic decrease in the retention potential of the stationary phase.

7.3.3
Operating Techniques

Once the solvent system has been chosen, there are basically three methods for loading CPC instruments, as follows:

a) The chromatograph is first filled with stationary phase. Then (under rotation) the mobile phase is pumped into the apparatus. In the case of the P.C. Inc. instruments, when the mobile phase is the lower phase of a two-phase system, solvent is pumped into the "head" end of the multilayer coil and when the mobile phase is the upper phase, it is pumped into the "tail" end. When conditions are stable and only mobile phase exits the apparatus, the sample is injected via a sample loop. The sample is best introduced as a solution (filtered) in a mixture of the two solvent phases. Detection of eluted compounds is by means of a UV detector or, for non-UV-active compounds, by means of TLC. At the end of the separation, the stationary phase can be blown out by nitrogen or washed out with methanol.
b) The chromatograph (without rotation) is first filled with stationary phase. Then the sample is injected, and only at the end is rotation started, followed by pumping of mobile phase. This "stationary" or "sandwich" injection has the disadvantage over "dynamic" injection (in a) above) in that there is no preliminary stabilization of mobile phase flow and a noisy baseline is observed when the mobile phase front appears (Conway 1990).
c) By using separate pumps for mobile and stationary phases, the two phases can be introduced simultaneously into the instrument (Fig. 7.15) (Slacanin et al. 1989). This has the double advantage of saving time and also of preselecting the proportions of the two phases. When the instrument is full, rotation and pumping of mobile phase is commenced. Just a few minutes are required to reach stable conditions and sample can then be introduced.

With two pumps it is possible to elute the sample under *gradient* conditions (Slacanin et al. 1989). By changing the proportions of the two phases in the chromatograph, sample retention times can obviously be modified – either to speed up or slow down a separation.

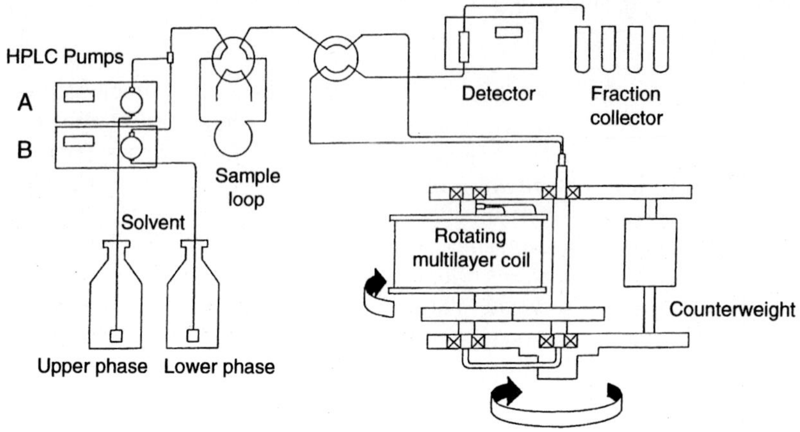

Fig. 7.15. Two-pump modification for the multilayer countercurrent chromatograph. (Reprinted with permission from Slacanin et al. 1989)

Both the Sanki and multilayer coil systems are capable of operating in the *"reversed-phase"* mode: at any time during the separation the flow direction can be reversed, provided elution is changed to the stationary phase. For example, in the coil system with two pumps, pumping with mobile phase is stopped, the "head" and "tail" entry/exit tubes are switched and pumping is re-started with the stationary phase layer (Slacanin et al. 1989).

7.3.4 Applications

CPC is now widely accepted as a routine preparative technique in both industrial and university laboratories. In the field of natural products, crude extracts and semi-pure fractions can be successfully chromatographed, with sample sizes ranging from milligrams to grams. Aqueous and non-aqueous solvent systems are used and the separation of compounds with a wide range of polarities is possible. CPC can also be exploited as an extraction technique – for the concentration of trace compounds from a large volume of solvent, for example. In this case, the solvent in which the compound is dissolved is employed as mobile phase and a stationary phase is chosen which has a high affinity for the product in question.

An extensive range of separations by CPC has been published (Hostettmann et al. 1986; Mandava and Ito 1988; Conway 1990; Marston and Hostettmann 1994; Foucault 1995; Ito and Conway 1996). The examples presented here have been selected either because they illustrate a particular technique or substance class, or they represent recent applications. Table 7.7 gives a number of separations of plant-derived natural products which involve at least one CPC step in the isolation procedure. Consultation of this table will give an idea of the solvent systems which can be applied to the different classes of natural products.

Table 7.7. Separations of plant-derived natural products by centrifugal partition chromatography

Sample	Instrument[a]	Solvent system	Reference
Flavonoids	MLCCC	$CHCl_3$-MeOH-H_2O 4:3:2	Zhang et al. 1988
Flavonol glycosides	CCC-1000	$CHCl_3$-MeOH-*n*BuOH-H_2O 7:6:3:4	Miserez et al. 1996
Biflavonoids	CCC-2000	Hexane-EtOAc-MeOH-H_2O 2:8:5:5	Kapadia et al. 1994
Xanthones	CCC-1000	Petrol ether-EtOAc-EtOH-H_2O 9:3:7:2	Rath et al. 1996
		Petrol ether-EtOAc-MeOH-H_2O 1:1:1:1	
		Cyclohexane-EtOAc-MeOH-H_2O 3:2:2:2	
	CCC-1000	Cyclohexane-EtOAc-MeOH-H_2O 4:5:3:3	Terreaux et al. 1995
Anthocyanidins	MLCCC	Hexane-EtOAc-*n*BuOH-HOAc-HCl 1% 2:1:3:1:5	Kunz et al. 1994
Proanthocyanidins	MLCCC	EtOAc-*n*PrOH-H_2O 35:2:2	De Mello et al. 1996a
	MLCCC	EtOAc-*n*PrOH-H_2O 140:8:80	De Mello et al. 1996b
	MLCCC	Hexane-EtOAc-MeOH-H_2O 8:16:7:10	Drewes and Taylor 1994
Tannins	Sanki L-90	*n*BuOH-*n*PrOH-H_2O 4:1:5	Hatano et al. 1992
	Sanki L-90	*n*BuOH-*n*PrOH-H_2O 2:1:3	Hatano et al. 1991
Lignans	Cross-axis CPC	Hexane-EtOAc-MeOH-0.5% NaCl 6:4:5:5	Gnabre et al. 1996
	MLCCC	Hexane-EtOAc-MeOH-0.5% NaCl 7:3:5:5	
	Sanki LLN	$CHCl_3$-MeOH-H_2O 10:10:6, 10:10:5, 13:7:8	Lee et al. 1996a
		EtOAc-EtOH-H_2O 2:1:2	
Lignan glycosides	MLCCC	Hexane-CH_2Cl_2-MeOH-H_2O 1:4:5:3, 2:5:4:3	Pettit and Schaufelberger 1988
Neolignans		Hexane-EtOAc-MeCN-H_2O 8:5:7:1	Nitao et al. 1991
Norlignans	MLCCC	Cyclohexane-EtOAc-Me_2CO-H_2O 5:3:4:5	Chifundera et al. 1991
		Cyclohexane-Me_2CO-EtOH-H_2O 7:6:1:3	
Stilbenes	Beijing	$CHCl_3$-MeOH-H_2O 4:3:2	Zhang and Fang 1994
Coumarin glycosides	MLCCC	$CHCl_3$-MeOH-H_2O 13:23:13	Van Wagenen et al. 1988
Naphthoquinones	Sanki NMF	Hexane-EtOAc-MeOH-H_2O 17:7:13:3	Dai et al. 1994
Anthraquinones	Sanki LLN	Hexane-EtOAc-MeOH-H_2O 9:1:5:5	Hermans-Lokkerbol et al. 1993
		$CHCl_3$-MeOH-H_2O-HOAc 5:6:4:0.05	

Table 7.7 (continued)

Sample	Instrument[a]	Solvent system	Reference
Phloroglucinols	CCC-1000	Petrol ether-EtOAc-MeOH-H_2O 10:5:5:1	Rocha et al. 1996
Cinnamic acid esters	MLCCC	Hexane-EtOAc-*n*PrOH-H_2O (pH 2) 0.5:7:0.4:4	Hahn and Nahrstedt 1993
Gingerols	MLCCC	Hexane-EtOAc-MeOH-H_2O 3:2:3:2	Farthing and O'Neill 1990
Furanone glucosides	MLCCC	$CHCl_3$-MeOH-H_2O 7:13:8	Roscher et al. 1996
		EtOAc-*n*BuOH-H_2O 2:2:5	
Theaspirones	MLCCC	Pentane-TBME-MeOH-H_2O 10:1:10:1	Herion et al. 1993
Iridoid glucosides	MLCCC	$CHCl_3$-MeOH-*n*PrOH-H_2O 5:6:1:4	Potterat et al. 1991
Hemiterpene glycosides	MLCCC	EtOAc-*n*BuOH-H_2O 3:2:5	Messerer and Winterhalter 1995
Norisoprenoid glycosides	MLCCC	$CHCl_3$-MeOH-H_2O 7:13:8	Skouroumounis and Winterhalter 1994
		EtOAc-*n*BuOH-H_2O 3:2:5	
		EtOAc-*n*BuOH-H_2O 4:1:5	
Ionyl glucosides	MLCCC	EtOAc-*n*BuOH-H_2O 3:2:5	Dietz and Winterhalter 1996
Abscisic alcohols	MLCCC	Hexane-EtOAc-MeOH-H_2O 3:7:5:5	Lutz and Winterhalter 1994
Sesquiterpenes	Sanki NMF	Hexane-EtOAc-MeOH-H_2O 4:1:4:1	Ying et al. 1995
Clerodane diterpenes	MLCCC	Hexane-EtOH-H_2O 2:1:1	Leitao et al. 1992
Labdane diterpenes	Sanki LLN	Hexane-CH_2Cl_2-MeOH-H_2O 5:20:17:8	Kagawa et al. 1993
Quassinoids	MLCCC	$CHCl_3$-MeOH-H_2O 5:6:4	Jaziri et al. 1991
Taxanes	MLCCC	Petrol ether-EtOAc-MeOH-H_2O 4:12:4:5	Guo et al. 1996
		CCl_4-CH_2Cl_2-MeOH-H_2O 5:5:6:4	Gunawardana et al. 1992
		Isooctane-EtOAc-MeOH-H_2O 7:3:6:4	
	Beijing	Hexane-EtOAc-EtOH-H_2O 6:3:2:5	Gu et al. 1994
Triterpenes	Sanki LLN	$CHCl_3$-MeOH-1% HOAc 2:2:1	Lee et al. 1996b
	MLCCC	Hexane-EtOAc-MeOH-H_2O 1:2:1:1	Fullas et al. 1996
	MLCCC	Hexane-EtOAc-MeCN-MeOH 5:2:5:4	Abbott et al. 1989
Saponins	CCC-1000	$CHCl_3$-MeOH-*n*PrOH-H_2O 5:6:1:4	Verotta et al.1996
	MLCCC	$CHCl_3$-MeOH-*i*PrOH-H_2O 5:6:1:4	Potterat et al. 1992
	MLCCC	$CHCl_3$-MeOH-*i*BuOH-H_2O 7:6:3:4	Diallo et al. 1991

Saponins (continued)	MLCCC	$CHCl_3$-MeOH-H_2O 7:13:8	Fullas et al. 1990
	Sanki HPCPC	EtOAc-*n*BuOH-H_2O gradient	Le Men-Olivier et al. 1995
Carotenoids	MLCCC	CCl_4-MeOH-H_2O 5:4:1	Diallo and Vanhaelen 1988
Polyacetylenes	MLCCC	Hexane-MeCN-TBME 10:10:1	Nitz et al. 1990
Ergot alkaloids	CCC-2000	$CHCl_3$-MeOH-H_2O 5:4:3	Petroski et al. 1992
Pyrrolizidine alkaloids	MLCCC	$CHCl_3$-0.2 mol/l potassium phosphate buffer	Cooper et al. 1996
Indole alkaloids	MLCCC	Me_2CO-*n*BuOH-H_2O 1:8:10	Quetin-Leclercq et al. 1988
Bis-indole alkaloids	MLCCC	EtOAc-MeOH-H_2O 4:1:3	Quetin-Leclercq et al. 1995
Aporphine alkaloids	Sanki LLN	$CHCl_3$-MeOH-0.5% HOAc 5:5:3	Lee et al. 1996c
Naphthyltetrahydro-isoquinoline alkaloids	Sanki NMF	$CHCl_3$-MeOH-0.5% HBr 5:5:3	Hallock et al. 1994
Diterpene alkaloids	Sanki HPCPC	C_6H_6-$CHCl_3$-MeOH-H_2O 5:5:7:2	Venkateswarlu et al. 1995
Cephalotaxus alkaloids	Beijing	$CHCl_3$-0.07 mol/l sodium phosphate buffer 1:1 (pH 5)	Cai et al. 1992
Camptothecin	CCC-1000	CCl_4-$CHCl_3$-MeOH-H_2O 2:2:3:1	Broglia et al. 1994
		CH_2Cl_2-MeOH-H_2O 5:3:1	
Vinca alkalolids	Beijing	Hexane-EtOH-H_2O 6:5:5	Zhou et al. 1990
Flavonoid alkaloids	Sanki NMF	$CHCl_3$-MeOH-0.5% HCl 5:5:3	Beutler et al. 1992
Cyanogenic glycosides	MLCCC	EtOAc-*n*PrOH-H_2O 35:2:20	Nahrstedt et al. 1993
Amino acids	MLCCC	$CHCl_3$-MeOH-H_2O 4:4:2	Bejar et al. 1995
Prostaglandins	Sanki NMF	Hexane-Et_2O-MeCN 5:2:5	Jirousek and Salomon 1988

[a] MLCCC = multilayer countercurrent chromatograph (PC Inc.); CCC-1000 = Pharma-Tech; Beijing = instrument constructed by Beijing Institute of New Technology Applications.

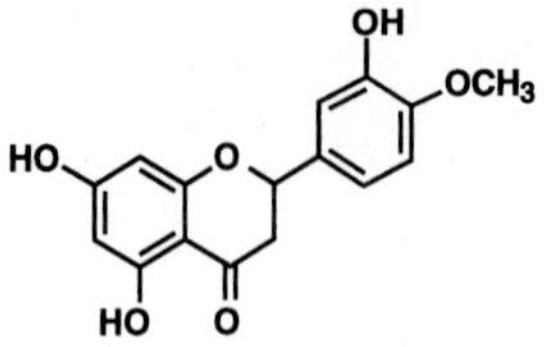

32

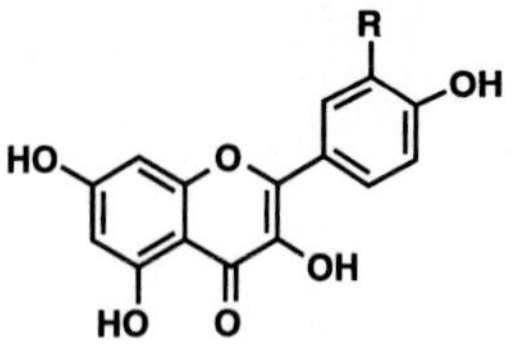

33 R = H
34 R = OH

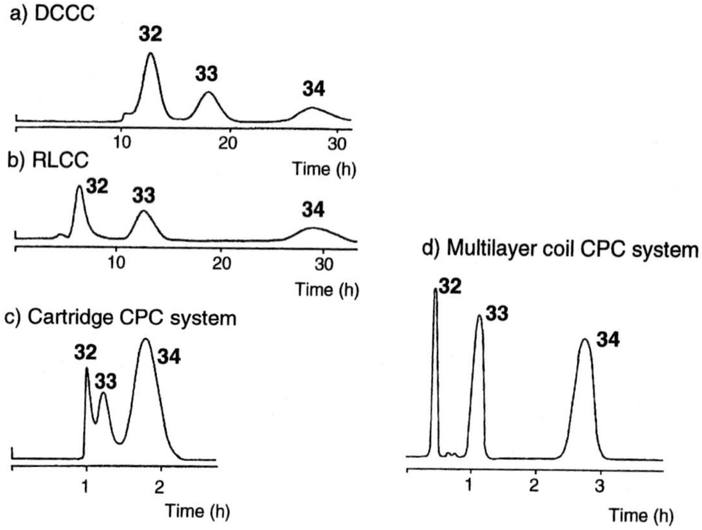

Fig. 7.16a–d. Separation of hesperetin (**32**), kaempferol (**33**) and quercetin (**34**) by different CCC methods; solvent system: $CHCl_3$-MeOH-H_2O 5:6:4; mobile phase: lower phase; detection: 254 nm: **a** flow rate 48 ml/h; **b** flow rate 18 ml/h; **c** flow rate 2 ml/min; rotational speed 600 rpm; 6 cartridges; **d** flow rate 3 ml/min; rotational speed 700 rpm. (Reprinted with permission from Marston et al. 1990)

A comparative separation of the flavonone hesperetin (**32**) and the flavonols kaempferol (**33**) and quercetin (**34**) by DCCC, RLCC and two CPC methods is shown in Fig. 7.16 (Marston et al. 1990). Elution was by order of increasing polarity when the lower phase was used as the mobile phase. Whereas DCCC and RLCC required more than 30 h for complete separation, the two CPC methods took approximately 3 h. Solvent consumption for RLCC was ca. 1500 ml, for DCCC and the CPC methods ca. 550 ml.

When eluting with the upper phase (mobile phase), separation times were longer (Fig. 7.17a). However, CPC is amenable to reversed-phase operation: during the separation the elution mode (head → tail or tail → head) can be changed, as long as the mobile phase is changed (Fig. 7.17b). By eluting first with upper phase and then changing to lower phase at 70 min, the separation is speeded up considerably.

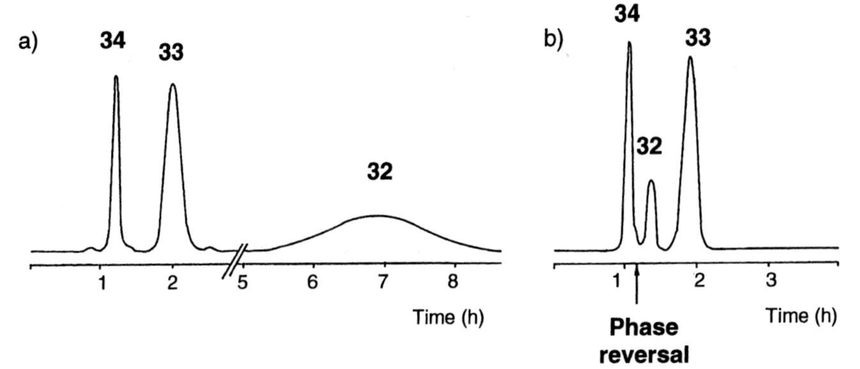

Fig. 7.17 a, b. Separation of hesperetin (**32**), kaempferol (**33**) and quercetin (**34**) with phase reversal; instrument:multilayer countercurrent chromatograph (PC Inc.); solvent system: $CHCl_3$-MeOH-H_2O 5:6:4; detection: 254 nm; **a** mobile phase:upper phase; flow rate 3 ml/min; rotational speed 700 rpm; **b** mobile phase:upper phase to 70 min, then lower phase; flow rate 3 ml/min; rotational speed 700 rpm. (Reprinted with permission from Marston et al. 1990)

7.3.4.1
Flavonoids

Many CPC separations of natural products involve polyphenols. The reason for this fact is that there is a tendency to "tail" or even to adsorb irreversibly on conventional chromatographic supports (silica gel, polyamide etc.). This problem does not occur with all-liquid separation techniques and quantitative recovery of injected sample is achieved.

An example of the separation of flavonol glycosides is shown in Fig. 7.18. This was performed on a Quattro instrument (630 ml column capacity) with the solvent system chloroform-ethanol-methanol-water 5:3:3:4. The column was filled with 50% upper phase and 50% lower phase by simultaneously pumping with the two-pump system shown in Fig. 7.15. The leaves of the African plant *Tephrosia vogelii* (Leguminosae) were first extracted with dichloromethane and then with methanol. A part of the methanol extract (500 mg) was dissolved in a mixture of 10 ml of upper phase and 10 ml of lower phase and then injected via a 60 ml sample loop. For elution, upper phase was chosen as the mobile phase (tail → head orientation). Rotation speed was 800 rpm and flow rate 3 ml/min. The most polar glycoside was the first to elute and all three glycosides were separated within 3 h.

CPC was employed as a key step in the purification of 26 phenolic compounds from the needles of Norway spruce (*Picea abies*, Pinaceae). Separations were performed on a Pharma-Tech CCC-1000 instrument (total capacity 660 ml) with the solvent system $CHCl_3$-MeOH-*i*PrOH-H_2O 5:6:1:4, using the lower phase as initial mobile phase and then switching to upper phase as mobile phase. In one run, an aqueous methanol extract of the mature needles (5.45 g) was injected into the instrument. Final purification by gel filtration and semi-preparative HPLC gave flavonol glycosides, stilbenes and catechins (Slimestad et al. 1996).

Fig. 7.18. Separation of flavonol glycosides **35–37** from *Tephrosia vogelii* (Leguminosae) by CPC; instrument: Quattro; solvent system: $CHCl_3$-MeOH-EtOH-H_2O 5:3:3:4; detection: 254 nm; mobile phase: upper phase; flow rate 3 ml/min; rotational speed 800 rpm; sample 500 mg

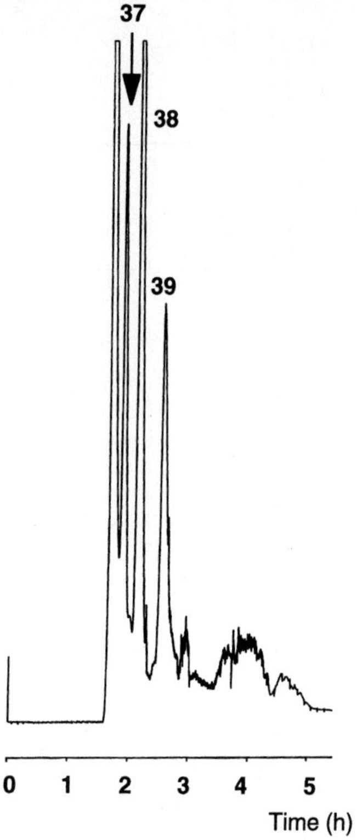

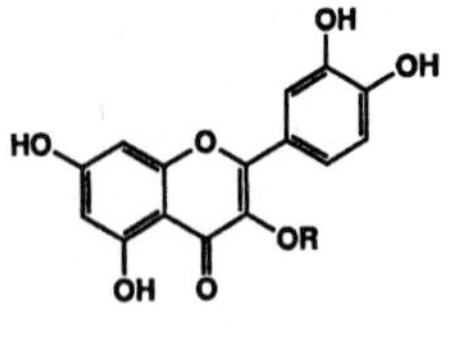

35 R = -Ara
36 R = -Glc
37 R = -Glc^6-Rha

Extracts of *Gingko biloba* (Gingkoaceae) leaves are important therapeutically in certain geriatric problems, including treatment of circulatory disorders of the brain. European sales of these preparations represent millions of dollars annually. Flavonoid glycosides from the leaves have been isolated by a combination of CPC and semi-preparative HPLC. The liquid-liquid step involved *gradient elution*, starting with water as stationary phase and eluting with ethyl acetate. Increasing amounts of isobutanol were then added to the ethyl acetate until the proportions ethyl acetate-isobutanol 6:4 were reached at the end of the elution. Seven flavonol glycosides were obtained from 500 mg of leaf extract (Vanhaelen and Vanhaelen-Fastré 1988).

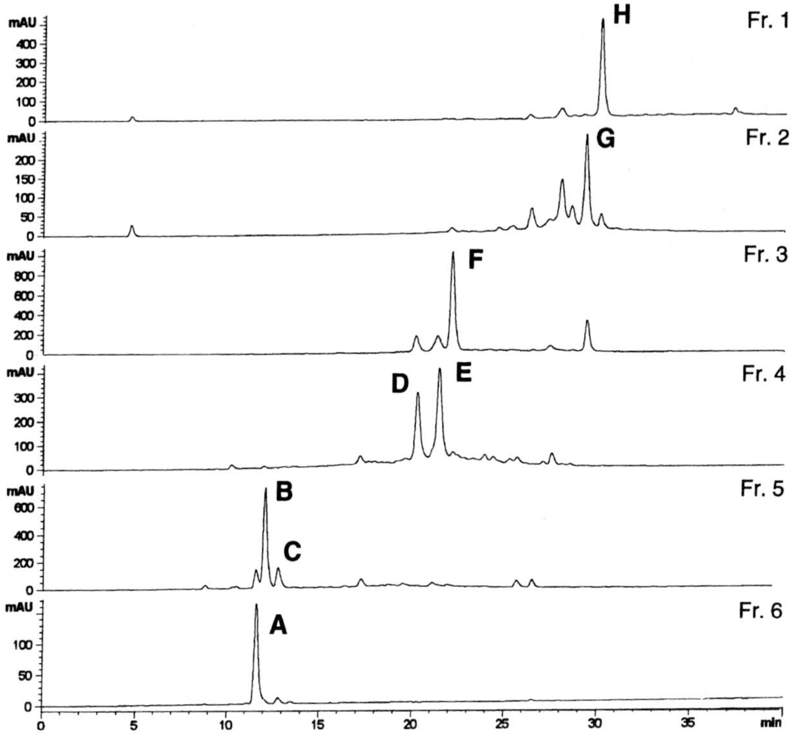

Fig. 7.19. CPC separation of xanthones from *Hypericum roeperanum* (Guttiferae). HPLC monitoring of fractions: Column Macherey-Nagel C-18 (4 × 250 mm; 5 μm); solvent MeOH-H_2O (+0.05% TFA) 55:45 → 100:0 over 40 min; flow rate 1 ml/min; detection 254 nm

7.3.4.2
Xanthones

This is another class of polyphenolics which lends itself well to liquid-liquid chromatography. Certain xanthones exhibit inhibition of monoaminooxidases (MAO). Inhibitors of MAO (in particular the A-type isoenzyme) have a potential as antidepressive drugs since they increase both noradrenaline and serotonin levels in the brain.

In one example, extensive use of CPC was made for the separation of xanthones from a Zimbabwe plant, *Hypericum roeperanum* (Guttiferae). A dichloromethane extract of the roots gave a fraction (363 mg) rich in xanthones after initial chromatography. Separation of the individual xanthones from this fraction by other methods, including semi-preparative HPLC, proved very difficult. However, CPC with a Pharma-Tech CCC-1000 instrument gave six fractions (1–6), which after a final purification step in each case, provided eight xanthones (A-H) (Fig. 7.19). For the CPC separation, the solvent hexane-EtOAc-MeOH-H_2O 1:1:1:1 was employed, with the upper phase as the mobile phase (Rath et al. 1996).

Xanthone H (5-*O*-demethyl-paxanthonin, **38**) and xanthone G (roeperanone, **39**) were new xanthones which could only be satisfactorily separated by CPC.

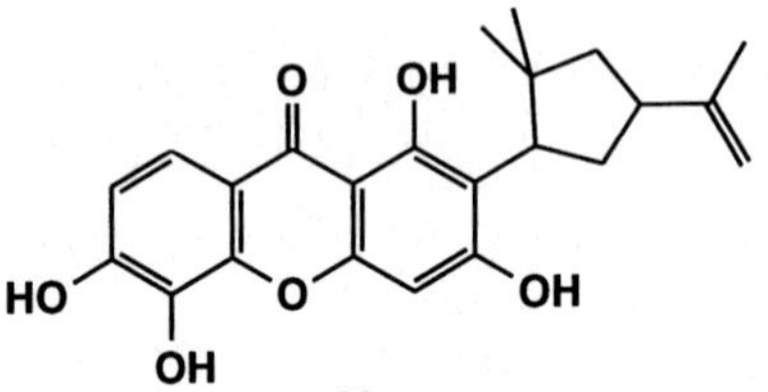

38

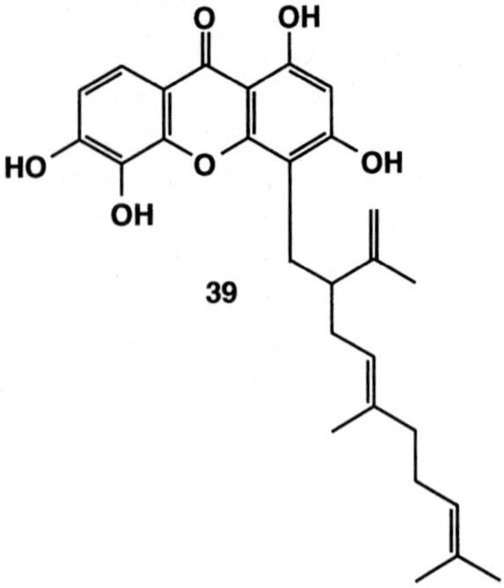

39

7.3.4.3
Chalcone Derivatives

Initial work on a tree named *Cordia goetzei* (Boraginaceae) from Tanzania showed the presence of antifungal compounds in the yellow inner stem bark (Marston et al. 1988 a). The substances responsible for this activity were isolated by DCCC and C-18 LPLC: they were potent inhibitors of the growth of the plant pathogenic fungus *Cladosporium cucumerinum*. Subsequent investigation (Marston et al. 1996) showed that the identity of the tree was in fact *Brackenridgea zanguebarica*, a member of the Ochnaceae family.

Direct fractionation of 20 g of a methanol extract of the yellow bark layer on a Pharma-Tech CCC-1000 instrument was possible with the solvent system cyclohexane-ethyl acetate-methanol-water 8:8:6:6 (upper layer as mobile phase). The extract was dissolved in 20 ml of each phase and introduced via a 60 ml sample loop. The separation column (660 ml) was first filled with 50% of each phase and then eluted with mobile phase. When steady run conditions had been obtained, the sample was injected. Seven fractions were obtained (A – G; Fig. 7.20), three of these giving the antifungal polyphenols **40**–**43**. Compound **40** crystallized from fraction B, while compounds **41**–**43** were purified by a final LPLC step on a C-18 support. A novel, inactive, spiro derivative (**44**) was isolated from fraction E after LPLC on C-18 (Marston et al. 1996).

This example illustrates the utility of the CPC method for handling of large quantities of crude extract. Furthermore, all five polyphenols were obtained by a two-step fractionation procedure.

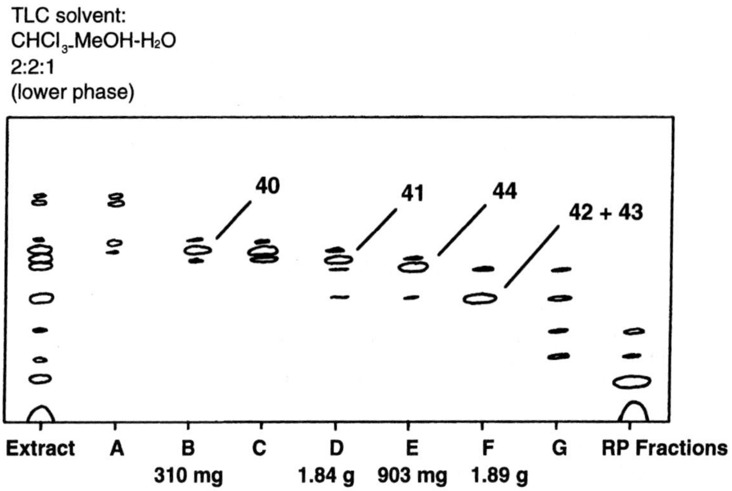

Fig. 7.20. TLC monitoring of fractions from the CCC-1000 CPC separation of *Brackenridgea zanguebarica* stem bark methanol extract. CPC solvent system:cyclohexane-EtOAc-MeOH-H_2O 8:8:6.6 (mobile phase: upper phase); flow rate 3 ml/min; rotational speed 1000 rpm; detection 254 nm; sample 20 g

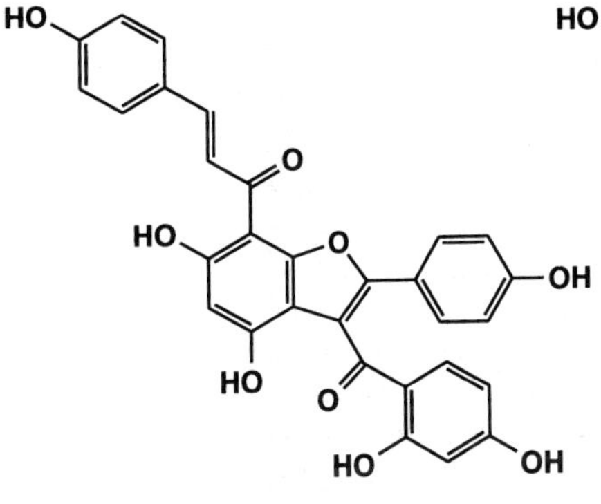

40

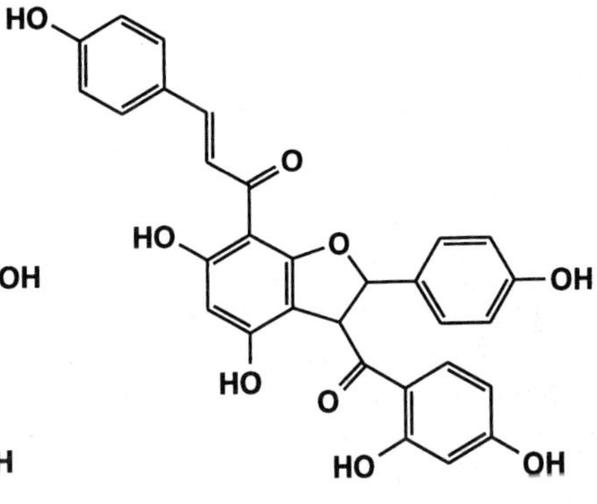

41

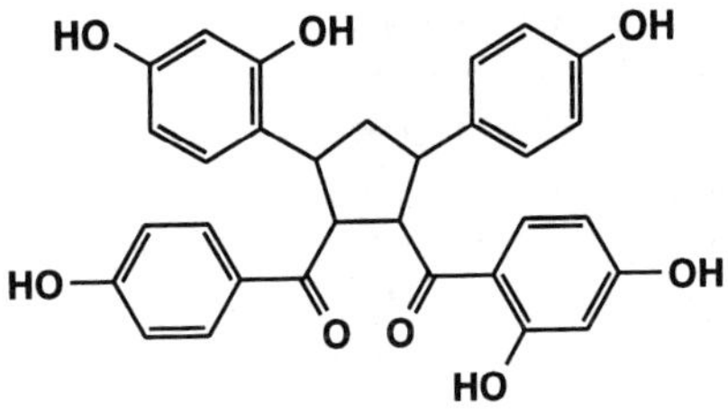

42

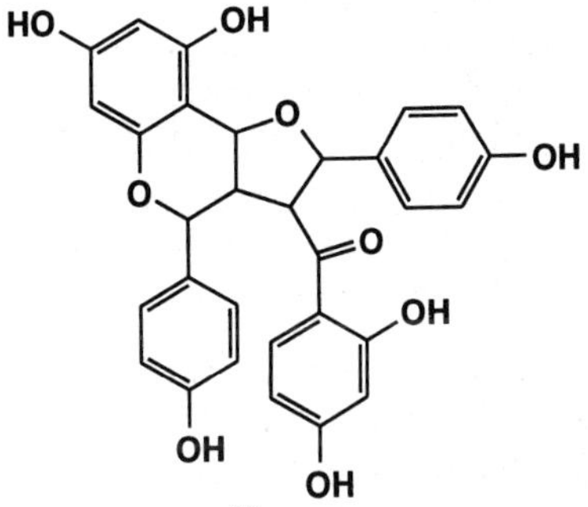

43

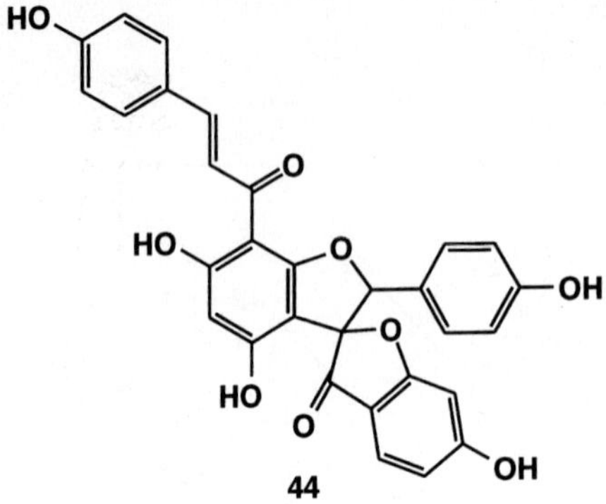

44

7.3.4.4
Phenylpropanoids

It is possible to isolate natural products by CPC alone, as shown by the purification of phenylpropanoids and a furanocoumarin from a dichloromethane extract of the leaves of *Diplolophium buchanani* (Apiaceae). Initial fractionation (Fig. 7.21) on a Pharma-Tech instrument gave semi-pure products (Fig. 7.22). A subsequent liquid-liquid step, using a different non-aqueous solvent system in each case, gave pure myristicin (**45**) and a mixture of elemicin (**46**) and trans-isoelemicin (**47**). The furanocoumarin oxypeucedanin (**48**) was obtained by simple crystallization of the corresponding CPC fraction. All four isolated compounds had both antifungal and larvicidal activities (Marston et al. 1995).

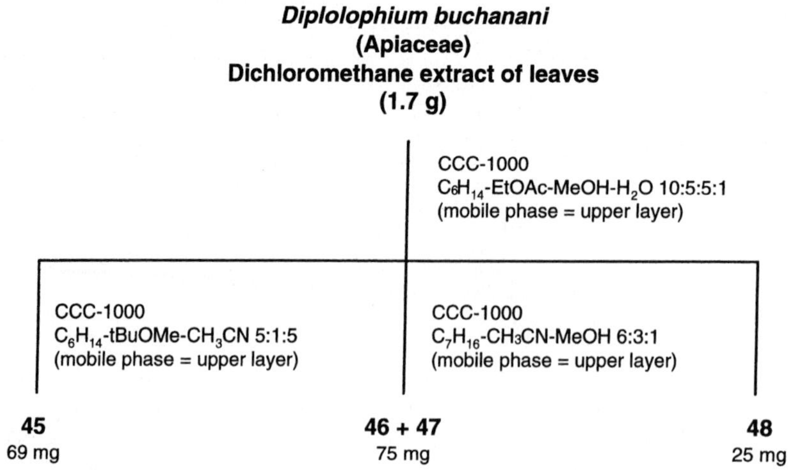

Fig. 7.21. Isolation of phenylpropanoids (**45**–**47**) and a furanocoumarin (**48**) from *Diplolophium buchanani*

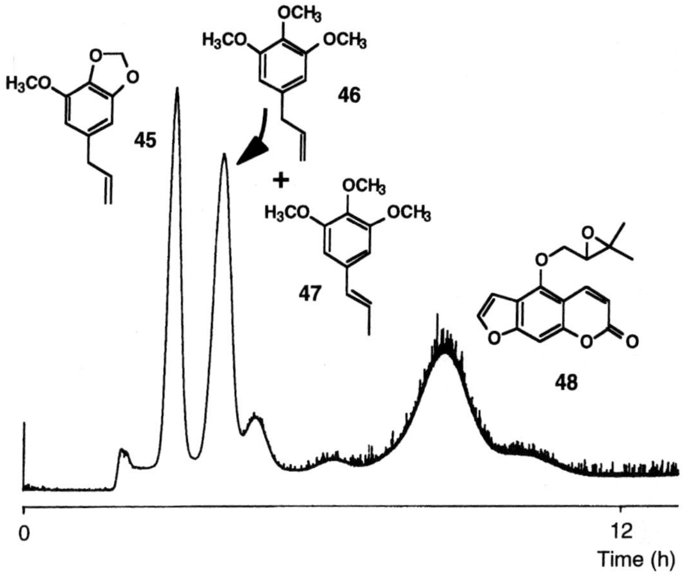

Fig. 7.22. Pharma-Tech CCC-1000 initial fractionation of *Diplolophium buchanani* leaves (dichloromethane extract). Solvent: hexane-EtOAc-MeOH-H_2O 10:5:5:1 (upper layer as mobile phase); flow-rate 3 ml/min; detection 254 nm; sample 1.7 g

7.3.4.5
Lignans

Initial fractionation of insecticidal neolignans from *Magnolia virginiana* (Magnoliaceae) by CPC was claimed to be better, less expensive and more efficient than traditional open-column or more recent flash chromatographic methods (Nitao et al. 1991). A hexane extract of the leaves was chromatographed with the lower layer of the solvent system hexane-acetonitrile-ethyl acetate-water 8:7:5:1 as mobile phase. This solvent contains only a small proportion of water, probably to provide compatibility with the very lipophilic extract. Subsequent purification of the fractions provided a biphenyl ether (**49**) and two biphenyls (**50, 51**), which were not only insecticidal to *Aedes aegypti* (the vector

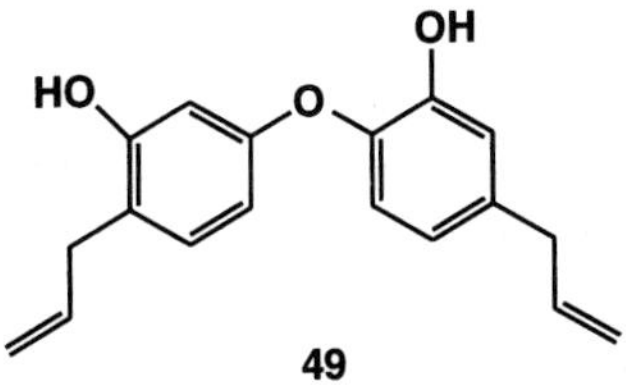

49

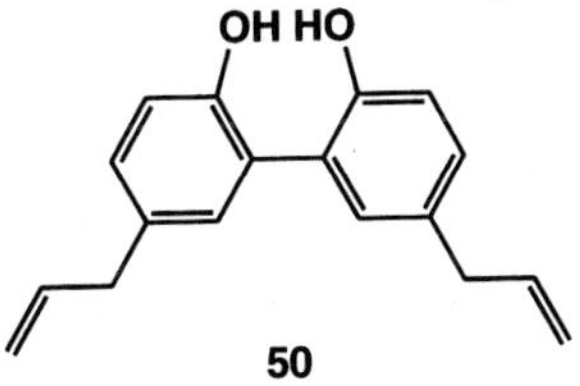

50

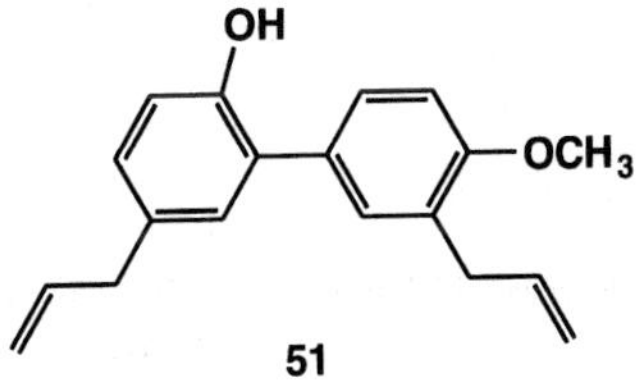

51

of yellow fever) but also fungicidal, bactericidal, and toxic to brine shrimp (Nitao et al. 1991).

During the investigation of anti-HIV-1 lignans from creosote bush, *Larrea tridentata* (Zygophyllaceae), Gnabre et al. (1996) employed a cross-axis coil planet centrifuge with aqueous phases containing *NaCl*, e.g. hexane-ethyl acetate-methanol-0.5% NaCl 6:4:5.5. The presence of salt reduced emulsification and gave improved stationary phase retention.

7.3.4.6
Proanthocyanidins

A combination of gel filtration, CPC and semi-preparative HPLC has been reported for the isolation of eight dimeric prorobinetinidins (proanthocyanidins) of general structure **52** from the stem bark of *Stryphnodendron adstringens* (Leguminosae). The CPC step involved separation on an MLCCC instrument with the upper layer of EtOAc-nPrOH-H_2O 35:2:2 as mobile phase (De Mello et al. 1996a).

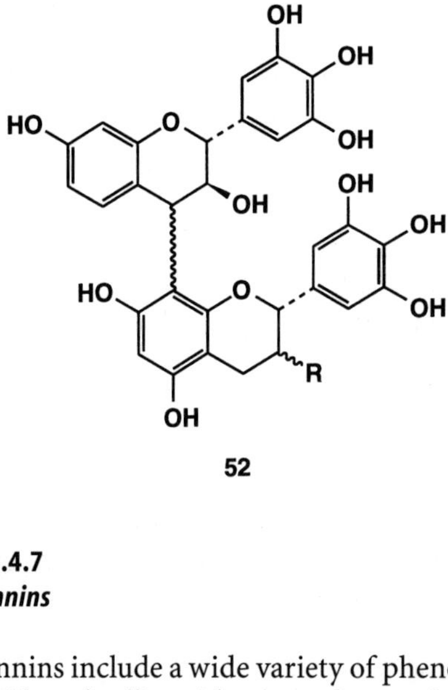

52

7.3.4.7
Tannins

Tannins include a wide variety of phenolic compounds, ranging from single glycosides of gallic acid to complex condensed and polymerized derivatives of catechin, epicatechin and related compounds. Their separation poses special problems since there is often irreversible adsorption and even hydrolysis on solid supports (Okuda et al. 1988). Preparative HPLC is also accompanied by sample loss and deterioration or contamination of the column (Yoshida et al. 1989). CPC has proved to be an ideal technique for the resolution of these particular problems. Okuda's group in Japan has been especially active in the application of Sanki CPC instruments to the purification of tannins.

Among the examples of successful separations performed on Sanki cartridge instruments is the resolution of two diastereomers, castalgin (**53**) and ve-

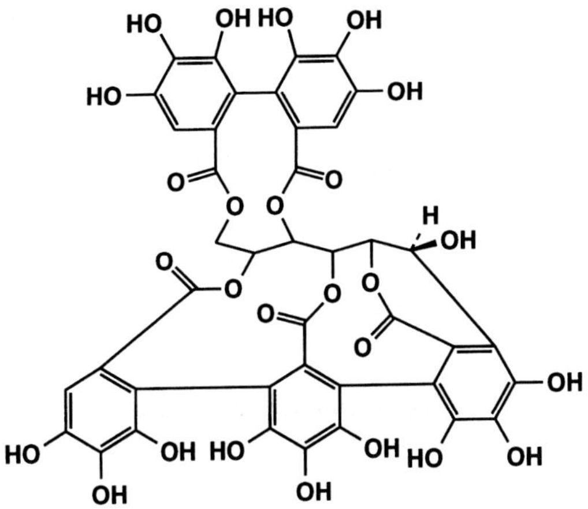

53

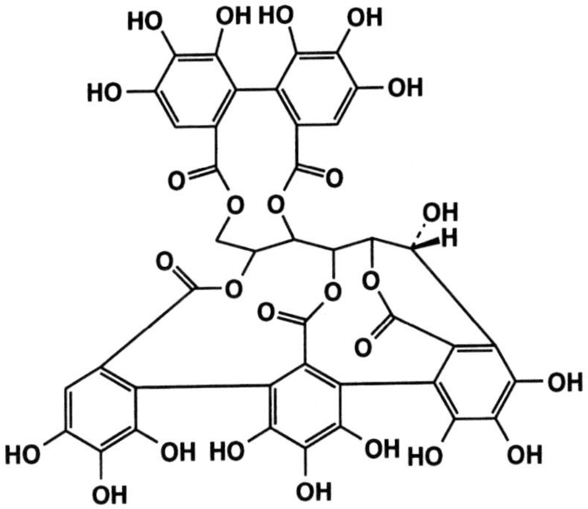

54

scalagin (**54**), which differ only in the configuration of a single hydroxyl group. They were extracted from *Lythrum anceps* (Lythraceae) leaves and chromatographed with the solvent system *n*-butanol-*n*-propanol-water (4:1:5), using the upper phase as the mobile phase (Okuda et al. 1986).

The same solvent system was used to obtain a trimeric (nobotanin J; MW 2764) and a tetrameric (nobotanin K; MW 3742) hydrolyzable tannin from the leaves of *Heterocentron roseum* (Melastomataceae). The isolation procedure (typical for Okuda's work on tannins) involved first a Diaion HP-20 step, then chromatography on Toyopearl HW-40 (methanol-acetone-water gradient). Final purification of the two tannins was by Sanki CPC on an L-90 instrument with 12 cartridges (total capacity 240 ml; 700 rpm; flow rate 3 ml/min) (Okuda et al. 1995).

Sorghum grain tannins have been separated by CPC using *n*-butanol-0.1 mol/l NaCl 1:1 as solvent system (with the lower phase as mobile phase). Since sodium chloride is insoluble in methanol, once the fractions have been collected they can be dried, tannin being dissolved in methanol and separated from the salt (Putman and Butler 1985).

7.3.4.8
Anthracene derivatives

Lipophilic root bark extracts of the African medicinal plant *Psorospermum febrifugum* (Guttiferae) exhibit strong growth inhibition of cancer cells and have antimalarial activity. Separation of the active anthranoid pigments by flash chromatography and low-pressure liquid chromatography resulted in considerable material losses, owing to irreversible adsorption on the supports. However, in a *single* CPC step (Sanki cartridge system), three pure compounds (**55**–**57**) and a mixture of a fourth anthranoid pigment (**58**) with an unidentified constituent were obtained without loss of product (Fig. 7.23). A non-aqueous solvent system was used for the separation (Marston et al. 1988b). Increasing the number of cartridges from six to twelve permitted a better resolution of the peaks but required a longer separation time. With twelve cartridges, the separation could be scaled up to a 500 mg sample size (Marston et al. 1990).

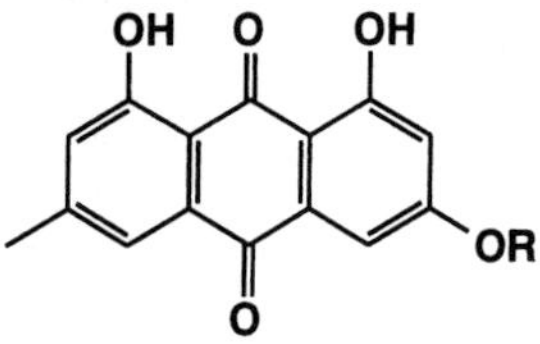

55

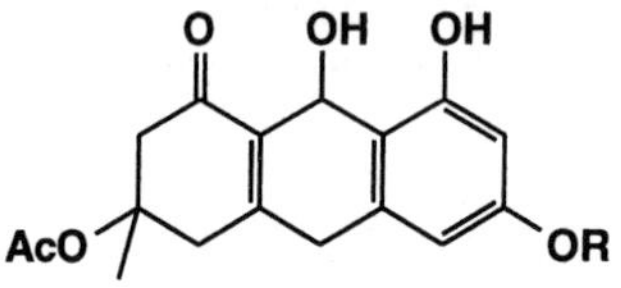

56

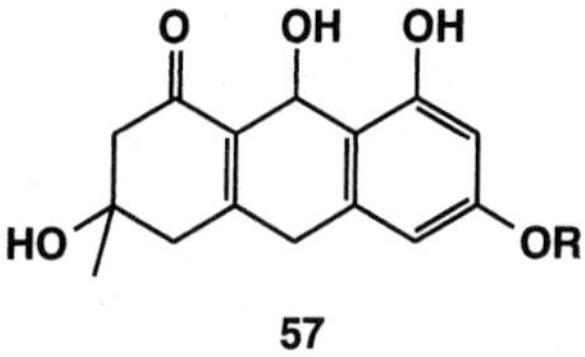

57

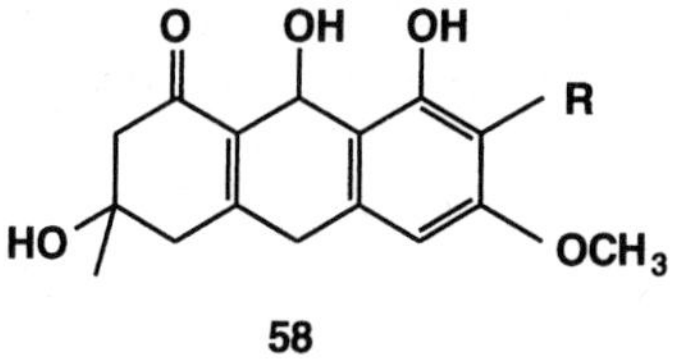

58

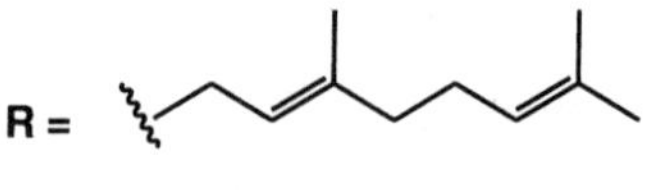

7.3.4.9
Triterpenes

Abbott et al. (1989) have studied the separation of triterpene acetates from *Asclepias linaria* (Asclepiadaceae) by three preparative chromatography techniques and came to the conclusion that CPC gave a less complete purification than silver ion exchange resin HPLC or non-aqueous reversed phase HPLC. However, they considered that the wider range of possible solvents for CPC tended to outweigh the less favorable separation capacity.

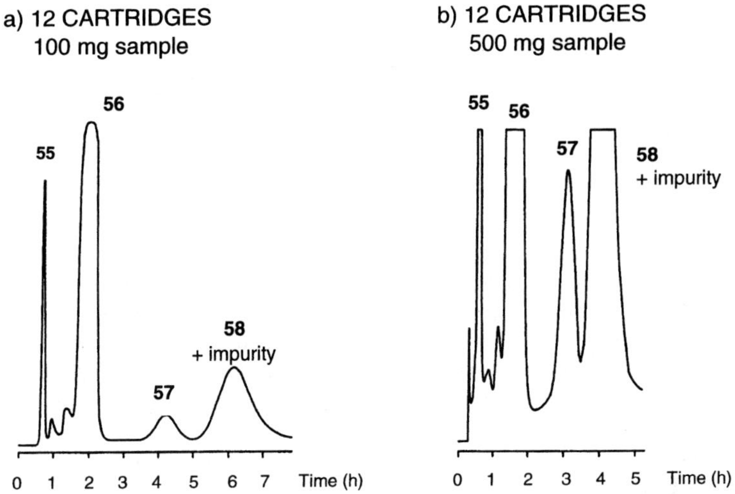

Fig. 7.23 a, b. Separation of a light petroleum extract of *Psorospermum febrifugum* (Guttiferae) root bark with a Sanki LLN cartridge instrument. Solvent system: *n*-hexane-acetonitrile-methanol 8:5:2 (upper layer as mobile phase); flow rate 6.5 ml/min; 800 rpm; detection 254 nm. (Reprinted with permission from Marston et al. 1990)

7.3.4.10
Saponins

Like DCCC, centrifugal partition chromatography is a useful technique for the separation of saponins. For example, two triterpene glycosides, asiaticoside (**59**) and madecassoside (**60**), which have a difference of only one hydroxyl group, were isolated from an extract of the medicinal plant *Centella asiatica* (Apiaceae) with an MLCCC instrument. The two-phase solvent system was chloroform-methanol-isobutanol-water 7:6:3:4, with the lower phase as mobile phase, at a flow-rate of 4 ml/min. A 400 mg sample was injected. Detection of the non-UV-active saponins was achieved by direct coupling to TLC (Diallo et al. 1991).

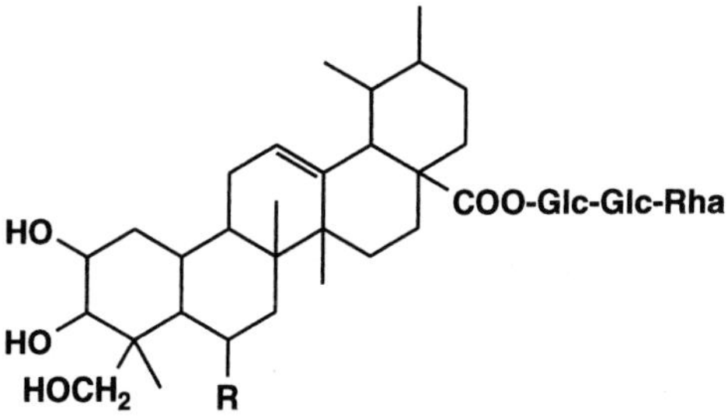

59 R = H (Asiaticoside)
60 R = OH (Madecassoside)

7.3.4.11
Alkaloids

The utility of CPC in separating alkaloids from gummy or tarry matrices has been shown in the preparative separation of pyrrolizidine alkaloids from different plant sources. After classical alkaloid extraction procedures, batches of up to 800 mg extract could be chromatographed on an MLCCC apparatus (380 ml capacity). Potassium phosphate buffer (0.2 mol/l) at an appropriate pH was used as stationary phase, while the mobile phase was chloroform. With the buffered stationary phase, good solute resolution can be obtained since structurally similar pyrrolizidine alkaloids differ in their pKa values (Cooper et al. 1996).

CPC with a Pharma-Tech CCC-1000 instrument has been employed for the separation of the antitumour drug camptothecin (**61**). From sources such as *Nothapodytes foetida* (Icacinaceae), camptothecin is often mixed with 9-methoxycamptothecin (**62**). These can be easily separated with the solvent systems

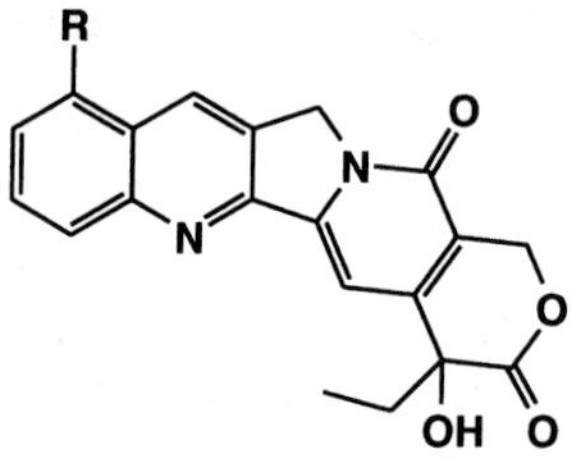

61 R = H (Camptothecin)
62 R = OCH_3 (9-Methoxycamptothecin)

$CHCl_3$-CCl_4-MeOH-H_2O 2:2:3:1 or CH_2Cl_2-MeOH-H_2O 5:3:1. The low solubility of the samples poses a problem with traditional chromatographic techniques. However, for the CPC separation, 500 mg of sample can be dissolved in 70 ml lower phase and 10 ml of upper phase (i.e. a *large volume*) for injection via a sample loop (Broglia et al. 1994).

Repetitive sample injections are possible for the separation of close-running compounds on rotating coil instruments. This has been shown for the separation of vincamine (**63**) and vincine (**64**) from *Vinca minor* (Apocynaceae)

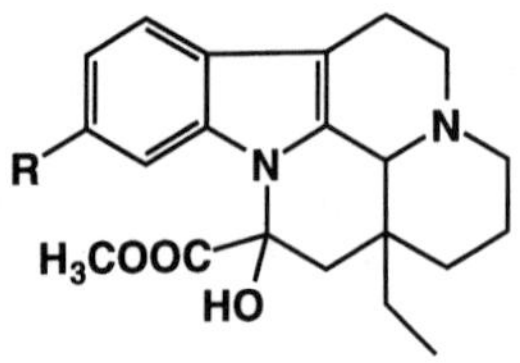

63 R = H (Vincamine)
64 R = OCH_3 (Vincine)

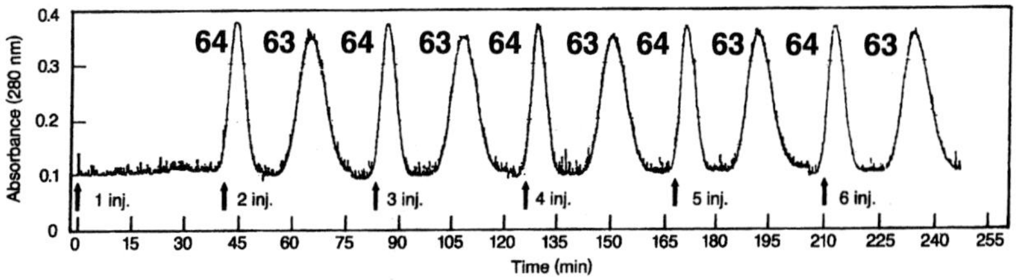

Fig. 7.24. CPC separation of vincamine (**63**) and vincine (**64**) by multiple sample injection. Solvent: hexane-EtOH-H_2O 6:5:5 (lower layer as mobile phase); flow-rate 2.6 ml/min; detection 280 nm. (Reprinted with permission from Zhou et al. 1990)

(Fig. 7.24). After twenty successive injections (at 42-minute intervals), each of 1.7 mg sample mixture, 16.5 mg of **63** and 14 mg of **64** were obtained on a Chinese 230-ml instrument. The solvent system was *n*-hexane-ethanol-water (6:5:5, lower phase as mobile phase). The resolution of the MLCCC system was not changed when the instrument was shut down overnight and re-started the next day with the same stationary phase in the column (Zhou et al. 1990).

7.3.4.12
Marine Natural Products

The mild conditions achieved with CPC and the rapidity of the method are ideal for the separation of delicate marine natural products (Table 7.8). Attempts at purification of antitumour ecteinascidins from the tunicate *Ecteinascidia turbinata*, for example, by normal- or reversed-phase chromatography led to extensive loss of activity. CPC, however, proved to be a very effective means of separating these light- and acid-sensitive alkaloids (Rinehart 1990).

The pyrroloquinoline alkaloids isobatzellines A (**65**), B (**66**) and C (**67**) have been isolated from a *Batzella* sponge. They exhibited in vitro cytotoxicity

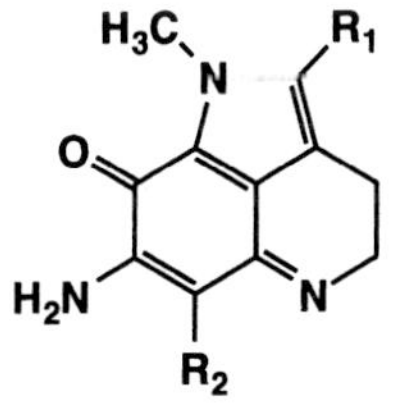

65 $R_1 = SCH_3$, $R_2 = Cl$
66 $R_1 = SCH_3$, $R_2 = H$
67 $R_1 = H$, $R_2 = Cl$

against the P-388 leukaemia cell line and antifungal activity against *Candida albicans*. CPC with an MLCCC apparatus was used in the purification, following extraction and solvent partitioning. Elution was performed with the upper phase of the solvent system heptane-chloroform-methanol-water (2:7:6:3) (Sun et al. 1990).

The non-aqueous solvent system heptane (or hexane)-dichloromethane-acetonitrile (10:3:7) has been applied to the separation of a variety of marine natural products (Table 7.8), including a long-chain methoxylamine pyridine,

Table 7.8. Separations of marine natural products by centrifugal partition chromatography

Organism	Class of compound	Solvent system	Reference
Sponge *Tedania ignis*	Diketopiperazine	$CHCl_3$-MeOH-H_2O 25:34:20	Dillman and Cardellina 1991
Sponge *Calyx podatypa*	*N*-Methylpyridinium salts	Heptane-MeCN-CH_2Cl_2 10:7:3	Stierle and Faulkner 1991
Sponge *Xestospongia wiedenmayeri*	Methoxylaminopyridines	Heptane-MeCN-CH_2Cl_2 10:6:3	Sakemi et al. 1990
Sponge *Bazella* sp.	Pyrroloquinoline alkaloids	Heptane-$CHCl_3$-MeOH-H_2O 2:7:6:3	Sun et al. 1990
		Heptane-EtOAc-MeOH-H_2O 4:7:4:3	Sun et al. 1990
		$CHCl_3$-*i*Pr_2NH-MeOH-H_2O 7:1:6:4	Sakemi et al. 1989
Sponges *Axinella* sp. and *Hymeniacidon* sp.	Pyrrololactams	*n*BuOH-(0.01 mol/l K_3PO_4-0.01 mol/l K_2HPO_4) 1:1	Schaufelberger and Pettit 1989
Sponges *Dercitus* sp. and *Stellatta* sp.	Acridine alkaloids	CH_2Cl_2-MeOH-H_2O 5:5:3	Gunawardana et al. 1989
Sponge *Pachypellina* sp.	*β*-Carboline alkaloid	Hexane-MeCN-CH_2Cl_2 10:7:3	Ichiba et al. 1994
Sponge *Discodermia polydiscus*	(Aminoimidazolinyl)indole	$CHCl_3$-MeOH-H_2O 5:10:6	Sun and Sakemi 1991
Sponge *Psammaplysilla purpurea*	Bastadin	Heptane-EtOAc-MeOH-H_2O 4:7:4:3	Carney et al. 1993
Sponge *Theonella* sp.	Cyclic peptide	Hexane-EtOAc-MeOH-H_2O 3:7:5:5	Fusetani et al. 1991
Sponge *Plakortis lita*	Cyclic peroxides	Heptane-CH_2Cl_2-MeCN 5:1:4	Sakemi et al. 1987
Bryozoan *Bugula neritina*	Bryostatin (macrocyclic lactone)	Hexane-EtOAc-MeOH-H_2O 14:6:10:7	Schaufelberger et al. 1991
Ascidian *Amphicarpa meridiana*	Alkaloids	$CHCl_3$-MeOH-5% NH_3 5:5:3	Schmitz et al. 1991
Tunicate *Clavelina picta*	Quinolizidines	Hexane-MeCN-CH_2Cl_2 10:7:3	Raub et al. 1991
Tunicate *Ecteinascidia turbinata*	Ecteinascidins	$C_6H_5CH_3$-Et_2O-MeOH-H_2O 2:2:2:1	Sakai et al. 1996
Fish oil	Fatty acid ethyl esters	Hexane-MeCN-CH_2Cl_2 5:4:1	Du et al. 1996

xestamine A (**68**), isolated from the sponge *Xestospongia wiedenmayeri* (Sakemi et al. 1990).

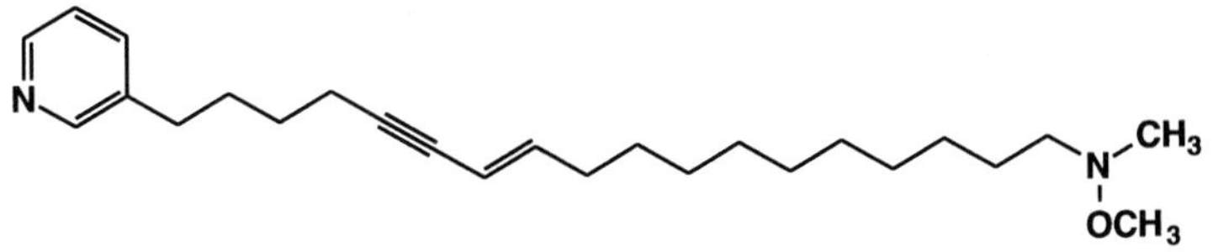

68

7.3.4.13
Antibiotics and Fungal Toxins

Since its inception, CPC has been associated with the field of antibiotics. Liquid-liquid partition techniques are particularly suitable for their separation because these bioactive metabolites are often produced in very small amounts and have to be removed from other secondary metabolites and non-metabolized media ingredients. Antibiotics are normally biosynthesized as mixtures of closely related congeners and many are labile molecules, thus requiring mild separation techniques with a high resolution capacity.

The application of CPC to antibiotics is not immediately evident from the published literature because many of the separations are from industrial laboratories or development plants, details of which are kept secret for obvious reasons. This is a pity because certain of the applications involve special large-scale apparatus – up to kilogram quantities are supposedly involved, notably in Japan.

To illustrate the approach involving antibiotics, one example is provided by the sporavidins, which are water-soluble basic glycoside antibiotics with complex structures. These are unstable under basic conditions and exist as mixtures of closely related compounds. A sample comprising six sporavidins was resolved on a Shimadzu prototype countercurrent chromatographic instrument (total capacity 325 ml; 800 rpm). Selection of the solvent system was based on partition coefficient data from chloroform-methanol-water, chloroform-ethanol-methanol-water and *n*-butanol-diethyl ether-water mixtures. After HPLC analysis, the final system adopted was *n*-butanol-diethyl ether-water (5:2:6). Sample was introduced by the so-called "sandwich" technique, in which sample was injected after filling with stationary phase and before mobile phase elution. The six components were separated within 3.5 h, employing a total elution volume of 500 ml (Harada et al. 1990).

Other examples are given in Table 7.9. *Streptomyces lusitanus* produces the hydroquinone derivative **69** (cyanocycline C) which is extremely unstable and cannot be purified by semi-preparative HPLC. However, isolation of this antibacterial compound was successfully performed on an MLCCC apparatus with the solvent system $CHCl_3$-MeOH-*i*PrOH-H_2O 3:10:10:10 (Gould et al. 1993).

The ivermectins B1 are broad-spectrum antiparasitic agents derived from the avermectins B1, produced by *Streptomyces avermitilis*. In order to obtain pure ivermectin reference compounds for analytical studies, a commercially

Table 7.9. Separations of antibiotics and fungal metabolites by centrifugal partition chromatography

Sample	Instrument[a]	Solvent system	Reference
Trichothecenes	CCC-1000	CCl_4-$CHCl_3$-MeOH-H_2O gradient CCl_4-CH_2Cl_2-MeOH-H_2O gradient Hexane-$CHCl_3$-MeOH-H_2O 1:3:3:2, 1.2:3:3:2, 8:12:15:5, 3:7:5:5, 1:1:1:1 Hexane-CH_2Cl_2-CCl_4-MeOH-H_2O 2:3:5:6:4 CCl_4-CH_2Cl_2-MeOH-H_2O 3:2:3:2 Hexane-EtOAc-MeOH-H_2O 1:1:1:1	Jarvis 1992 Jarvis et al. 1992 Jarvis et al. 1996
Stachybotrys fungal toxins (trichothecenes, phenylspirodrimanes, cyclosporin)	CCC-1000	Hexane-CH_2Cl_2-CCl_4-MeOH-H_2O 2:1:7:6:4 Hexane-CCl_4-MeOH-H_2O 3:7:3:2 Hexane-EtOAc-MeOH-H_2O 3:2:2:2	Jarvis et al. 1995
Macrolide antibiotics (RS-22A, B, C)	Sanki LLN	BuOH-EtOH-H_2O 4:1:4	Ubukata et al. 1995
Aldecalmycin	Sanki LLN	$CHCl_3$-MeOH-H_2O 5:6:4	Sawa et al. 1994
Benarthin	Sanki NMF	*n*BuOH-HOAc-H_2O 15:1:15	Aoyagi et al. 1992
Cyclic hexadepsipeptide	Sanki	$CHCl_3$-MeOH-H_2O 2:2:1 Hexane-EtOAc-MeCN 7:2:3	Ueno et al. 1993
Glycoside antibiotics	Shimadzu CCC	*n*BuOH-EtOH-H_2O 5:2:1	Harada et al. 1990
Pristinamycins	SFCC CPHV 2000	$CHCl_3$-EtOAc-MeOH-H_2O-HCOOH 12:8:15:10:2	Drogue et al. 1992
Dunaimycins (Macrolide antibiotics)	MLCCC	Hexane-EtOAc-MeOH-H_2O 70:30:15:6, 8:2:10:5, 8:2:5:5	Hochlowski et al. 1991
Tiacumicin	MLCCC	CCl_4-$CHCl_3$-MeOH-H_2O 7:3:7:3	Chen et al. 1988
Ivermectins	Shimadzu CCC	Hexane-EtOAc-MeOH-H_2O 19:1:10:10	Oka et al. 1996
Cyanocyclines	MLCCC	$CHCl_3$-MeOH-*i*PrOH-H_2O 3:10:10:10	Gould et al. 1993
Australifungin	MLCCC	Hexane-EtOAc-MeOH-25 mmol/l K_2HPO_4 pH 6.9 7:3:5:5	Mandala et al. 1995

[a] MLCCC = multilayer countercurrent chromatograph (PC Inc.); CCC-1000 = Pharma-Tech; SFCC = instrument no longer available.

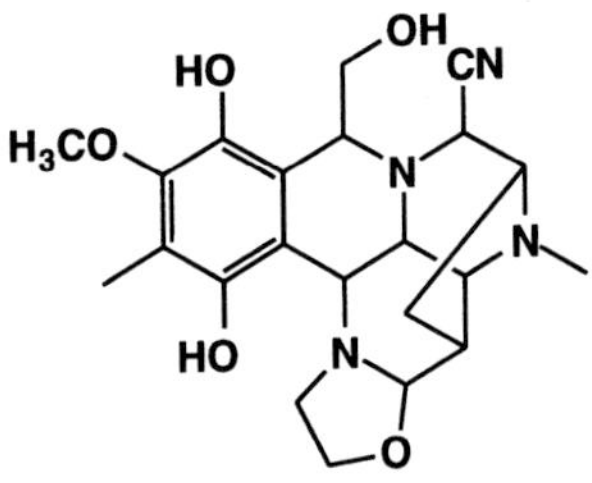

69

available ivermectin mixture was resolved into its components by CPC on a Shimadzu HSCCC-1A multilayer coil-planet centrifuge (capacity 300 ml). With the solvent system hexane-ethyl acetate-methanol-water 19:1:10:10 (lower phase as mobile phase), 99.9% pure ivermectin B1a was obtained, together with ivermectin B1b and avermectin B1a (Oka et al. 1996).

Two species of the genus *Baccharis* (Asteraceae) contain highly cytotoxic trichothecenes similar to those found in the common soil fungi *Myrothecium* (Jarvis et al. 1992). They are only present in small amounts and their isolation presents a number of problems which are not resolved by TLC and HPLC (Jarvis 1992). The application of CPC has provided the possibility of overcoming some of these difficulties. Using a Pharma-Tech CCC-1000 instrument with interchangeable columns, it was possible first to determine the separation conditions on 55 ml analytical coils. By changing to 350 ml semi-preparative coils, the transposition of the conditions enabled difficult separations to be performed. Thus, trichoverrins of the series A, B and C, very closely related 2′,3′-*trans*-trichoverrins and isomeric 2′,3′-*cis*-trichoverrins could all be separated from one another. Aliquots (400 mg) of crude trichoverrin mixture were injected into the CCC-1000 instrument and chromatographed with the solvent system hexane-chloroform-methanol-water 1:3:3:2 (lower phase as mobile phase). A second separation using gradient elution (dichloromethane-carbon tetrachloride-methanol-water 2:3:3:2 → 5:2:3:2; lower phase as mobile phase) completed the procedure. Jarvis (1992) stated that the relative retention times in CPC were not easy to predict by TLC or HPLC. On silica gel, the relative order of retention times was roridin A > roridin D > roridin E > verrucarin A. On a reversed-phase column the order was roridin E > roridin D > roridin A > verrucarin A. In contrast, the order for CPC with the solvent system carbon tetrachloride-methanol-water 5:3:2 was roridin A > roridin D > verrucarin A > roridin E. This example shows a) that analytical CPC is a good method for selecting suitable solvent systems and b) that CPC can provide a different selectivity for a separation that is not possible by conventional liquid-solid chromatography.

7.3.4.14
pH-Zone-Refining

This is a recently developed preparative technique which provides important advantages over the conventional CPC method – including a ten-fold (and greater) increase in sample loading capacity, high concentration of fractions and concentration of minor impurities. The method uses a retainer acid (or base) in the stationary phase to retain the sample constituents in the column and an eluent base (or acid) to elute these components according to their pKa values and hydrophobicities (Weisz et al. 1994). It produces a succession of highly concentrated rectangular peaks with minimum overlap, similar to those observed in displacement chromatography. The mechanism and mathematical treatment of the technique can be found in the monograph by Ito and Conway (1996) and in a paper by Ito and Ma (1996).

Major areas of application include separations of amino acids and peptides (Ma and Ito 1995). Alkaloids are potentially candidates for pH-zone-refining CCC and three alkaloids from *Crinum moorei* (Liliaceae) have been purified by this means (Ma et al. 1994).

The main drawback with pH-zone-refining CCC is that it is presently only applicable to ionizable compounds. This, of course, considerably limits its use in the natural products field. Removal of acid or base from the separated products may also be inconvenient.

7.3.5 Perspectives

Centrifugal partition chromatography has been the subject of important developments and improvements over the last few years. On the mechanical side there is still room for much modification and innovation but the basic technique is of great utility for the separation of natural products. In addition to the general advantages mentioned for liquid-liquid chromatography, CPC has the following attributes:

- high reproducibility (no disruption by solid media),
- good resolution (350–1000 theoretical plates),
- gradient operation possible,
- large sample volumes possible,
- reversed-phase operation,
- repetitive injection possible.

Centrifugal partition chromatographic techniques have fast separation speeds and high partition efficiencies which, in certain cases, approach those of HPLC. For compounds which have α (selectivity) = 1.2 and K (partition coefficient) = 1, in HPLC 185,000 theoretical plates are needed to achieve baseline resolution (Rs = 1.5), whereas only 2200 theoretical plates are required in CPC (Foucault 1991). The volume ratio of the stationary phase to the total column capacity is much higher than that of reversed-phase HPLC, giving increased sample loading. Furthermore, the solvent consumption is less: up to a factor of 10 in some cases. However, there are disadvantages (low flow rates etc.) and CPC cannot be considered a competitive method to preparative-scale HPLC but rather a *complementary* technique. This complementarity can also be exploited when other chromatographic techniques fail to separate a sample: CPC can be tried since its selectivity may be different.

Once the partition coefficients of the solutes are known, it is theoretically possible to predict the location of eluted solute peaks for CPC (Ito and Bowman 1971). Virtually any pair of immiscible liquids can be employed, including non-aqueous systems and aqueous polymer systems. The mild separation conditions generated by these completely liquid-liquid systems mean that biological samples can be handled with ease, e.g. the separation of erythrocytes (Ito 1981). The purification of water-sensitive prostaglandin H_2 was possible by CPC with a non-aqueous solvent system (Jirousek and Salomon 1988).

The applications of CPC are numerous and not limited to the separation of natural products: separations of synthetic intermediates, inorganic elements and rare earth elements are all possible. Even the fractionation of whole cells has been performed (Sutherland et al. 1987). Recently, the technique has also been applied as an enzymatic reactor for organic synthesis (Bousquet et al. 1995).

Future developments could include the coupling of small CPC units in parallel or series to give larger sample capacity. Other avenues which may be worth exploring include CPC with supercritical fluid carbon dioxide as a mobile phase (Yu et al. 1996) and co-current chromatography for speeding up separations (Lucy and Hausermann 1995).

7.4 References

Abbott TP, Kleiman R (1991) J Chromatogr 538:109

Abbott T, Peterson R, McAlpine J, Tjarks L, Bagby M (1989) J Liq Chromatogr 12:2281

Abe F, Chen R, Yamauchi T (1988) Chem Pharm Bull 36:2784

Adinolfi M, Corsaro MM, Lanzetta R, Parrilli M, Scopa A (1989) Phytochemistry 28:284

Andersen OM, Aksnes DW, Nerdal W, Johansen OP (1991) Phytochem Anal 2:175

Aoyagi T, Hatsu M, Kojima F, Hayashi C, Hamada M, Takeuchi T (1992) J Antibiot 45:1079

Aquino R, Peluso G, De Tommasi N, De Simone F, Pizza C (1996) J Nat Prod 59:555

Arisawa M, Fujita A, Morita N, Cox PJ, Howie RA, Cordell GA (1987) Phytochemistry 26:3301

Aritomi M, Kumori T, Kawasaki T (1985) Phytochemistry 24:2438

Bakhtiar A, Gleye J, Moulis C, Fouraste I (1994) Phytochemistry 35:1593

Baldé AM, Pieters LA, Gergely A, Kolodziej H, Claeys M, Vlietinck AJ (1991) Phytochemistry 30:337

Bejar E, Amarquaye A, Che CT, Malone MH, Fong HHS (1995) Int J Pharmacog 33:25

Berthod A (1995) Instrumentation Science and Technology 23:75

Berthod A, Armstrong DW (1988) J Liq Chromatogr 11:547

Beutler JA, Cardellina JH, McMahon JB, Boyd MR, Cragg GM (1992) J Nat Prod 55:207

Bifulco G, Bruno I, Riccio R, Lavayre J, Bourdy G (1995) J Nat Prod 58:1254

Bousquet OR, Braun J, Le Goffic F (1995) Tetrahedron Lett 36:8195

Brill GM, McAlpine JB, Hochlowski JE (1985) J Liq Chromatogr 8:2259

Broglia S, Verotta L, Battistini M (1994) Fitoterapia 65:520

Cai DG, Gu MJ, Zhu GP, Zhang JD, Zhang TY, Ito Y (1992) J Liq Chromatogr 15:2873

Cardenas LC, Rodriguez J, Riguera R, Chamy MC (1992) Liebigs Ann 665

Carney JR, Scheuer PJ, Kelly-Borges M (1993) J Nat Prod 56:153

Casapullo A, Minale L, Zollo F, Lavayre J (1994) J Nat Prod 57:1227

Chaurasia N, Wichtl M (1986) Dtsch Apoth Ztg 126:1559

Chaurasia N, Wichtl M (1987) J Nat Prod 50:881

Chen RH, Hochlowski JE, McAlpine JB, Rasmussen RR (1988) J Liq Chromatogr 11:191

Chifundera K, Messana I, Galeffi C, De Vicente Y (1991) Tetrahedron 47:4369

Choi Y, Pezzuto JM, Kinghorn AD, Farnsworth NR (1988) J Nat Prod 51:110
Conway WD (1990) Countercurrent chromatography:apparatus, theory and applications. VCH, Weinheim
Conway WD, Ito Y (1984) J Liq Chromatogr 7:291
Conway WD, Petroski RJ (eds.) (1995) Modern countercurrent chromatography. American Chemical Society, Washington, DC
Conway WD, Hammond RL, Sarlo AM (1988) J Liq Chromatogr 11:107
Cooper RA, Bowers RJ, Beckham CJ, Huxtable RJ (1996) J Chromatogr A 732:43
Corthout J, Pieters LA, Claeys M, Van den Berghe DA, Vlietinck AJ (1991) Phytochemistry 30:1129
Craig LC (1944) J Biol Chem 155:519
Craig LC, Craig D (1956) In: Weissberger A (ed) Technique of organic chemistry, 2nd edn, vol 3. Interscience, New York, NY, pp. 149–174
Dai J, Decosterd LA, Gustafson KR, Cardellina JH, Gray GN, Boyd MR (1994) J Nat Prod 57:1511
D'Auria MV, Minale L, Riccio R (1993) Chem Rev 93:1839
Decosterd LA, Hoffmann E, Kyburz R, Bray D, Hostettmann K (1991) Planta Med 57:548
De Rosa S, Di Vincenzo G (1992) Phytochemistry 31:1085
De Tommasi N, De Simone F, Aquino R, Pizza C, Liang ZZ (1990) J Nat Prod 53:810
De Tommasi N, Aquino R, De Simone F, Pizza C (1992) J Nat Prod 55:1025
Della Greca M, Fiorentino A, Monaco P, Previtera L (1994) Phytochemistry 36:1479
Della Greca M, Fiorentino A, Monaco P, Previtera L, Zarrelli A (1995) Phytochemistry 40:533
De Mello JP, Petereit F, Nahrstedt A (1996a) Phytochemistry 42:857
De Mello JP, Petereit F, Nahrstedt A (1996b) Phytochemistry 41:807
Diallo B, Vanhaelen M (1988) J Liq Chromatogr 11:227
Diallo B, Vanhaelen-Fastré R, Vanhaelen M (1991) J Chromatogr 558:446
Dietz H, Winterhalter P (1996) Phytochemistry 42:1005
Dillman RL, Cardellina JH (1991) J Nat Prod 54:1159
Do T, Popov S, Marekov N, Trifonov A (1987) Planta Med 53:580
Domon B, Hostettmann K (1984) Helv Chim Acta 67:1310
Domon B, Hostettmann M, Hostettmann K (1982) J Chromatogr 246:133
Drewes SE, Taylor CW (1994) Phytochemistry 37:551
Drogue S, Rolet MC, Thiébaut D, Rosset R (1992) J Chromatogr 593:363
Du Q, Shu A, Ito Y (1996) J Liq Chromatogr 19:1451
Farthing JE, O'Neill MJ (1990) J Liq Chromatogr 13:941
Foucault AP (1991) Anal Chem 63:569A
Foucault AP (ed) (1995) Centrifugal partition chromatography. Marcel Dekker, New York, NY
Franke A, Rimpler H (1987) Phytochemistry 26:3015
Fukuda N, Yonemitsu M, Kimura T (1993) Liebigs Ann 491
Fullas F, Choi Y, Kinghorn AD, Bunyapraphatsara N (1990) Planta Med 56:332
Fullas F, Wani MC, Wall ME, Tucker JC, Beecher CWW, Kinghorn AD (1996) Phytochemistry 43:1303
Fusetani N, Sugawara T, Matsunaga S, Hirota H (1991) J Am Chem Soc 113:7811
Gafner F, Chapuis JC, Msonthi JD, Hostettmann K (1987) Phytochemistry 26:2501
Gerhardt G, Sinnwell, V, Kraus L (1989) Planta Med 55:200
Gering-Ward B, Junior P (1989) Planta Med 55:75
Ghannamy U, Kopp B, Robien W, Kubelka W (1987) Planta Med 53:172
Girault JP, Bathori M, Kalasz H, Mathé I, Lafont R (1996) J Nat Prod 59:522
Gnabre JN, Ito Y, Ma Y, Huang RC (1996) J Chromatogr A 719:353
Gould SJ, He W, Cone MC (1993) J Nat Prod 56:1239
Gu J, Tong X, Ding Y, Fang Q (1994) Fitoterapia 65:149
Gunawardana GP, Kohmoto S, Burres NS (1989) Tetrahedron Lett 30:4359
Gunawardana GP, Premachandran U, Burres NS, Whittern DN, Henry R, Spanton S, McAlpine JB (1992) J Nat Prod 55:1686
Guo Y, Diallo B, Jaziri M, Vanhaelen-Fastré, Vanhaelen M (1996) J Nat Prod 59:169

Hahn R, Nahrstedt A (1993) Planta Med 59:71
Hallock YF, Dai J, Bokesch HR, Dillah KB, Manfredi KP, Cardellina JH, Boyd MR (1994) J Chromatogr A 688:83
Harada K, Kimura I, Yoshikawa M, Suzuki H, Kakazawa S, Hattori K, Komori K, Ito Y (1990) J Liq Chromatogr 13:2373
Hatano T, Yazaki K, Okonogi A, Okuda T (1991) Chem Pharm Bull 39:1689
Hatano T, Yoshihara R, Hattori S, Yoshizaki M, Shingu T, Okuda T (1992) Chem Pharm Bull 40:1703
Hecker E (1955) Verteilungsverfahren im Laboratorium. Verlag Chemie, Weinheim
Hefendehl FW, Kuhne W (1984) Farmaceutisch Tijdschrift Belgie 61:353
Herderich M, Feser W, Schreier P (1992) Phytochemistry 31:895
Herion P, Full G, Winterhalter P, Schreier P, Bicchi C (1993) Phytochem Anal 4:235
Hermans-Lokkerbol A, Verpoorte R (1986) Planta Med 52:299
Hermans-Lokkerbol A, Verpoorte R (1987) Planta Med 53:546
Hermans-Lokkerbol AC, Van der Heijden R, Verpoorte R (1993) J Liq Chromatogr 16:1433
Hochlowski JE, Mullally MM, Brill GM, Whittern DN, Buko AM, Hill P, McAlpine JB (1991) J Antibiot 44:1318
Hoffmann B, Hölzl J (1988) Phytochemistry 27:3700
Hoffmann B, Hölzl J (1989) Phytochemistry 28:247
Hostettmann K (1980) Planta Med 39:1
Hostettmann K, Hostettmann M (1981) GIT Fachz. Lab., Supplement Chromatographie 22
Hostettmann K, Hostettmann M (1982) In: Harborne JB, Mabry TJ (eds) The flavonoids: advances in research. Chapman and Hall, London, p 1
Hostettmann K, Marston A (1995) Saponins. Cambridge University Press, Cambridge
Hostettmann K, Hostettmann-Kaldas M, Sticher O (1979) Helv Chim Acta 62:2079
Hostettmann K, Appolonia C, Domon B, Hostettmann M (1984a) J Liq Chromatogr 7:231
Hostettmann K, Hostettmann M, Marston A (1984b) Nat Prod Rep 1:471
Hostettmann K, Hostettmann M, Marston A (1986) Preparative chromatography techniques: applications in natural product isolation. Springer, Berlin Heidelberg New York
Houghton PJ, Hairong Y (1987) Planta Med 53:262
Humpf HU, Wintoch H, Schreier P (1992) J Agric Food Chem 40:2060
Ichiba T, Corgiat JM, Scheuer PJ, Kelly-Borges M (1994) J Nat Prod 57:168
Iorizzi M, De Riccardis F, Minale L, Riccio R (1993) J Nat Prod 56:2149
Ito Y (1980a) J Chromatogr 188:33
Ito Y (1980b) J Chromatogr 192:75
Ito Y (1981) J Biochem Biophys Methods 5:105
Ito Y (1984) J Chromatogr 301:377
Ito Y (1987) Sep Sci Tech 22:1971
Ito Y, Bhatnagar R (1981) J Chromatogr 207:171
Ito Y, Bowman RL (1970a) Science 167:281
Ito Y, Bowman RL (1970b) J Chromatogr Sci 8:315
Ito Y, Bowman RL (1971) Anal Chem 43:70A
Ito Y, Bowman RL (1977) Anal Biochem 82:63
Ito Y, Bowman RL (1978) Anal Biochem 85:614
Ito Y, Chou FE (1988) J Chromatogr 454:382
Ito Y, Conway WD (1984) Anal Chem 56:534A
Ito Y, Conway WD (eds.) (1996) High-speed countercurrent chromatography. John Wiley, New York
Ito Y, Ma Y (1996) J Chromatogr A 753:1
Ito Y, Zhang TY (1988) J Chromatogr 449:135
Ito Y, Oka H, Slemp JL (1989) J Chromatogr 475:219
Ito Y, Kitazume E, Slemp JL (1991) J Chromatogr 538:81
Itoh A, Tanahashi T, Nagakura N (1991) Phytochemistry 30:3117
Jarvis BB (1992) Phytochem Anal 3:241
Jarvis BB (1995) Nat Toxins 3:10

Jarvis BB, DeSilva T, McAlpine JB, Swanson SJ, Whittern DN (1992) J Nat Prod 55:1441
Jarvis BB, Wang S, Ammon HL (1996) J Nat Prod 59:254
Jaziri M, Diallo B, Vanhaelen M (1991) J Chromatogr 538:227
Jimenez C, Villaverde MC, Riguera R, Castedo L, Stermitz FR (1987) Phytochemistry 26:1805
Jirousek MR, Salomon RG (1988) J Liq Chromatogr 11:2507
Kagawa K, Tokura K, Uchida K, Kakushi H, Shike T, Kikuchi J, Nakai H, Dorji P, Subedi L (1993) Chem Pharm Bull 41:1604
Kapadia GJ, Oguntimein B, Shukla YN (1994) J Chromatogr A 673:142
Kery A, Turiak G, Tetenyi P (1988) J Chromatogr 446:157
Khan IA, Erdelmeier CAJ, Sticher O, Rali T (1992) J Nat Prod 55:1270
Kijima H, Ide T, Otsuka H, Takeda Y (1995) J Nat Prod 58:1753
Kissmer B, Wichtl M (1987) Arch Pharm 320:541
Komori T, Krebs HC, Itakura Y, Higuchi R, Sakamoto K, Taguchi S, Kawasaki T (1983) Liebigs Ann 2092
Krajewski D, Toth G, Schreier P (1996) Phytochemistry 43:141
Kubo I (1991) J Chromatogr 538:187
Kubo I, Matsumoto A, Kozuka M, Wood WF (1985) Chem Pharm Bull 33:3821
Kubo I, Hanke FJ, Marshall GT (1988) J Liq Chromatogr 11:173
Kubo I, Murai Y, Soediro I, Soetarno S, Sastrodihardjo S (1991) J Nat Prod 54:1115
Kunz S, Burkhardt G, Becker H (1994) Phytochemistry 35:233
Le Men-Olivier L, Renault JH, Thepenier P, Jacquier MJ, Zeches-Hanrot M, Foucault AP (1995) J Liq Chromatogr 18:1655
Lee SS, Doskotch RW (1996) J Nat Prod 59:738
Lee YW, Cook CE, Ito Y (1988) J Liq Chromatogr 11:37
Lee YW, Fang QC, Ito Y, Cook CE (1989) J Nat Prod 52:706
Lee S, Lin M, Liu C, Lin Y, Chen Liu KCS (1996a) J Nat Prod 59:1061
Lee S, Lin B, Liu KC (1996b) Phytochemistry 43:847
Lee S, Chen C, Chen I, Chen C (1996c) J Nat Prod 59:55
Leitao GG, Kaplan MAC, Galeffi C (1992) Phytochemistry 31:3277
Lichius JJ, El Khyari D, Wichtl M (1991) Planta Med 57:159
Lucy CA, Hausermann BP (1995) Anal Chim Acta 307:173
Lutz A, Winterhalter P (1994) Phytochemistry 36:811
Ma Y, Ito Y (1995) J Chromatogr A 702:197
Ma Y, Ito Y, Sokoloski E, Fales HM (1994) J Chromatogr A 685:259
Mahran GEH, Glombitza KW, Mirhom YW, Hartmann R, Michel CG (1996) Planta Med 62:163
Malterud KE, Hanche-Olsen IM, Smith-Kielland I (1989) Planta Med 55:569
Mandala SM, Thornton RA, Frommer BR, Curotto JE, Rozdilsky W, Kurtz MB, Giacobbe RA, Bills GF, Cabello MA, Martin I, Pelaez F, Harris GH (1995) J Antibiot 48:349
Mandava NB, Ito Y (eds) (1988) Countercurrent chromatography:theory and practice. Marcel Dekker, New York
Marcos M, Villaverde MC, Riguera R, Castedo L, Stermitz FR (1990) Phytochemistry 29:2315
Markham KR, Andersen OM, Viotto ES (1988) Phytochemistry 27:1745
Marston A, Hostettmann K (1991) Nat Prod Rep 8:391
Marston A, Hostettmann K (1994) J Chromatogr A 658:315
Marston A, Zagorski MG, Hostettmann K (1988a) Helv Chim Acta 71:1210
Marston A, Borel C, Hostettmann K (1988b) J Chromatogr 450:91
Marston A, Slacanin I, Hostettmann K (1990) Phytochem Anal 1:3
Marston A, Hostettmann K, Msonthi JD (1995) J Nat Prod 58:128
Marston A, Msonthi JD, Hostettmann K (1996) In: Hostettmann K, Chinyanganya F, Maillard M, Wolfender JL (eds) Chemistry, biological and pharmacological properties of African medicinal plants. University of Zimbabwe Publications, Harare, Zimbabwe, p. 253
Menet JM, Thiebaut D, Rosset R (1994) J Chromatogr A 659:3
Merfort I (1992) Phytochemistry 31:2111
Merfort I, Wendisch D (1987) Planta Med 53:434

Merfort I, Wendisch D (1988) Planta Med 54:247
Messerer M, Winterhalter P (1995) Nat Prod Letters 5:241
Miething H, Rauwald HW (1990) J Chromatogr 498:303
Miething H, Seger V (1989) J Chromatogr 478:433
Minale L, Riccio R, Zollo F (1993) Prog Chem Org Nat Prod 62:75
Miserez F, Potterat O, Marston A, Mungai GM, Hostettmann K (1996) Phytochemistry 43:283
Murayama W, Kobayashi T, Kosuge Y, Yano H, Nunogaki Y, Nunogaki K (1982) J Chromatogr 239:643
Nahrstedt A, Rockenbach J (1993) Phytochemistry 34:433
Nahrstedt A, Wray V, Engel B, Reinhard E (1985) Planta Med 51:517
Nahrstedt A, Lechtenberg M, Brinkler A, Seigler DS, Hegnauer R (1993) Phytochemistry 33:847
Nahrstedt A, Rockenbach J, Wray V (1995) Phytochemistry 39:375
Neto JO, Agostinho SMM, Da Silva MF, Vieira PC, Fernandes JB, Pinheiro AL, Vilela EF (1995) Phytochemistry 38:397
Nitao JK, Nair MG, Thorogood DL, Johnson KS, Scriber JM (1991) Phytochemistry 30:2193
Nitz S, Spraul MH, Drawert F (1990) J Agric Food Chem 38:1445
Ohsaki A, Kasetani Y, Asaka Y, Shibata K, Tokoroyama T, Kubota T (1991) Phytochemistry 30:4075
Oka F, Oka H, Ito Y (1991) J Chromatogr 538:99
Oka H, Ikai Y, Hayakawa J, Harada K, Suzuki M, Shimizu A, Hayashi T, Takeba K, Nakazawa H, Ito Y (1996) J Chromatogr A 723:61
Okuda T, Yoshida T, Hatano T, Yazaki K, Kira R, Ikeda Y (1986) J Chromatogr 362:375
Okuda T, Yoshida T, Hatano T (1988) J Liq Chromatogr 11:2447
Okuda T, Haba K, Arata R, Nakata F, Shingu T, Okuda T (1995) Chem Pharm Bull 43:1101
Otsuka H (1995) Phytochemistry 39:1111
Otsuka H, Yamasaki K, Yamauchi T (1989) Phytochemistry 28:3197
Otsuka H, Kubo N, Sasaki Y, Yamasaki K, Takeda Y, Seki T (1991) Phytochemistry 30:1917
Otsuka H, Kamada, K, Ogimi C, Hirata E, Takushi A, Takeda Y (1994) Phytochemistry 35:1331
Otsuka H, Kamada K, Yao M, Yuasa K, Kida I, Takeda Y (1995) Phytochemistry 38:1431
Otsuka H, Kashima N, Nakamoto K (1996a) Phytochemistry 42:1435
Otsuka H, Yao M, Hirata E, Takushi A, Takeda Y (1996b) Phytochemistry 41:1351
Parada F, Krajewski D, Duque C, Jäger E, Herderich M, Schreier P (1996) Phytochemistry 42:871
Passreiter CM (1992) Phytochemistry 31:4135
Passreiter CM, Willuhn G, Röder E (1992) Planta Med 58:557
Petroski RJ, Powell RG, Clay K (1992) Nat Toxins 1:84
Pettit GR, Schaufelberger DE (1988) J Nat Prod 51:1104
Piacente S, Pizza C, De Tommasi N, Mahmood N (1996) J Nat Prod 59:565
Pieters LA, Vlietinck AJ (1986) Planta Med 52:465
Potterat O, Stoeckli-Evans H, Msonthi JD, Hostettmann K (1987) Helv Chim Acta 70:1551
Potterat O, Msonthi JD, Hostettmann K (1988) Phytochemistry 27:2677
Potterat O, Saadou M, Hostettmann K (1991) Phytochemistry 30:889
Potterat O, Hostettmann K, Stoeckli-Evans H, Saadou M (1992) Helv Chim Acta 75:833
Putman LJ, Butler LG (1985) J Chromatogr 318:85
Quetin-Leclercq J, Angenot L, Dupont L, Bisset NG (1988) Phytochemistry 27:4002
Quetin-Leclercq J, Llabres G, Warin R, Belem-Pinheiro ML, Mavar-Manga H, Angenot L (1995) Phytochemistry 40:1557
Rath G, Potterat O, Mavi S, Hostettmann K (1996) Phytochemistry 43:513
Raub MF, Cardellina JH, Choudhary MI, Ni C, Clardy J, Alley MC (1991) J Am Chem Soc 113:3178
Rauwald HW, Lohse K (1992) Planta Med 58:259
Riccio R, D'Auria MV, Minale L (1985) Tetrahedron 41:6041
Riccio R, Greco OS, Minale L, Laurent D, Duhet D (1986) J Chem Soc Perkin Trans I 665
Rinehart KL (1990) J Nat Prod 53:771

Rocha L, Marston A, Potterat O, Kaplan MAC, Hostettmann K (1996) Phytochemistry 42:185
Röder E, Breitmaier E, Birecka H, Frohlich MW, Badzies-Crombach A (1991) Phytochemistry 30:1703
Röder E, Liang XT, Kabus KJ (1992) Planta Med 58:283
Romussi G, Parodi B, Caviglioli G (1991) Pharmazie 46:679
Roscher R, Herderich M, Steffen JP, Schreier P, Schwab W (1996) Phytochemistry 43:155
Sakai R, Jares-Erijman EA, Manzanares I, Elipe MVS, Rinehart KL (1996) J Am Chem Soc 118:9017
Sakemi S, Higa T, Anthoni U, Christophersen C (1987) Tetrahedron 43:263
Sakemi S, Sun HH, Jefford CW, Bernardinelli G (1989) Tetrahedron Lett 30:2517
Sakemi S, Tottori LE, Sun HH (1990) J Nat Prod 53:995
Sawa R, Takahashi Y, Itoh S, Shimanaka K, Kinoshita N, Homma Y, Hamada M, Sawa T, Naganawa H, Takeuchi T (1994) J Antibiot 47:1266
Schaufelberger DE, Pettit GR (1989) J Liq Chromatogr 12:1909
Schaufelberger DE, Chmurny GN, Beutler JA, Koleck MP, Alvarado AB, Schaufelberger BW, Muschik GM (1991) J Org Chem 56:2895
Schmitz FJ, DeGuzman FS, Hossain HB, van der Helm D (1991) J Org Chem 56:804
Schübel H, Treiber A, Stöckigt J (1984) Helv Chim Acta 67:2078
Schwab W, Schreier P (1990) Phytochemistry 29:161
Signer R, Allemann K, Koehli E, Lehmann W, Mayer H, Ritschard W (1956) Dechema Monograph 27:36
Skouroumounis GK, Winterhalter P (1994) J Agric Food Chem 42:1068
Slacanin I, Marston A, Hostettmann K (1989) J Chromatogr 482:234
Slimestad R, Andersen OM, Francis GW (1994) Phytochemistry 35:550
Slimestad R, Marston A, Hostettmann K (1996) J Chromatogr A 719:438
Snyder JK, Nakanishi K, Hostettmann K, Hostettmann M (1984) J Liq Chromatogr 7:243
Soicke H, Leng-Peschlow E (1987) Planta Med 53:37
Sorensen JM, Arlt W (1979) Liquid-liquid equilibrium data collection. Dechema, Frankfurt
Speranza G, Manitto P, Cassara P, Monti D, de Castri D, Chialva F (1993) J Nat Prod 56:1089
Stierle DB, Faulkner DJ (1991) J Nat Prod 54:1134
Sun HH, Sakemi S (1991) J Org Chem 56:4307
Sun HH, Sakemi S, Burres N, McCarthy P (1990) J Org Chem 55:4964
Sutherland IA, Heywood-Waddington D, Ito Y (1987) J Chromatogr 384:197
Szendrei K, Varga E, Hajdu Z, Herke I, Lafont R, Girault JP (1988) J Nat Prod 51:993
Tanahashi T, Nagakura N, Kuwajima H, Takaishi K, Inoue K, Inouye H (1989) Phytochemistry 28:1413
Tani C, Nagakura N, Sugiyama N (1975) Yakugaku Zasshi 95:838
Tanimura T, Pisano JJ, Ito Y, Bowman RL (1970) Science 169:54
Terreaux C, Maillard M, Gupta MP, Hostettmann K (1995) Phytochemistry 40:1791
Ubukata M, Shiraishi N, Kobinata K, Kudo T, Yamaguchi I, Osada H, Shen Y, Isono K (1995) J Antibiot 48:289
Ueno M, Amemiya M, Someno T, Masuda T, Iinuma H, Naganawa H, Hamada M, Ishizuka M, Takeuchi T (1993) J Antibiot 46:1658
Van der Heijden R, Hermans-Lokkerbol A, Verpoorte R, Baerheim-Svendsen A (1987) J Chromatogr 396:410
Vanhaelen M, Vanhaelen-Fastré R (1988) J Liq Chromatogr 11:2969
Van Wagenen BC, Huddleston J, Cardellina JH (1988) J Nat Prod 51:136
Venkateswarlu V, Srivastava SK, Joshi BS, Desai HK, Pelletier SW (1995) J Nat Prod 58:1527
Verotta L, Caldiroli S, Gariboldi P, Tato M (1996) Phytochem Anal 7:245
Vilegas JHY, Gottlieb OR, Kaplan MAC, Gottlieb HE (1989) Phytochemistry 28:3577
Waldmann D, Winterhalter P (1992) Vitis 31:169
Weisz A, Scher AL, Shinomiya K, Fales HM, Ito Y (1994) J Am Chem Soc 116:704
Winterhalter P, Sefton MA, Williams PJ (1990) J Agric Food Chem 38:1041
Winterhalter P, Harmsen S, Trani F (1991) Phytochemistry 30:3021
Ying B, Peiser G, Ji Y, Mathias K, Tutko D, Hwang Y (1995) Phytochemistry 38:909

Yonemitsu M, Fukuda N, Kimura T, Komori T (1987) Liebigs Ann 193
Yoshida T, Hatano T, Okuda T (1989) J Chromatogr 467:139
Yu T, Li S, Chen Y, Wang HP (1996) J Chromatogr A 724:91
Zalkow LH, Asibal CF, Glinski JA, Bonetti SJ, Gelbaum LT, VanDerveer D, Powis G (1988) J Nat Prod 51:690
Zhang J, Fang Q (1994) Planta Med 60:190
Zhang TY, Hua X, Xiao R, Kong S (1988) J Liq Chromatogr 11:233
Zhang M, Stout MJ, Kubo I (1992) Phytochemistry 31:247
Zhou J, Fang Q, Lee Y (1990) Phytochem Anal 1:74

CHAPTER 8

Separation of Macromolecules

The chromatography of macromolecules (often termed "biochromatography") involves principally the separation of proteins (and enzymes), peptides, carbohydrates, oligonucleotides and nucleic acids.

The separation of macromolecules will assume increasing importance as more companies produce carbohydrates and nucleic acid products such as oligonucleotides for antisense therapeutics and gene therapy. Extensive use of chromatography is essential for the large-scale production of pharmaceutical-grade bioactive proteins. Proteins (including enzymes) produced by recombinant technology also require specific processing and purification methodologies.

Unlike substances normally separated by classical chromatography, the biopolymers to be separated are often ionic in nature, having both hydrophobic and hydrophilic characteristics and a high molecular weight. They are easily denatured and involved in unspecific interactions with surfaces. The kinetics of adsorption and desorption processes at the stationary surface are slow. Solvation, dissociation and conformational changes can also occur. Consequently, for chromatography, these factors have to be taken into account. Because of the high molecular weight and the large solvating envelope, diffusion processes tend to be very slow. To enable mass transfer to occur at a suitable rate, macroporous supports are needed. Supports should have sufficiently large pores and small particles so that the stationary surface is fully accessible to the solute (Gooding and Regnier 1990). For HPLC, columns can basically be classified into two categories: those with small pore size (60–100 Å) and those with wide pore size (>150 Å). Sorbents with small pores have a larger surface area (> 200 m^2/g) and a better separation can potentially be achieved, as is the case with small organic molecules. For macromolecules, wide-pore sorbents are necessary, many with pore size 300 Å (if conventional packings are used, there is irreversible retention of proteins in the small pores). There is also a new trend to go toward even wider pores (>1000 Å). The other alternative is the use of non-porous materials, e.g. TSK.

Polymeric columns are becoming more and more popular (Hatano 1985). Many of these are based on a cross-linked polystyrene-divinylbenzene backbone, which can be sulphonated or aminated to form ion-exchange functional groups or can be used without functionalization for non-ion-exchange applications (Benson and Woo 1984). Also available are acrylates, which contain ester linkages. These are more stable than silica but less stable than polymers derived from styrene.

The separation of peptides and proteins is currently recognized as one of the most challenging areas of chromatography (Guiochon 1993). Samples commonly involve complex mixtures of very similar compounds. The constituents of the mixtures are often delicate and biological activity is easily lost. It may be necessary to remove trace contaminants for adequate analytical investigation. This is the case, for example, with X-ray crystallography which cannot be applied without pure protein crystals.

An advantage of preparative protein chromatography is that the amount of sample is sometimes very small. Polypeptide hormones and enzymes of pharmaceutical interest can be effective in microgram dosages. Separation on analytical instruments is thus possible.

As most of the recent advances in this particular area of separation technology involve high- and medium-pressure liquid chromatography, emphasis will be placed upon these applications.

Another domain which requires the application of similar techniques (ion exchange chromatography, size exclusion chromatography etc.) is the isolation of polar and/or water-soluble compounds from marine sources. This has been well summarized by Shimizu (1985).

8.1 Size Exclusion Chromatography

Size exclusion chromatography (SEC) is a more accurate term for the technique which is often referred to as gel filtration chromatography (GFC) or gel permeation chromatography (GPC).

Solutes are selectively separated in order of decreasing molecular size on porous hydrophilic support materials (polymer gels or modified silica gels) by the process of steric exclusion. In theory, SEC relies on two effects: different rates of diffusion for dissolved particles as a function of their size and shape, and the grading effect achieved as a function of the pore size of the packing material. Any solute molecules in the mixture which are larger than the pores of the support material are excluded and migrate with the eluent. The method allows high solute recoveries to be achieved.

While SEC is a relatively low resolution technique and has low sample capacity, this chromatography is simple, rapid and is compatible with a broad range of mobile phases.

Although SEC is supposed to produce separation by molecular size alone, solute-support interactions have been observed in every support material that has ever been introduced for SEC. The support matrix is neutral and hydrophilic to minimise this non-size exclusion partitioning.

The most widely used materials are cross-linked dextran or agarose (Sephadex/Sepharose), polyacrylamide beads (Bio-Gel) and dextran derivatives (Sephacryl).

Semi-rigid, porous, hydrophilic gels have been developed by the Tosoh Corporation (TSK-Gel) in Japan for medium-pressure LC. These are known as the Toyopearl HW series (also marketed by TosoHaas in the USA and Europe and by E. Merck as Fractogel TSK) and consist of a matrix copolymerized from oligoethyleneglycol, glycidylmethacrylate and pentaerythrodimethacrylate. They are available in three different spherical particle granulometries (S: 20–40 μm, F: 30–60 μm, C: 50–100 μm). The type designations are as follows: HW-40, HW-50, HW-55, HW-65 and HW-75. These are classified by their molecular weight fractionation ranges (HW-40 covers proteins of molecular weights 100–10,000, while HW-75 is used for molecular weight 500,000–50,000,000 proteins). The gel media have high chemical stability and are compatible with organic solvents.

Classical hydrophilic gel supports have very poor mechanical stability under pressure and cannot be effectively used for high performance SEC (HPSEC). For this reason, two major types of HPSEC supports were introduced: *surface-modified controlled-porosity inorganic materials* and *controlled-porosity organic packings* (polymeric sorbents of the methacrylate or vinyl type). Controlled-porosity glass and silica inorganic SEC supports are essentially pure silica and behave as weak acids. They are, therefore, cation exchangers. Other limitations include an inability to work above pH 8, the leaching of organic bonded phases and the residual surface silanol groups. The porous silica carries covalently bonded inert phases (usually glycerylpropyl) which deactivate the adsorption sites on the silica surface. Organic supports have different limitations. They have poor mechanical strength but rigidity can be improved by increasing the degree of crosslinking. However, this increases the *hydrophobicity* and decreases the pore volume. The slight hydrophobic character of organic bonded phases may be used to retain proteins in a reversed phase mode.

The organic supports for SEC include the rigid TSK H-type (porous polystyrene divinylbenzene resin) and PW gels of Tosoh in Japan and the semirigid highly crosslinked agarose Superose (Hjerten et al. 1984) (Table 8.1).

Microparticulate sorbents (particle diameter between 5 and 13 μm) available for HPSEC come in prepacked columns of length 25 or 30 cm, with i.d.

Table 8.1. Typical stationary phases for HPSEC (Churms 1996)[a]

Sorbent	Matrix	Particle diameter (μm)	Supplier
LiChrospher DIOL (100, 500, 1000 Å)	Silica (glycerylpropyl-bonded)	10	Merck
SynChropak (100, 500, 1000, 4000 Å)	Silica (glycerylpropyl-bonded)	10	SynChrom
TSK-Gel SW	Silica	10–13	Tosoh
TSK-Gel PW	Methacrylate	10–17	Tosoh
TSK-Gel H	Polystyrene divinyl benzene	10–13	Tosoh
Ultrahydrogel	Methacrylate	6–13	Waters
Toyopearl	Methacrylate copolymer	25–40	Tosoh
Superose	Agarose		Pharmacia
12HR		10	
6HR		13	

[a] See also Unger and Janzen (1986).

4–8 mm. Such columns have plate numbers of 10,000–50,000 per metre; the optimal length of 50–100 cm is achieved by coupling two or more columns in series. If the sample has a wide distribution of molecular weight, the range of pore sizes necessary for fractionation can also be provided by coupling columns which contain stationary phases of the same type but with differing porosity.

Silica-based stationary phases have a lower pore volume and their resolution is generally inferior to polymeric sorbents. Since polymeric sorbents are also more stable at elevated temperatures and at high pH, these stationary phases are being increasingly preferred in HPSEC.

Bacteria can grow in most of the buffers used in HPSEC. Columns, therefore, have to be stored in a solvent which is not conducive to bacterial growth, e.g. 10% isopropanol in water.

SEC has been widely used in the fractionation of polysaccharides of industrial or biochemical importance (Churms 1996). Stationary phases of lower porosity are employed in the separation and analysis of mixtures of oligosaccharides, such as those obtained on hydrolysis of starch (maltodextrins) or the oligogalacturonic acids produced by enzymatic digestion of pectins. Examples are as follows:

- conventional SEC, on gels of small pore size (e.g. Bio-Gel P-2, P-4 or P-6, Bio-rad, Richmond, CA, USA) gives high resolution of maltodextrins (Churms 1996),
- semi-rigid gels, which are capable of withstanding higher pressures than the older dextran and polyacrylamide gels, have been used in MPLC separations of oligosaccharides. Chromatography on Trisacryl GF05 (Pharmacia, Bromma, Sweden) resolved degradation products of a glucuronomannoglycan from a plant gum (Churms and Stephen 1987), while Toyopearl HW-40S (Tosoh, Tokyo, Japan) provided a separation of oligosaccharides from the degradation of mucins (Goso and Hotta 1990). In both cases, the volatile buffer 0.1 mol/l pyridinium acetate (pH 5.0) was used as mobile phase. Simple evaporation of the solvent provided the oligosaccharides,
- fractionation of carrageenans and alginates has been performed on Toyopearl HW-55S and HW-75S columns by HPSEC. Two columns (10 cm i.d.) of different pore sizes were coupled and the eluent was 0.1 mol/l sodium nitrate. The polysaccharide components were removed by precipitation with i-propanol. About 1.25 g sample could be fractionated on each injection (Lecacheux and Brigand 1988).

A water-soluble glucan polysaccharide was obtained from fruiting bodies of the mushroom *Boletus erythropus* (Basidiomycetes) by a two-step procedure involving initial ion exchange chromatography on a DEAE Trisacryl M column and a final SEC fractionation on a S400 HR column (90×2.6 cm) in 20 mmol/l Tris-HCl buffer (pH 7.2) with 8 mol/l urea to reduce interactions (Chaveau et al. 1996).

In another application, the main allergen from rye grass (*Lolium perenne*, Gramineae) pollen was purified by HPSEC. The pollen (1 g) was extracted with 10 ml of 0.15 mol/l phosphate-buffered saline (pH 7.4). Batches of up 100 μl extract were injected onto a Protein Pak 125 column (Waters) and eluted with

0.15 mol/l phosphate-buffered saline (pH 7.2) at 1 ml/min. After dialysis and lyophilisation, the sample was run again on the same column. The PBS eluent hindered formation of aggregates and minimized ionic interactions between the proteins and the silica gel stationary phase of the column. Yield of the purified allergen was about 300 mg/g (Brieva and Rubio 1986).

An SEC step was employed in the purification of α-glucosidase from millet seeds (*Panicum miliaceum*, Gramineae). The crude enzyme was first fractionated on a CM-cellulofine column (17.5×3.2 cm) and then by SEC on a Sephadex G-100 column (Pharmacia). Final purification was achieved by *preparative disc gel electrophoresis* (HSI GT Tube Gel Electrophoresis Unit, pH 4). Various intermediate steps were undertaken, including *ultrafiltration* (Amicon, PM-10 membrane), *preparative isoelectric focusing* (Rotfor TM Preparative IEF Cell, pH 6.5–9.0) and dialysis (Yamasaki et al. 1996).

8.2 Ion Exchange Chromatography

This is the most widely used technique in the separation of biomolecules.

In the case of proteins, their amphoteric character means they can exist in either cationic or anionic forms, depending on the pH. The isoelectric point determines the pH of the medium and the choice of the type of ion exchanger. The strength of binding is determined by the number of cooperatively effective ionic groups on the protein, the charge density of the ion exchanger and the ionic strength of the mobile phase. Thus, elution of the proteins can be brought about by a gradient of increasing ionic strength, a pH gradient or a combination of both. There is a wide range of possibilities available for influencing retention and selectivity.

Ion exchange chromatography separates proteins by charge under near physiological and non-denaturing conditions and the resins generally have a high loading capacity. The capacity of an ion-exchange column depends on the exchange capacity of the resin. This can be as high as 5 meq/g.

When nucleic acids and proteins have a predominant negative charge at physiological pH values, anion exchange chromatography is used. Proteins are eluted from the column by a rising salt gradient (e.g. NaCl). The greater the charge density of the molecule being separated, the greater the ionic strength of the eluent required. The problem with anion exchange chromatography is that it is performed at high pH, an aspect which can affect the stability of silica-based supports. This complication is eliminated with copolymeric ion exchange resins, such as Vydac columns.

If the peptide or protein of interest is basic, cation exchange chromatography is the method of choice. For cation exchange, phosphate (pk_{a1} = 2.1), formate (pk_a = 3.7) and acetate (pk_a = 4.8) buffers can be explored.

The most common functional groups on ion exchangers are as follows:

- TMAE (trimethylammoniumethyl) $-CH_2CH_2-N^+(CH_3)_3$; strongly basic; pK > 13,
- DMAE (dimethylaminoethyl) $-CH_2CH_2-N(CH_3)_2$; weakly basic; pK 8–9,

- DEAE (diethylaminoethyl) $-CH_2CH_2-N(CH_2CH_3)_2$; weakly basic; pK 9.5–11,
- COO^- (carboxymethyl) $-CH_2COOH$; weakly acidic; pK 4.5,
- SO_3^- (sulpho) $-CH_2CH_2SO_3^-$; strongly acidic; pK < 1.

Some commercially available ion exchangers suitable for the chromatography of proteins are shown in Table 8.2.

Ion exchange chromatography is suitably used as a first separation step in combination with RPC.

Separation of carbohydrates is conveniently performed by high pressure anion-exchange chromatography (HPAEC). Pellicular resin columns, such as CarboPac PA-1, rely on the ionic attachment of small anion-exchange resin spheres to a larger cation-exchange resin sphere. The support gives excellent resolution for oligo- and polysaccharides. Desalting of the separated saccharides can be achieved by dialysis, ion exchange or gel filtration.

A Hitachi gel 3019-s (30–40 µm) cation exchanger has been used for the separation of acidic and neutral saccharides. Mono- and oligosaccharides were chromatographed on a 600×22 mm column at 8 bar, with 0.5% formic acid as eluent (Kumanotani et al. 1979).

Polysaccharides showing complement activation have been isolated from the leaves of *Plantago major* (Plantaginaceae) by a combination of ion-exchange chromatography and SEC. Preliminary extraction of the leaves with 80% ethanol was followed by extraction first with water at 50 °C, then water at 100 °C and finally DMSO. Subsequently, the three extracts were subjected to ion-

Table 8.2. Commercially available ion exchange supports for protein separations (see also Unger and Janzen 1986)

Support matrix[a]	Anion exchanger[a]	Cation exchanger[a]	Supplier
Cellulose	DEAE	CM, phospho	Bio-Rad Whatman
	TEAE, QAE		Bio-Rad
Sephacel (spherical cellulose beads)	DEAE		Pharmacia
Sephadex (dextran beads)	DEAE, QAE	CM, sulphopropyl	Pharmacia
Agarose-based	DEAE	CM	Pharmacia Bio-Rad
	PEI		Pharmacia
Synthetic polymer-based (Trisacryl)	DEAE	CM, sulphopropyl	IBF
Polymer-based	DEAE	CM, sulphopropyl	TosoHaas
Polymer-based (PS-DVB)	DEAE, TEAE	sulphopropyl	The Separations Group (Vydac)

[a] DEAE, diethylaminoethyl; TEAE, triethylaminoethyl; QAE, diethyl-(2-hydroxy-propyl) aminoethyl; PEI, polyethylene imino; CM, carboxymethyl; PS-DVB, polystyrene-divinylbenzene.

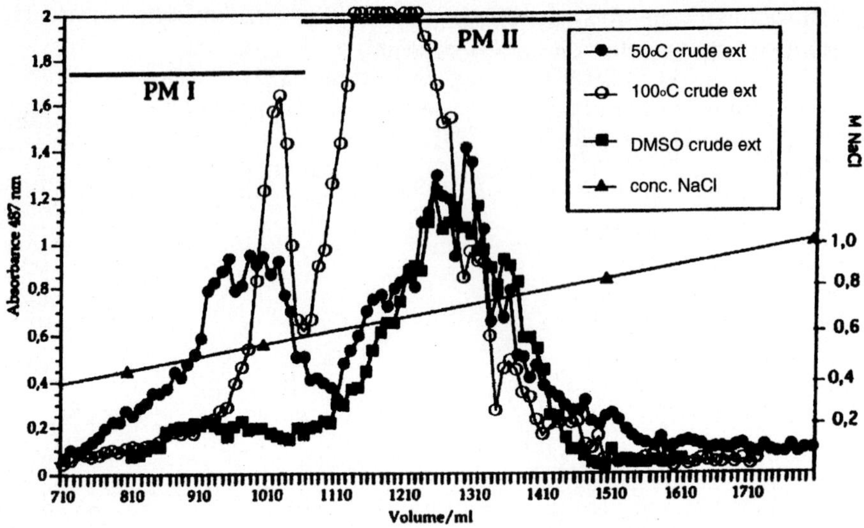

Fig. 8.1. Separation of polysaccharides from *Plantago major* leaf extracts on a DEAE Sepharose Fast Flow column. Fractions PM I were eluted with 0.4–0.6 mol/l NaCl and fractions PM II with 0.6–0.8 mol/l NaCl. (Reprinted with permission from Samuelsen et al. 1995)

exchange chromatography on a DEAE Sepharose Fast Flow column (750 ml). Elution was begun with water at 1 ml/min and then a gradient of sodium chloride (Fig. 8.1). The active PM I fraction of the 50 °C extract was fractionated by Superose 6HR 10/30 SEC. Elution with 10 mmol/l sodium chloride gave an immunostimulating arabinogalactan (Samuelsen et al. 1995).

An α-amylase has been purified from poplar leaves, using a mixture of anion exchange resins, affinity chromatography and gel filtration. An initial DEAE-cellulose step with 0.1 mol/l Tris/HCl, pH 8, 5 mmol/l $MgCl_2$ and 10 mmol/l 2-mercaptoethanol was followed by separation on Q-Sepharose Fast Flow (15×1.6 cm) (Pharmacia). After equilibration with the same sovent as that used on DEAE-cellulose, the eluted protein was precipitated by addition of solid ammonium sulphate to 50% saturation. Following affinity chromatography the sample was ultrafiltered (YM-10 membranes, Amicon) and then purified by gel filtration over Sephacryl S-300-HR (Pharmacia). A 17,700-fold purification was achieved in total (Witt and Sauter 1996).

Anion exchange chromatography with a Q-Sepharose column was also an intermediate step in the partial purfication of squalene synthase (SQS) from cultured cells of dandelion (*Taraxacum officinale*, Asteraceae). Following solubilization of SQS activity from microsomes and triple-joint-column chromatography (SP-Sepharose, Q-Sepharose and Bio-Gel HTP), a Bio-Gel HTP (hydroxyapatite) step was performed. Subsequent Q-Sepharose chromatography gave a further 25-fold purification of SQS. A 1×3 cm column was used, first equilibrating with 50 mmol/l K-Pi (pH 7.0) buffer and then eluting with a linear gradient of 50–500 mmol/l of the same buffer (0.5 ml/min) (Komine et al. 1996).

8.3 Hydrophobic Interaction Chromatography

Solute retention in both hydrophobic interaction chromatography (HIC) and reversed-phase chromatography (RPC) is based on hydrophobic interactions between the sample and the stationary phase. The main difference between the two methods is that RPC uses packings that have much denser and/or more hydrophobic bonded stationary phase ligands. These hydrophobic surfaces strongly absorb solutes such as proteins and require the use of organic solvents in order to achieve elution. The biopolymers attach spontaneously to the hydrophobic ligands and, to achieve elution, the surface tension of the eluent must be decreased by addition of the organic solvent. The combination of harsh mobile phases and strong surface adsorption in RPC can lead to denaturation of the protein. On the other hand, HIC uses bonded stationary phase ligands of lower density or decreased hydrophobicity (short alkyl or small aryl functions), resulting in mild adsorption of proteins (Kato 1987). Retention is achieved using aqueous solutions of high ionic strength in the absence of organic solvents. Under these aqueous conditions, accessible hydrophobic residues on the native protein interact with the column support. The protein can be chromatographed with its conformational structure intact, thereby allowing a high recovery of biological activity. Inevitably, therefore, HIC has almost exclusively been used for the separation of proteins.

Most of the stationary phases for high pressure HIC are silica-based (Table 8.3). More recently, polymer-based supports have been introduced. These provide greater chemical stability over a pH range of 2–12 and they allow column regeneration at high pH.

Proteins are adsorbed onto hydrophobic supports at a high ionic concentration (salt concentration 1–3 mol/l; usually 1.2–1.8 mol/l) and subsequently

Table 8.3. Stationary phases for high pressure HIC (Cooke et al. 1990)

Ligand	Support matrix	Supplier
Polyether (e.g. Spherogel CAA-HIC)	5 μm spherical silica	Beckman
Polyethylene glycol	12 μm spherical silica	Rainin
Salt-modified diol	5 μm spherical silica	Supelco
Short alkyl chain	5 μm polyaspartamide-coated silica	Poly LC
Propyl	5 μm silica	J.T. Baker
Short alkyl chain (e.g. Synchropack-Propyl)	5 μm PEI-coated silica	Synchrom
Butyl	10 μm hydrophilic polymer	Interaction
Phenyl, ether, butyl (e.g. TSK Phenyl-5PW)	10 μm hydrophilic polymer	TosoHaas [a]
Phenyl, alkyl	10 μm Sepharose	Pharmacia

[a] TosoHaas is a joint venture of the Tosoh Corporation from Japan and the Rohm and Haas Company from the USA.

eluted using a falling salt gradient when the ionic concentration is lower, i.e. when the hydrophobic interactions become weaker. The procedure is employed for samples containing high levels of salt and as a routine means of separating water-soluble proteins. It can also be used to remove strongly hydrophobic constituents from protein solutions.

The salt most commonly used in HIC is ammonium sulphate and Tris/HCl (tris(hydroxymethyl)aminomethane/hydrochloric acid) as gradient former. Other possibilities for salts are potassium phosphate, sodium chloride, sodium sulphate or potassium acetate.

HIC is commonly used in combination with other chromatographic techniques. For instance, recombinant human granulocyte-macrophage colony stimulating factor (rhGM-CSF) was purified to homogeneity by a three-step chromatographic procedure which included HIC on a Phenyl-Sepharose fast-flow column (Belew et al. 1994).

In another example, the enzyme phenylalanine ammonia lyase (PAL) was purified from the yeast *Rhodotorula glutinis*. The enzyme is widely distributed in higher plants and is the first enzyme of the phenylpropanoid pathway. A crude extract of the yeast was salt-precipitated and subjected to gel filtration on Sephacryl S-400 (Pharmacia) with 50 mmol/l Tris-HCl buffer, pH 8.5, containing 25% glycerol. Pure enzyme was then obtained by HIC on Phenyl Sepharose CL-4B (20 × 0.8 cm column, total bed volume 40 ml) (Pharmacia). After introduction of the crude fraction onto the column, it was washed with 50 mmol/l Tris-HCl buffer, pH 7.5, containing 1 mol/l ammonium sulphate. Finally, PAL was eluted with a double gradient of decreasing ammonium sulphate concentration (1 $\rightarrow$ 0 mol/l) and increasing glycerol concentration (0 $\rightarrow$ 20%) in Tris-HCl buffer, pH 7.5. An overall yield of 33% and a 195-fold purification were obtained (D'Cunha et al. 1996).

8.4 Reversed-Phase Chromatography

Reversed-phase chromatography (RPC) is the most important technique for peptide purification in preparative/process chromatography, with resolution higher than ion exchange or hydrophobic interaction chromatographies. Pharmacia has also introduced the concept of "fast protein liquid chromatography" (FPLC) for their system dedicated to the separation of proteins. The wide-scale application of RPC in industrial protein purification is still hampered by the denaturing conditions intrinsic to the process and other problems associated with the use of organic solvents. However, peptides with molecular weights of less than 25,000 can usually be restored to the native conformation following separation and this technique is usually the first choice for their chromatography. Ion exchange and hydrophobic interaction chromatography often provide good recoveries (> 90%) of biologically active material and these methods are preferred for larger molecular weight samples. In those cases of separation of proteins by RPC, the procedure is achieved at low pH in the presence of small amounts of TFA or phosphoric acid in the eluent. Under these conditions, the surface silanols are not ionized and the protein is thought to form an ion-pair with the acid, becoming an almost neutral entity.

Unbuffered hydro-organic solvent mixtures are employed as mobile phases. The elution strength of the commonly-used organic modifiers is as follows: methanol ≤ acetonitrile < ethanol ~ acetone ~ dioxan < iso-propanol ≤ tetrahydrofuran (Snyder et al. 1979). In the RPC of peptides, the most commonly used solvent is acetonitrile which has a low viscosity, thus giving high efficiencies. Iso-propanol is an alternative solvent and has a different selectivity but its viscosity is high, producing low column efficiencies. Another possibility is to use a mixture of acetonitrile and iso-propanol. The most widely used mobile phase additive is TFA. However, 13 mmol/l TFA is very aggressive to columns, which has meant that columns more resistant to low pH hydrolysis have had to be introduced. Another important factor for good column efficiencies and recoveries is the pore size – this should be much larger than the size of the solutes (100–150 Å for small carbohydrates and glycoconjugates, 300 Å for peptides, proteins and polysaccharides). In summary, therefore, a 300-Å packing material with RP-8 or RP-18 stationary phase and low silanol interactions is a good starting point. In problematic cases, less hydrophobic phases such as CN and phenyl are alternatives, or materials based on organic polymers, when silanol interactions preponderate.

While C-18 silica has proved most useful for the separation of carbohydrates, C-4 and, to a lesser extent, C-8 silica are used for the separation of proteins and glycoproteins because these packings provide less retention of the macromolecules. Phenyl silica is also suited to protein RPC.

The hydrolytic instability of silica-based stationary phases at high pH is a major problem in the large-scale purification of pharmaceutical proteins, most of which are glycosylated. Columns cannot be cleaned with sodium hydroxide solution, which is the most effective agent for desorbing residual proteins from the surface of the stationary phase before re-use. Other options are therefore necessary. One of these is the use of polymeric columns, such as PLRP-S 300 Å (Polymer Laboratories). This particular packing has a polystyrene/divinylbenzene-based matrix which has retention characteristics analogous to alkyl-bonded silica gel but is stable in the pH range 1–14.

The RPC separation of oligosaccharides is illustrated by the isolation of tetrasaccharides containing (2 → 1)- and (2 → 6)-linked β-D-fructofuranose residues from the red squill (*Urginea maritima*, Liliaceae). In this particular example, slices of the bulbs were extracted by 80% methanol. Size exclusion chromatography of the extract on Biogel P2 and P4 columns in sequence gave fractions which were further purified by RPC on a 7 µm 120 Å pore size C-18 column (300 × 10 mm). Elution was with pure water and a differential refractometer was used for detection (Praznik and Spies 1993).

The application of RPC to the preparative purification of proteins and synthetic peptides has been frequently described (e.g. Seipke et al. 1986). This is illustrated by the large scale purification (in gram amounts) of gonadotropin releasing hormone analogues and amidated human pancreatic tumour growth hormone releasing factor, both produced by solid phase synthesis. For this purpose, Rivier and co-workers used C-18, C-4 or diphenyl derivatized silicas (pore

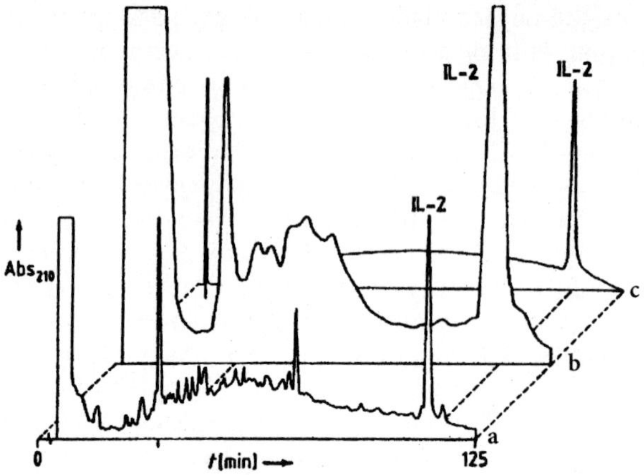

Fig. 8.2 a–c. Analytical and preparative HPLC separation of recombinant IL-2 from *E. coli*: **a** analytical HPLC; column: Hipore C18, 25 cm×4.6 mm; eluent A: 0.1% TFA in water; eluent B: 0.1% TFA/80% acetonitrile in water; gradient: 35% to 85% B in 60 min; flow-rate: 1.5 ml/min; detection 210 nm; **b** preparative HPLC of 60 mg crude IL-2; column: Prep Pak radially compressed cartridge, 30 cm×5.7 cm (Waters) with Vydac TP 218 (totally porous C-18), 15–20 μm, 300 Å; gradient as for a; flow-rate: 50 ml/min; detection with 1/50 split; **c** analytical HPLC of material obtained in b; HPLC conditions as in a. (Reprinted with permission from Seipke et al. 1986)

size 300 Å) of different particle sizes (10–20 μm). Mobile phases included triethylammonium phosphate pH 2.25 or 6.5, 0.1% trifluoroacetic acid, ammonium acetate pH 6.5 and acetonitrile (Rivier et al. 1984). *Desalting* of the purified peptides was achieved by diluting with water and running the octadecylsilyl column in 0.1% TFA (Rivier et al. 1984). Alternatively, if a volatile mobile phase (containing, e.g. ammonium bicarbonate) is used, the pure peptide can be obtained directly by lyophilisation (Knighton et al. 1982).

Recombinant interleukin 2 (IL-2) from *E. coli* can be purified by semi-preparative HPLC. After work-up, the raw material contains up to 40% IL-2, and, after optimization by analytical HPLC, the separation conditions can be transferred to semi-preparative HPLC (Fig. 8.2). The protein has 133 amino acid residues (MW about 15,000) and, due to its considerable hydrophobicity, IL-2 is only eluted after all impurities. In contrast to C-18 packing material, a C-4 support is relatively insensitive to overloading and is better for preparative work (the recovery of hydrophobic proteins is increased when the silica gel is coated with shorter alkyl chains) (Seipke et al. 1986).

8.4.1
Ion-Pair RPC

The column packing is usually the same as in RPC and the mobile phase is a buffered hydro-organic eluent containing an ion-pairing agent. Some of the mobile phase additives normally used in RPC, e.g. trifluoroacetic acid or phosphoric acid, may undergo ion-pair formation with the solute (protein, peptide, glycoproteins, acidic carbohydrates). Examples of cationic ion-pairing agents are tetrabutylammonium or alkyl-trimethylammonium salts. There are two different theories for retention (Hearn 1985): i) formation of an ion-pair between the solute and ion-pair reagent in the mobile phase – this is then retained by the stationary phase; ii) formation of an in situ ion-exchange column – this involves initial retention of the ion-pairing agent followed by ion exchange between the charged solute and the mobile phase counter-ions.

8.5 Affinity Chromatography

Affinity chromatography is a selective, non-destructive method which relies on biospecific interactions (Cuatrecasas 1972). It can be used to concentrate very dilute solutions. While ion-exchange chromatography gives a 5- to 10-fold enrichment of the sample, affinity chromatography can give over 100-fold enrichment. Another advantage is that affinity chromatography can stabilize proteins once they are bound onto the column. A disadvantage lies in the low capacity for the sample and, although it is the most selective technique, it may also be the most expensive chromatographic technique. The high selectivity advantage of affinity chromatography means that suitable adsorbents are often not commercially available. Thus the adsorbent has to be synthesized (at high cost). Alternatively, antibody-based adsorbents can be developed but their production and immobilization on a support matrix is similarly not without problems.

A list of commercial affinity packings is given by Unger and Janzen (1986). This review also includes activated packings which can be coupled to the required ligands.

The ideal adsorbent requires elements of both non-selectivity and selectivity (i.e. adsorbents that are generally applicable and stable), requirements which have been met by *synthetic dye ligands*. The most widely used synthetic affinity ligands, therefore, are the Colour Index (C.I.) Reactive Blue 2 dyes, to which Cibacron Blue F3G-A belongs (Clonis et al. 1987). Cibacron Blue is a triazinyl dye whose three-dimensional structure resembles NAD^+ and acts as an ADP-ribose analogue. Most enzymes that bind purine nucleotides show affinity for the dye. In one example, a phosphonic acid group, a known inhibitor of alkaline phosphatase, has been incorporated into the terminal aminobenzene ring of C.I. Reactive Blue 2, providing in one step a 330-fold purification of calf intestinal alkaline phosphatase from crude intestinal extract (Lindner et al. 1989).

Separation occurs through specific interactions (e.g. electrostatic interactions between charged groups, non-polar interactions and hydrogen bond formation). Supports used include porous hydrophilic organic gels (e.g. agarose) and modified silica gels with spacer-bound ligands. Certain conditions are conducive to specific, irreversible adsorption of the biomolecules being separated, while at the same time any impurities pass through the column without being retained. The product of interest, adsorbed on the column, can be eluted by modifying the mobile phase, either with a salt or through a pH gradient (non-specific elution). Alternatively, elution of a tightly adsorbed biomolecule, such as an enzyme from an enzyme-ligand complex, can be achieved by using a solution containing a high concentration either of the ligand or a ligand analogue which can compete with the covalently attached ligand for the enzyme (specific elution). For example, the regulatory subunit of cAMP-dependent protein kinase can be eluted from an affinity column packed with cAMP-Sepharose by the ligand analogue 1,N^6-ethenoadenosine 3′,5′-cyclic monophosphate (Charlton et al. 1983). However, for this technique a large volume of the ligand analogue-containing eluent is sometimes needed. To avoid this problem, the eluent from the column can be passed through a filter (e.g. microporous filter YM-10, Amicon). Enzyme is retained on the filter and the filtrate, together with the small molecules of ligand analogue, is recycled back to the inlet of the affinity column (Charlton 1985).

In the purification of chloroplast adenylate kinase (AK) from tobacco, Blue Sepharose affinity chromatography involving specific elution with the AK inhibitor diadenosine pentaphosphate was used. The five-step procedure also included an initial ammonium sulphate precipitation, DEAE-Sephacel anion exchange chromatography (0–0.3 mol/l NaCl), gel filtration over Sephadex G-100 and FPLC with a Mono Q HR 5/5 anion exchange column (0–0.3 mol/l NaCl in Q buffer: 20 mmol/l ethanolamine-HCl, pH 9.5, 14 mmol/l 2-mercaptoethanol), giving a 58-fold enrichment (Schlattner et al. 1996).

8.6 Metal Interaction Chromatography

This is an extension of the affinity technique, commonly known as immobilized-metal-ion affinity chromatography (IMAC) (Porath et al. 1975), in which metal ions are bound to a chelator (such as iminodiacetate) attached to the matrix. The matrices are of inorganic (silica gel), biological (agarose) or synthetic organic origin. Immobilised metal ions on the support are an effective way of binding peptides and proteins. When the adsorbed sample is to be eluted, an imidazole gradient can be used.

As an example of IMAC with a synthetic polymer, TSK-gel chelate-5PWE, a hydrophilic resin-based material, has been used to separate peptides (Porath 1988).

Figueroa et al. (1986) have employed iminodiacetate chelator coupled to a silica gel support, in the presence of Cu^{II}, to fractionate a crude sample of dog myoglobin.

High resolution of mixtures is possible by this technique and, in addition, 10^3- to 10^4-fold purifications of proteins have been achieved (Porath 1988); the future potential of IMAC is therefore quite considerable.

8.7 Tentacle Supports

In order to avoid irreversible conformational changes of biopolymer caused by multifunctional interactions between the stationary phase and the sample,

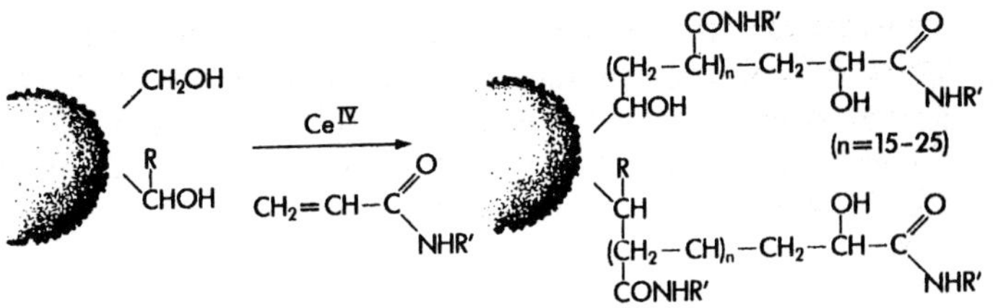

Fig. 8.3. Formation of "tentacle" supports

E. Merck has developed "tentacle" technology, in which there is a flexible arrangement of the chromatographically active groups on the surface of the stationary phase (Müller 1990). Mass transfer between the stationary and mobile phases is also accelerated, resulting in a markedly increased chromatographic resolution. The chromatographic supports are modified by linear polymerisation so that the functional groups are no longer bound to the matrix by short spacer groups but sit on flexible "tentacle"-like polymer chains. An inorganic (e.g. 5 µm and 1000 Å pore size LiChrospher diol) or polymeric (e.g. Fractogel) matrix can be used for attachment of the polymer chain. As an example, Fig. 8.3 shows the graft polymerisation of acrylamide derivatives on a hydrophilic support for an ion exchanger. The Ce(IV)-catalyzed reaction produces chains of 15–25 monomer units with functional groups located along their length. The polymer chains are able to move freely in space and are thus flexible enough to adapt to the macromolecule conformation. Due to the hydrophilic character of the tentacles, possible interference of proteins with the matrix is minimized.

Other advantages of "tentacle" chromatography are as follows:

- higher capacity than conventional supports,
- recovery, both in terms of mass and biological activity, is high,
- high selectivity,
- high efficiency,
- high flow rates are possible because there is fast adsorption of macromolecules to the functional groups.

"Tentacle" technology has been applied for the preparation of ion-exchange, HIC, SEC and affinity supports.

8.8 The Influence of Buffers

In the case of proteins, there is usually a specific conformation responsible for the biological function of the molecule under physiological conditions. During separation, this structure should remain intact in order to conserve bioactivity. To avoid protein denaturation, therefore, the presence of aqueous buffer systems is necessary. Extremely pure buffer must often be used. Such purity is required for electrophoresis and HPLC. In electrophoresis, very high quality urea, formaldehyde and acrylamide are needed. Reagents not prepared specifically for electrophoresis may contain ions or impurities that may cause band broadening and high background noise if they are not removed before use. Buffers that contain heavy metal ions and UV-absorbing impurities may cause high background noise or unstable baselines with the sensitive detectors used in HPLC.

8.9 References

Belew M, Zhou Y, Wang S, Nyström LE, Janson JC (1994) J Chromatogr A 679:67
Benson JR, Woo DJ (1984) J Chromatogr Sci 22:386
Brieva A, Rubio N (1986) J Chromatogr 370:165
Charlton JP (1985) J Chromatogr 346:247
Charlton JP, Huang C, Huang LC (1983) Biochem J 209:581
Chaveau C, Talaga P, Wieruszeski JM, Strecker G, Chavant L (1996) Phytochemistry 43:413
Churms SC (1996) J Chromatogr A 720:151
Churms SC, Stephen AM (1987) Carbohydr Res 167:239
Clonis YD, Atkinson A, Bruton CJ, Lowe CR (1987) Reactive dyes in protein and enzyme technology. Stockton Press, New York
Cooke N, Shieh P, Miller N (1990) LC-GC Int 3:8
Cuatrecasas P (1972) Affinity chromatography of macromolecules. In: Meister A (ed) Advances in enzymology, vol. 36. John Wiley, New York
D'Cunha GB, Satyanarayan V, Nair PM (1996) Phytochemistry 42:17
Figueroa A, Corradini C, Feibush B, Karger BL (1986) J Chromatogr 371:335
Gooding KM, Regnier FE (1990) HPLC of biological macromolecules. Marcel Dekker, New York
Goso Y, Hotta K (1990) Anal Biochem 188:181
Guiochon GA (1993) Anal Chim Acta 283:309
Hatano H (1985) J Chromatogr 332:227
Hearn MTW (1985) Ion-pair chromatography. Marcel Dekker, New York
Hjerten S, Liu ZQ, Yang D (1984) J Chromatogr 296:115
Kato Y (1987) Adv Chromatogr 26:67
Knighton DR, Harding DRK, Napier JR, Hancock WS (1982) J Chromatogr 249:193
Komine H, Takahashi T, Ayabe S (1996) Phytochemistry 42:405
Kumanotani J, Oshima R, Yamauchi Y, Takai N, Kurosu Y (1979) J Chromatogr 176:462
Lecacheux D, Brigand G (1988) Carbohydr Polym 8:119
Lindner NM, Jeffcoat R, Lowe CR (1989) J Chromatogr 473:227
Müller W (1990) J Chromatogr 510:133
Porath J (1988) J Chromatogr 443:3
Porath J, Carlsson J, Olsson I, Belfrage G (1975) Nature (London) 258:598
Praznik W, Spies T (1993) Carbohyd Res 243:91
Rivier J, McClintock R, Galyean R, Anderson H (1984) J Chromatogr 288:303
Samuelson AB, Paulsen BS, Wold JK, Otsuka H, Yamada H, Espevik T (1995) Phytother Res 9:211
Schlattner U, Wagner E, Greppin H, Bonzon M (1996) Phytochemistry 42:589
Seipke G, Müllner H, Grau U (1986) Angew Chem Int Ed 25:535
Shimizu Y (1985) J Nat Prod 48:223
Snyder LR, Dolan JW, Gant JW (1979) J Chromatogr 165:3
Unger KK, Janzen R (1986) J Chromatogr 373:227
Witt W, Sauter JJ (1996) Phytochemistry 41:365
Yamasaki Y, Konno H, Masima H (1996) Phtochemistry 41:703

CHAPTER 9

Separation of Chiral Molecules

Optically active natural products generally occur in only one enantiomeric form and do not necessitate chiral separations for their isolation. On the other hand, synthetic, semi-synthetic or modified derivatives may require purification to obtain optically pure compounds.

There is an increasing demand for optically pure isomers because it is now apparent that many chiral drugs and agrochemicals display different activity and toxicity profiles with respect to their absolute configuration. This has been especially well studied for enzyme and receptor interactions at the molecular level. Data concerning the activities and toxicities of the individual pure enantiomers are now required by the relevant authorities for all new chiral drugs submitted for registration. Furthermore, chromatography of racemates can replace the often lengthy elaboration of an enantioselective synthesis and, at the same time, furnish both enantiomers required for comparative testing. The scale of separation normally lies in the 100–200 mg range since this quantity is generally sufficient for testing. However, applications involving amounts ranging from 5 to 100 mg have also been reported.

Chiral differentiation can be achieved by:

- a *chiral stationary phase* (CSP) (Francotte 1994),
- an achiral stationary phase and chiral mobile phase,
- an achiral stationary phase and chiral additive in the mobile phase.

The last two methods are rather more complex since there are difficulties in recovering the separated enantiomers, contaminated with mobile phase or additive.

The indirect separation of enantiomers by conversion to diastereoisomers with an optically active reagent is not covered here since the separation procedure is classical chromatography.

9.1 Chiral Separations by Medium- or High-Pressure Liquid Chromatography

The largest range of applications in this category comes from the pharmaceutical field: pure enantiomers are required for biological testing, toxicological studies and clinical testing. Enantiomers of labelled compounds are also conveniently obtained by liquid chromatography, e.g. the anticoagulant ^{14}C-warfarin (Fitos et al. 1990).

Another important area of applications is the food and drink industry. Flavours and fragrances are dependent on enantioseparation for the correct properties. The *R*-isomer of limonene, for example, has an orange odour, while the *S*-isomer has a lemon odour.

There are fundamentally two major classes of chiral stationary phases (CSP) for use in pressure liquid chromatography:

- chiral organic polymeric materials in the pure form or as a coating on a macroporous support. This class includes oligo- and polysaccharides, polyacrylamides, polyacryl esters and protein-based phases,
- materials obtained by chiral chemical modifications of the surface of the support (mostly silica gel). Modifying agents are amino-acid derivatives, crown ethers, cinchona alkaloids, carbohydrates, amines, tartaric acid derivatives, cyclodextrins and binaphthyl compounds. Silica gel functionalized with *quinine* and *quinidine*, for example, has been used for the resolution of racemic arylalkylcarbinols, 1,4-benzodiazepinones and binaphthols (Rosini et al. 1985; Salvadori et al. 1992; Lämmerhofer and Lindner 1996). Covalent binding of macrocyclic antibiotics provides many sites for interaction with the substrate. In this fashion, Astec have started to market a spherical 5 μm silica gel with the glycopeptide vancomycin attached (Chirobiotic V). Since *bovine serum albumin* (BSA) is a relatively cheap and readily available protein, there is considerable interest in preparative scale separations with BSA-bound silica gel (Jacobson et al. 1992). Using tryptophan as a model compound, 0.25 mg could be resolved in a single run on a 500×22 mm column packed with 20–45 μm BSA-modified silica (Erlandsson et al. 1986). Warfarin has been separated on Amicon BSA-spherical silica (pore size 300 Å, particle size 15 μm) with the solvent system 0.05 mol/l KH_2PO_4 (pH 6.8) – 8% isopropanol (Felix et al. 1995).

The disadvantage of chiral sorbents obtained by immobilizing chiral compounds on silica gel is that they only have moderate loading capacities (only part of the sorbent is actually capable of differentiating the enantiomers). This is exemplified by the Chiracel polysaccharide-based CSPs, in which the polymeric chiral sorbents are deposited on silica gel. For pure polymeric chiral phases (polysaccharides and poly(meth)acrylamide derivatives), on the other hand, the density of active sites is very high. On pure polymeric cellulose triacetate, for example, resolution of 40–150 g of racemate per run on a 20×100 cm column is possible (Francotte 1994).

Classical interaction forces such as ionic, hydrophobic, dipolar and π-π interactions and hydrogen bonding are involved during separation on CSP.

Chiral stationary phases can be divided into five types based on their solute-stationary phase bonding interactions:

- helical polymer phases,
- ligand exchange columns,
- protein phases,
- chiral cavity phases,
- donor-acceptor type columns.

Chiral packing materials with good mechanical properties are necessary for pressure chromatography techniques. Furthermore, enantiomeric separations require more stringent control of mobile phase composition, flow rate, temperature etc. than conventional separations.

In order to find suitable conditions for a preparative chiral separation, it is advisable to start with an analytical column. Choice of conditions and scale-up can be performed in a similar manner to that described in Chap. 5. Since the number of packing materials available is still limited, a screening of the different commercially available columns is the easiest approach. In view of the cost advantage, cellulose triacetate should be tried in the first instance. Recycling (Werner 1989) and peak shaving techniques can be employed, as in classical LC (Chap. 5). Polarimetric detection (with or without coupling to UV detection) is important for localizing separated enantiomers.

A certain number of CSPs are commercially available, a representative selection of which is shown in Table 9.1. Particle sizes in this list range from 15 to 40 μm. Numerous examples of applications are to be found in reviews by Francotte (1994) and Francotte and Junker-Buchheit (1992).

Table 9.1. Commercially available chiral stationary phases for preparative separations (Francotte and Junker-Buchheit 1992)

	Chiral unit	Supplier
Cyclodextrin phases		
ChiraDex	β-Cyclodextrin	Merck
Cyclobond I	β-Cyclodextrin	Astec
Cyclobond I acetylated	β-Cyclodextrin, acetylated	Astec
Cyclobond DMP	β-Cyclodextrin 3,5-dimethylphenylcarbamate	Astec
Cyclobond I PT	β-Cyclodextrin *p*-methylbenzoate	Astec
Cyclobond I RN	β-Cyclodextrin (*R*)-naphthylethylcarbamate	Astec
Cyclobond I SN	β-Cyclodextrin (*S*)-naphthylethylcarbamate	Astec
Cyclobond I SP	β-Cyclodextrin, (*S*)-hydroxypropyl	Astec
Cyclobond RSP	β-Cyclodextrin (*R*,*S*)-hydroxypropyl	Astec
Cyclobond II	γ-Cyclodextrin	Astec
Cyclobond II acetylated	γ-Cyclodextrin, acetylated	Astec
Cyclobond III	α-Cyclodextrin	Astec
Cyclobond III acetylated	α-Cyclodextrin, acetylated	Astec
Nucleodex β-PM	β-Cyclodextrin, permethylated	Macherey-Nagel
Si100 α-cyclodextrin	α-Cyclodextrin	Serva
Si100 β-cyclodextrin	β-Cyclodextrin	Serva
Si100 γ-cyclodextrin	γ-Cyclodextrin	Serva
Cellulose-based phases		
Chiralcel OA	Cellulose triacetate	Daicel
Chiralcel OB	Cellulose tribenzoate	Daicel
Chiralcel OC	Cellulose phenylcarbamate	Daicel
Chiralcel OD	Cellulose 3,5-dimethylphenylcarbamate	Daicel

Table 9.1 (Continued)

	Chiral unit	Supplier
Cellulose-based phases		
Chiralcel OF	Cellulose 4-chlorophenylcarbamate	Daicel
Chiralcel OG	Cellulose 4-methylphenylcarbamate	Daicel
Chiralcel OJ	Cellulose 4-methylbenzoate	Daicel
Chiralcel OK	Cellulose cinnamate	Daicel
Chiralcel CA-1	Cellulose triacetate	Daicel
Cellulose triacetate	Cellulose triacetate	Merck
CONBRIO-TAC	Cellulose triacetate	Perstorp Biolytica
Polyacrylamide phase		
ChiraSpher	Poly-[(*S*)-N-acryloylphenylalanine ethyl ester]	Merck
π-Acid and π-Base phases		
Bakerbond DNBPG (covalent)	Dinitrobenzoylphenylglycine	Baker
Bakerbond DNBLeu (covalent)	Dinitrobenzoylleucine	Baker
Pirkle Prep-DPG	Dinitrobenzoylphenylglycine	Regis
Pirkle Prep-LPG	Dinitrobenzoylleucine	Regis
Si100-DNB-Leu	Dinitrobenzoylleucine	Serva
Si100-DNB-PhGly	Dinitrobenzoylphenylglycine	Serva
SUMICHIRAL OA-1000	(*S*)-Naphthylethylamine	SCAS
SUMICHIRAL OA-2000	(*R*)-Phenylglycine dinitrobenzoylamide	SCAS
SUMICHIRAL OA-2100	(*R*)-Phenylglycine, (*S*)-chlorophenylisovaleric acid	SCAS
SUMICHIRAL OA-2200	(*R*)-Phenylglycine, (1*R*,3*R*)-chrysanthemic acid	SCAS
SUMICHIRAL OA-2500	(*R* or *S*)-Naphthylglycine dinitrobenzoylamide	SCAS
SUMICHIRAL OA-3000	(*S*)-Valine *t*-butylurea	SCAS
SUMICHIRAL OA-3100	(*S*)-Valine dinitrophenylurea	SCAS
SUMICHIRAL OA-3200	(*S*)-*t*-Leucine dinitrophenylurea	SCAS
SUMICHIRAL OA-3300	(*R*)-Phenylglycine dinitrophenylurea	SCAS
SUMICHIRAL OA-4000/4100	(*S*)-Valine, (*S* or *R*)-Naphthylethylamine	SCAS
SUMICHIRAL OA-4400/4500	(*S*)-Proline, (*S* or *R*)-Naphthylethylamine	SCAS
SUMICHIRAL OA-4600/4700	(*S*)-*t*-Leucine, (*S* or *R*)-Naphthylethylamine	SCAS
SUMICHIRAL OA-4800/4900	(*S*)-Indoline-2-carboxylic acid, (*S* or *R*)-Naphthylethylamine	SCAS
Spherisorb Chiral-1	(*R*)-Phenylethylamine urea	Phase Sep

9.1.1
Cellulose Derivatives

These derivatives are used as pure polymers or as coatings on inert achiral supports.

The most frequently used phase, in pure polymer form, is cellulose triacetate (inexpensive, wide applicability, high loading capacity) (Hesse and Hagel 1973; Lindner and Mannschreck 1980). An extensive description of this polymer can be found in a book by Subramanian (1994). For filling columns, pressures above 25 MPa (250 bar) should be avoided, meaning that the dynamic axial compression technique (Prochrom) cannot be applied. Columns equipped with a manually operated movable piston have to be used (Subramanian 1994). In cases where solubility is a problem or incomplete separations occur, recycling provides a useful solution: this has been shown in the separation of farnesiferol A and koladonin, which are diastereomeric natural products, one an axial and one an equatorial isomer (Schlögl and Widhalm 1984).

The disadvantages of cellulose triacetate are its low efficiency, poor reproducibility and need for slow flow rates.

Cellulose tribenzoate has been applied to the separation of racemic α-bisabolol (**1**). For this purpose, 500 mg of sample was injected onto a 200 × 10 mm column (Günther et al. 1993).

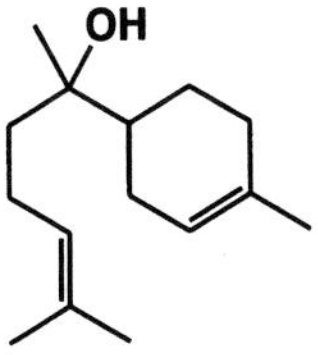

1

Chiralcel OD (cellulose 3-5-dimethylphenyl carbamate; Daicel) has been used in the separation of stereoisomeric isoflavanones. The racemic (at C-3) compounds ferreirin (**2**) and dihydrocajanin (**3**) were isolated from the heartwood of *Swartzia polyphylla* (Leguminosae) by silica gel open-column chromatography and C-18 semi-preparative HPLC. In order to obtain the optical isomers for biological testing, a Chiralcel OD column was employed with the solvent n-hexane-isopropanol-trifluoroacetic acid (90:10:0.5 for ferreirin and 80:20:0.5 for dihydrocajanin). There were no appreciable differences between

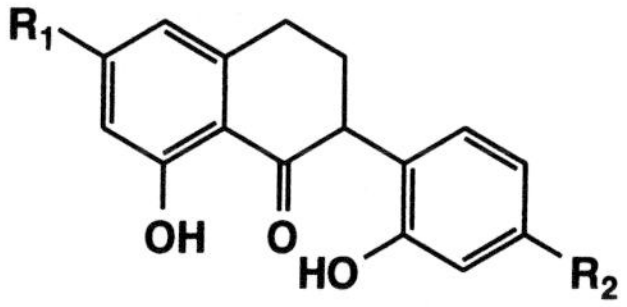

2 $R_1 = OH$, $R_2 = OCH_3$
3 $R_1 = OCH_3$, $R_2 = OH$

the stereoisomers in their antibacterial activity against cariogenic bacteria. For example, (*R*)-ferreirin was equipotent with (*S*)-ferreirin against *Streptococcus mutans* and *S. sobrinus* (Osawa et al. 1992).

Amylose derivatives, developed as a coating on silica gel, have also been investigated (Okamoto et al. 1990). The 3,5-dimethylphenyl carbamates of amylose (e.g. Chiralpak AD, Daicel) have given very effective separations.

9.1.2 Cyclodextrin Phases

Cyclodextrins are cyclic oligosaccharides which form stable inclusion complexes with a wide variety of molecules in their highly hydrophobic cavities. The size of the hydrophobic cavity, which differs for α-, β- and γ-cyclodextrins, and the substituent on the cyclodextrin are important for determining the ability of these oligosaccharides to complex a defined molecule. For preparative purposes, cyclodextrins can be cross-linked with bis(epoxypropyl)ethylene glycol (Zsadon et al. 1983) or immobilized on silica gel (e.g. Cyclobond I, Shaw et al. 1993).

The direct resolution of the endogenous plant growth regulator (–)-jasmonic acid from the corresponding racemate was achieved on a Nucleodex β-PM 5 μm permethylated cyclodextrin column with acetonitrile-0.1% triethylammonium acetate (pH 4) 30:70 (Kramell et al. 1996). The same column separated (–)- and (+)-*cis,trans*-abscisic acid, with methanol instead of acetonitrile (Kramell et al. 1996).

The separation of optically active flavanones has been compared on cyclodextrin (Cyclobond), cellulose-based and polyacrylamide phases. Although no chiral stationary phase was found for the resolution of all the flavanones, the individual supports each had their own merits (Krause and Galensa 1990). The separation of flavanone glycoside diastereomers has been reported on Cyclobond I in the reversed-phase mode (methanol-water solvent) (Krause and Galensa 1991).

9.1.3 Poly(meth)acrylamides

The gel structure of cross-linked, optically active polyacrylamides and polymethacrylamides prevents their utilization at high pressure. However, their mechanical performance can be improved by polymerization of the acrylic monomer on the surface of silica gel, giving a grafted polymer (Blaschke et al. 1986). Preparative separations are performed with (*S*)-phenylalanine ethyl ester, (*S*)-1-cyclohexylethylamine and menthylamine derivatives.

The 2,2′-spirobibenz[*e*]indan **4** was optically resolved from a racemate by liquid chromatography on a (+)-poly(triphenylmethylmethacrylate) chiral stationary phase (chirality here is a result of helicity alone). Absolute stereochemistry was determined by the c.d. exciton chirality method (Harada et al. 1985). The polymer is particularly adapted to the resolution of compounds having an axial, planar or helical chirality.

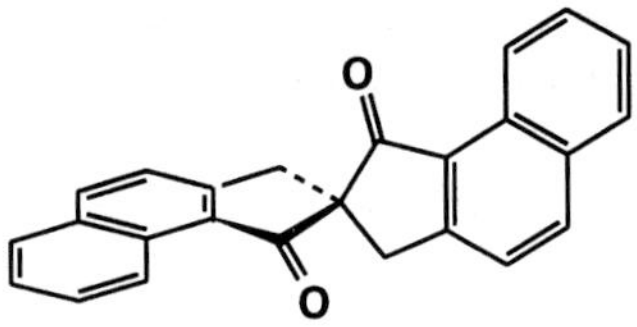

4

9.1.4
π-Acidic and π-Basic Phases

Much of the development work on these phases has been accomplished by Pirkle. The most frequently used π-acidic (π-acceptor) supports are derived from phenylglycine (DNBPG) and leucine (DNBLeu), covalently or ionically bonded to 3-aminopropyl silica gel (Fig. 9.1). Gram quantities of racemates have been resolved by the phase shown in this figure, including alcohols, lactams, lactones, sulphoxides, bi-*β*-naphthols and hydantoins (Pirkle and Finn 1982). For π-basic (π-donor) supports, naphthylalanine or naphthylethylamine derivatives can be employed.

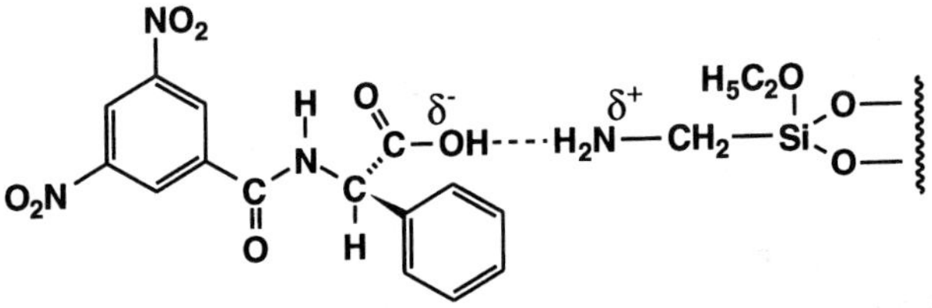

Fig. 9.1. Chiral stationary phase comprising (*R*)-*N*-3,5-dinitro-benzoylphenylglycine (DNBPG) ionically bonded to 40 μm aminopropyl silica

9.1.4.1
Gossypol

Gossypol (5) is found in the seeds of the cotton plant *Gossypium hirsutum* (Malvaceae) and in other species of the genus *Gossypium*. It is known for its ability to inhibit maturation of human sperm and is an effective male oral antifertility agent. Gossypol is a chiral molecule since it possesses restricted rotation about

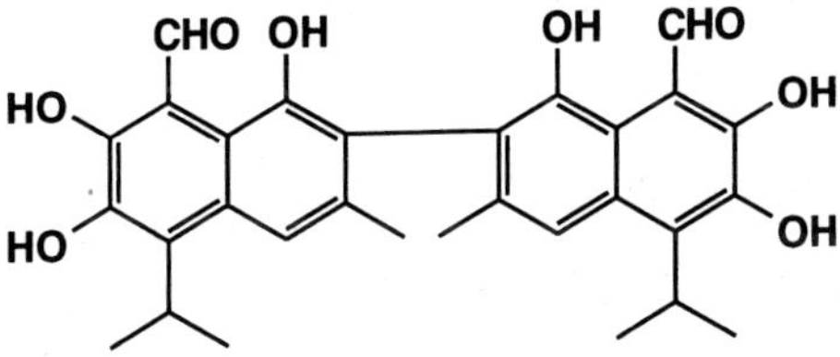

5

the inter-naphthyl bond (atropisomerism) and its enantiomers have been separated on Pirkle-type bonded phases. It was thought that gossypol occurs as a racemate in the cotton plant. However, *G. hirsutum* consistently yields a significant enantiomeric excess of (+)-gossypol (Dechary and Pradel 1971). *Thespesia populnea* (Malvaceae) produces (+)-gossypol and this enantiomer is without activity in laboratory animals (King and de Silva 1968). In order to isolate gram quantities of both enantiomers for antifertility studies, a Schiff's base (**6**) was prepared. Batches of 120 mg of this base were then separated on aminopropyl coated Hypersil (5 µm; 300×21 mm column) which had been converted to a salt with *N*-(3,5-dinitrobenzoyl)-D-(–)-phenylglycine. The solvent hexane-dichloromethane-acetonitrile-isopropanol 90:5:3:2 was effective for elution. Alkaline hydrolysis of the separated products gave pure enantiomers of gossypol (Matlin and Zhou 1984). Only the (–)-isomer was active as an oral antispermatogenic agent, thus confirming the above-mentioned results (Matlin et al. 1987).

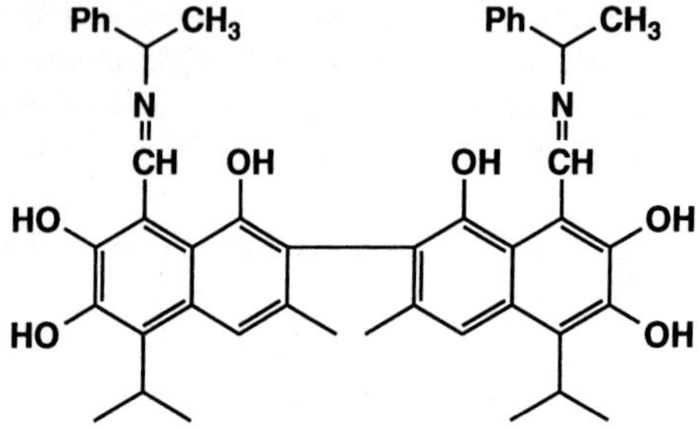

6

In the isolation of tremetone, the major toxic component of *Eupatorium rugosum* (Asteraceae), the final two steps involved first chromatography on a Pirkle-type column modified with D-phenylglycine and then on a 5-µm L-phenylglycine column (solvent: hexane-isopropanol 99.5:0.5). The chiral columns were effective for purifying the required benzofuran derivative (Beier et al. 1993).

9.1.5
Ligand-Exchange Chromatography

This is based on the reversible formation of complexes between metal ions (Cu^{2+} or Ni^{2+}) and chiral complexing agents (such as α-amino acids) carrying functional groups capable of interacting as ligands

9.2
Chiral Separations by Flash Chromatography

Chiral stationary phases of the type shown in Fig. 9.1 have been used to good effect in flash chromatography. Benzodiazepinone 7 (220 mg), for example, was completely resolved into its enantiomers within 20 min on a 150×40 mm bed of

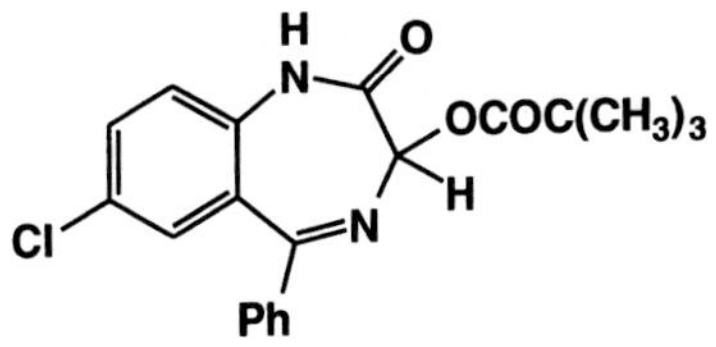

7

Pirkle derivatized aminopropylated silica, using a 2-propanol-hexane eluent (Pirkle et al. 1985). It should be mentioned, however, that the racemate had a large separation factor ($\alpha > 2$) and that racemates with smaller separation factors were only partially separated by this flash chromatographic method.

Recently, an alternative method has been demonstrated with cellulose tris(3,5-dimethylphenyl carbamate)-coated 40–63 μm flash chromatography silica (pore size 150 Å). Several applications have been demonstrated with this phase, including synthetic racemic compounds (e.g. oxprenolol, 2-phenoxypropionic acid) and flavanone. Separation factors observed on the flash column were very similar to those obtained on the equivalent coated silica HPLC column. The analytical HPLC column can therefore be used to establish optimum separation conditions. For a column of dimensions 25×2 cm, up to 100 mg of sample could be loaded. The same column could be reused many times without appreciable loss of efficiency (Matlin et al. 1995).

9.3 Chiral Separations by Gas-Liquid Chromatography

Relatively few publications deal with the preparative-scale separations of enantiomers by GC due to the very low sample capacity of open-tubular columns. However, chiral supports using modified cyclodextrins are available for chiral separations, e.g. octakis-(2,3-di-*O*-acetyl-6-*O*-*tert*-butyldimethylsilyl)-γ-cyclodextrin and heptakis (2,3-di-*O*-methyl-6-*O*-*tert*-butyldimethylsilyl)-γ-cyclodextrin.

Using packed columns, pure enantiomers can be obtained in milligram amounts on these supports. Hardt and König (1994) give an example in which 1.8 m stainless steel packed columns were prepared by coating Chromosorb W HP (Merck) with 5% (w/w) of a 1:1 (w/w) mixture of heptakis(2,6-di-*O*-methyl-3-*O*-pentyl)-γ-cyclodextrin and OV-1701. A carrier gas flow of 400 ml/min helium was employed and the fractions to be collected were trapped in a bath of liquid nitrogen. By this means, the separation of 2 mg racemic methyl jasmonate (injected at 120 °C) into (+)- and (–)-isomers was demonstrated.

The liverwort *Lophocolea bidentata* contains a highly fragrant sesquiterpene, **8**, which was isolated by supercritical fluid extraction of the whole plant with carbon dioxide, followed by preparative GC on a β-cyclodextrin column (Rieck et al. 1995). During synthesis of the same compound, it was separated from diastereoisomer **9** by another preparative GC step, using a stainless steel column (2.0 m × 5.3 mm) filled with 2.5% heptakis(6-*O*-dimethylthexylsilyl-

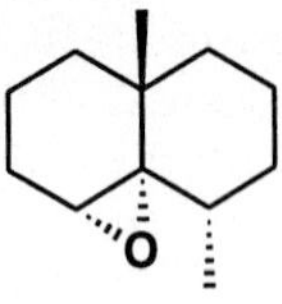

8

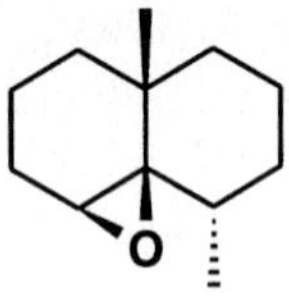

9

2,3-di-*O*-methyl)-*β*-cyclodextrin – SE-52 (20:80, w/w) on Chromosorb G-HP (110° isothermal) (Rieck et al. 1995).

9.4
Chiral Separations by Countercurrent Chromatography

The disadvantages of direct preparative separations of enantiomers by HPLC are the limited capacity of many columns and the high cost of solid stationary phases. Countercurrent chromatography, on the other hand, although lacking the resolution of HPLC methods, requires no solid support and separations can be achieved by the addition of a suitable chiral selector to the stationary liquid phase.

9.4.1
Droplet Countercurrent Chromatography

The complete resolution of DL-isoleucine into its enantiomers has been achieved in a two-phase buffered n-butanol-water system containing Cu^{2+} and *N*-n-dodecyl-L-proline (Takeuchi et al. 1984). Teflon tubes of i.d. 4 mm and a flow rate (descending mode) of 1 ml/min were employed. The copper ions were extracted into the organic phase as complexes with the long-chain proline derivative, thus facilitating the distribution of the enantiomers into the organic phase.

9.4.2
Rotation Locular Countercurrent Chromatography

Prelog and coworkers were the first to show that the separation of *α*-amino alcohol enantiomers was possible by their partition between an aqueous and an organic phase (Prelog et al. 1982). They made use of partition chromatography with an aqueous solution of hexafluorophosphate on Kieselguhr as the stationary phase and a solution of the chiral tartaric ester (di-non-5-yl tartrate) in 1,2-dichloroethane as the lipophilic phase. Extension of this method to support-free RLCC was achieved with a two-phase system consisting of 0.5 mol/l sodium hexafluorophosphate solution (pH adjusted to 4 with HCl) as stationary aqueous phase and a 0.3 mol/l solution of (*R,R*)-di-non-5-yl tartrate in 1,2-dichloroethane as lipophilic mobile phase. The 16-column RLCC instrument was run at 60–70 rpm, with a flow rate of 17–20 ml/h and a slope of 40°. For the separation of the enantiomers of (±)-norephedrine, best results were obtained at low temperature. i.e. 2–3 °C. The enantiomers of (±)-norephedrine probably form different diastereotopic complexes with the tartrate ester (**10, 11**) and these complexes are then partitioned differently between the two solvent phases. Very pure enantiomers (≥395%) could be obtained by this method, after introducing 200 mg of racemic norephedrine hydrochloride (Domon et al. 1982).

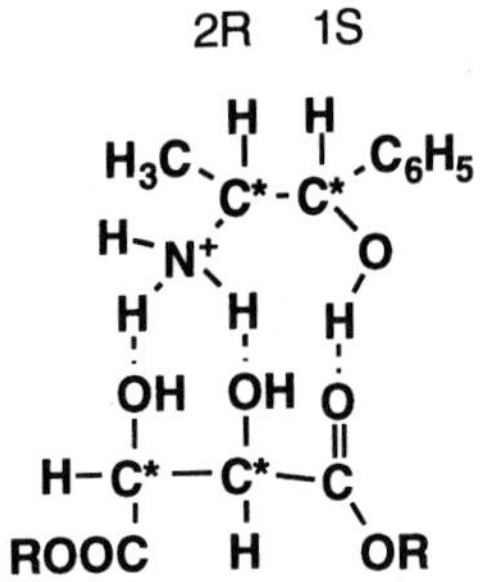

10

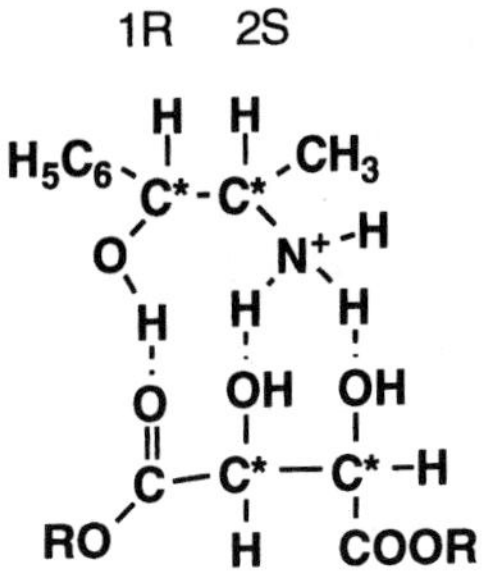

11

9.4.3
Centrifugal Partition Chromatography

Several chiral selectors have been tried for the separation of racemic compounds by CPC. Initial attempts with *β*-cyclodextrin, (2*R*,3*R*)-di-n-butyl tartrate and *N,N′*-bis-[*N*-(3, 5-dinitrobenzoyl)-(*S*)-phenylalanyl]-3,6-dioxa-1,8-octanediamine failed but *N*-dodecanoyl-L-proline-3,5-dimethylanilide (**12**) was successful for the separation of amino-acid derivatives (Oliveros et al. 1994). The molecule bears a π-basic 3,5-dimethylanilide group. When amino-acids are derivatized with a π-acid dinitrobenzoyl group, the π-π interactions are responsible for the chiral recognition. This was the same selector described by Pirkle and Murray (1993) for use in HPLC. The instrument used was a series 1000 chromatograph from Sanki Engineering, with a stacked circular partition disk rotor (total volume 240 ml, rotational speed 1200 rpm). With the solvent system heptane-ethyl acetate-methanol-water 3:1:3:1 (mobile phase = lower phase), the chiral selector was found preferentially in the upper, stationary phase and the pairs of racemic amino-acid derivatives *N*-(3,5-dinitrobenzoyl)-*tert*-butylvalinamide and *N*-(3,5-dinitrobenzoyl)-*tert*-butylleucinamide were

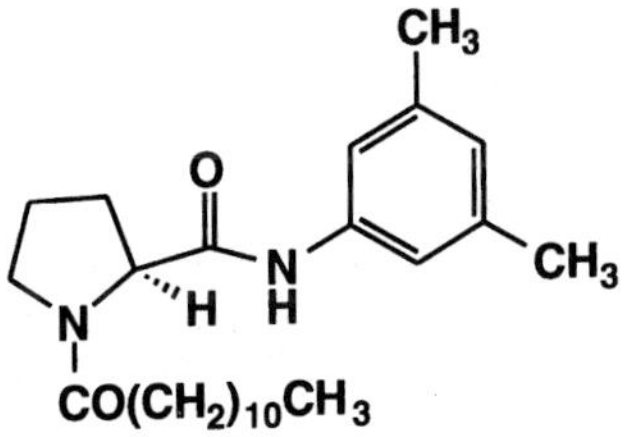

12

separated into their (+)- and (–)-enantiomers. Sample quantities were, however, rather small: 2 ml of a 0.001 mol/l solution of amino-acid derivative (Oliveros et al. 1994).

There were three other problems with this approach:

- free amino-acids were not separated,
- the pure enantiomers had to be separated from the chiral selector by conventional column chromatography,
- on-line UV detection was not possible because of the presence of the chiral selector. Monitoring was done by TLC.

Larger amounts of (±)-amino-acid derivatives have now been separated by CPC using the same chiral selector (Ma et al. 1995). About 150 ml of selector-free stationary phase was pumped into the column (330 ml capacity) of a P.C. Inc. instrument. Then 200 ml of stationary phase containing selector was introduced. In this way, the column had 130 ml of selector-free stationary phase at its outlet, ready to absorb any selector carried over in the mobile phase during the chromatographic run. This avoided contamination of the pure enantiomers with selector and consequently UV detection could be used. In the first experiment, up to 1 g of dinitrobenzoyl- (DNB-) leucine was resolved within 9 h using different concentrations of chiral selector (up to 60 mmol/l), in the presence of the solvent hexane-ethyl acetate-methanol-10 mmol/l HCl (6:4:5:5). The second part employed the pH-zone-refining technique with the same instrument. For these studies, a two-phase solvent system composed of methyl *tert*-butyl ether and water was used, with a retainer acid (TFA; 40 mmol/l) and chiral selector (40 mmol/l) added to the organic stationary phase and an eluent base (ammonia; 20 mmol/l) added to the aqueous mobile phase. Two grams of (±)-DNB-leucine were eluted (single rectangular UV peak) in about 3 h. The first zone (pH 6.5) was composed of (–)-DNB-leucine and the second zone (pH 6.8) of (+)-DNB-leucine (Ma et al. 1995).

Further investigations by conventional CPC with differing proportions of chiral selector and different compositions of the hexane-ethyl acetate-methanol-10 mmol/l HCl solvent system have shown that peak resolution was improved by increasing the amount of chiral selector in the organic stationary phase and by increasing the hydrophobicity of the solvent system (Ma and Ito 1995).

At the present time chiral separations have been limited to (±)-DNB-amino-acids but it should be possible to extend the applications to other racemic mixtures by the choice of appropriate chiral selectors.

9.5 References

Beier RC, Norman JO, Reagor JC, Rees MS, Mundy BP (1993) Nat Toxins 1:286
Blaschke G, Bröker W, Fraenkel W (1986) Angew Chem 98:808
Dechary JM, Pradel P (1971) J Am Oil Chem Soc 48:563
Domon B, Hostettmann K, Kovacevic K, Prelog V (1982) J Chromatogr 250:149
Erlandsson P, Hansson L, Isaksson R (1986) J Chromatogr 370:475
Felix G, Descorps V, Kopaciewicz W, Coryell B (1995) LC-GC Int. 8:396
Fitos I, Visy J, Magyar A, Katjar J, Simonyi M (1990) Chirality 2:161

Francotte E (1994) J Chromatogr 666:565
Francotte E, Junker-Buchheit A (1992) J Chromatogr 576:1
Günther K, Carle R, Fleischhauer I, Merget S (1993) Fres J Anal Chem 345:787
Harada N, Iwabuchi J, Yokota Y, Uda H, Okamoto Y, Yuki H, Kawada Y (1985) J Chem Soc Perkin Trans I 1845
Hardt I, König WA (1994) J Chromatogr A 666:611
Hesse G, Hagel R (1973) Chromatographia 6:277
Jacobson S, Felinger A, Guiochon G (1992) Biotechnol Bioeng 40:1210
King TJ, de Silva LB ((1968) Tetrahedron Lett 3:261
Kramell R, Schneider G, Miersch O (1996) Phytochem Anal 7:209
Krause M, Galensa R (1990) J Chromatogr 514:147
Krause M, Galensa R (1991) J Chromatogr 588:41
Lämmerhofer M, Lindner G (1996) J Chromatogr A 741:33
Lindner KR, Mannschreck A (1980) J Chromatogr 193:308
Ma Y, Ito Y (1995) Anal Chem 67:3069
Ma Y, Ito Y, Foucault A (1995) J Chromatogr 704:75
Matlin SA, Zhou R (1984) J High Res Chromatogr 7:629
Matlin SA, Belenguer A, Tyson RG, Brookes AN (1987) J High Res Chromatogr 10:86
Matlin SA, Grieb SJ, Belenguer AM (1995) J Chem Soc Chem Commun 301
Okamoto Y, Kaida Y, Hayashida H, Hatada K (1990) Chem Lett 909
Oliveros L, Franco Puertolas P, Minguillon C, Camacho-Frias E, Foucault A, Le Goffic F (1994) J Liq Chromatogr 17:2301
Osawa K, Yasuda H, Maruyama T, Morita H, Takeya K, Itokawa H (1992) Chem Pharm Bull 40:2970
Pirkle WH, Finn JM (1982) J Org Chem 47:4037
Pirkle WH, Murray PG (1993) J Chromatogr 641:11
Pirkle WH, Tsipouras A, Sowin TJ (1985) J Chromatogr 319:392
Prelog V, Stojanac Z, Kovacevic K (1982) Helv Chim Acta 65:377
Rieck A, Bülow N, König WA (1995) Phytochemistry 40:847
Rosini C, Bertucci C, Pini D, Altemura P, Salvadori P (1985) Tetrahedron Lett 26:3361
Salvadori P, Pini D, Rosini C, Bertucci C, Uccello-Barretta G (1992) Chirality 4:43
Schlögl K, Widhalm M (1984) Monatsh Chemie 115:1113
Shaw CJ, Sanfilippo PJ, McNally JJ, Park SA, Press JB (1993) J Chromatogr 631:173
Subramanian G (1994) A practical approach to chiral separations by liquid chromatography. VCH, Weinheim
Takeuchi T, Horikawa R, Tanimura T (1984) J Chromatogr 284:285
Werner A (1989) Kontakte (Darmstadt) 50
Zsadon B, Décsei L, Szilasi M, Tüdos F (1983) J Chromatogr 270:127

CHAPTER 10

Separation Strategy and Combination of Methods

A separation problem depends on the number and nature of components in a mixture. Obtaining a pure product from a synthetic reaction, for example, may require the removal of small amounts of a single by-product. In this situation, the problem could probably be resolved by an inexpensive method which is less time-consuming than chromatography – crystallisation etc. It may also be possible to separate a mixture by a single chromatographic step. In reality, however, the task is usually much more complex. For example, the isolation of a single bioactive compound from a plant extract containing several thousand components can be a daunting prospect and may involve many separation steps. The separation of cyanogenic constituents of *Xeranthemum cylindraceum* (Asteraceae) required LPLC, MPLC, HPLC, CTLC and DCCC for complete purification (Schwind et al. 1990). In these cases, a *combination* of techniques is the best approach.

The *strategy* of separation is the choice of the techniques to be employed and this depends on a number of factors:

- extraction method,
- complexity of extract or mixture,
- sample preparation,
- sample polarity,
- sample stability,
- sample solubility,
- sample size,
- complementarity of separation techniques.

When choosing a separation strategy, it is often useful to pick steps which differ as much as possible in selectivity. This can be achieved by varying the mode of separation. On the other hand, if only one stationary phase is used throughout the purification steps, selectivity is maximised by varying the eluent.

During an isolation procedure, the scale of the operation decreases: as the purity of the product increases, there is a corresponding diminution of sample quantity. This implies that the initial fractionation steps are those which can separate large amounts of material, e.g. column chromatography using relatively cheap stationary phases (silica, alumina, polyamide or XAD ion exchange resins), flash chromatography or countercurrent chromatography. Gel filtration is also becoming increasingly popular as a first purification step. Subsequent chromatographic steps on smaller quantities can be performed with more expensive column packings and equipment. Preparative HPLC is often reserved for final purification for several reasons: a) preliminary purification is necessary to remove elements which may irreversibly adsorb to the solid support; b) since much smaller sample quantities are available at the end of a separation scheme, the capacities of the HPLC columns are not exceeded; c) resolution is very high.

Reversed-phase packing material is much more expensive than normal silica gel for use as a first open-column or flash purification step. However, there is less irreversible adsorption to derivatized silica gel and it can be regenerated. For these reasons, reversed-phase supports have been used in conjunction with semi-preparative HPLC (final purification) for the separation of marine natural products. For initial fractionation, extracts were mixed with RP silica (32–63 μm) and loaded onto a flash column as either an aqueous slurry or a powder. By this means, loads of up to 20 g crude extract per 100 g support were possible (Blunt et al. 1987).

Selected strategies for the isolation of natural products are illustrated below. Although many different combinations of separation method are theoretically possible, certain strategies have proved their efficacy and are found very frequently.

10.1 Hydrophilic Compounds

The isolation of polar natural products presents a major challenge in separation science. Many polar compounds – polysaccharides, peptides, saponins etc. – possess unique biological activities and it is essential that straightforward and gentle techniques, or combinations of techniques, are developed for their isolation.

The chromatography of macromolecules (often termed "biochromatography") involves principally the separation of proteins, peptides, enzymes, carbohydrates, oligonucleotides and nucleic acids. As more companies produce therapeutics (including products for gene therapy) based on these macromolecules, their purification is assuming increasing importance. The separation of macromolecules may rely on a sequence of chromatographic modes such as size exclusion, ion exchange and affinity chromatographies. This aspect is discussed in Chap. 8.

Water-soluble marine natural products pose special problems and have been dealt with by Shimizu (1985).

Aqueous extracts of plants (often in the form of teas or infusions) are frequently found in traditional and folk medicine. The isolation of their active principles (many of which are extremely hydrophilic) is essential for the comprehension of their mode of action. The importance of not just the isolation procedure but also of the extraction procedure is well illustrated by the separation of saponins from the Ethiopian soap substitute plant *Phytolacca dodecandra* (Phytolaccaceae) (Domon and Hostettmann 1984). Maceration of the berries with solvents of increasing polarity (petrol ether, chloroform, methanol) and fractionation of the methanol extract gave bidesmosidic saponins which had absolutely no molluscicidal activity. However, direct water extraction of the berries gave monodesmosidic saponins with potent molluscicidal activity. Therefore, when attempting the isolation of polar natural products, it is important to plan carefully *all* stages of the separation procedure.

10.1.1
Combinations Involving Liquid-Liquid Partition and Liquid Chromatography

A very efficaceous strategy for the separation of natural products is the combination of an all-liquid chromatographic step with a liquid-solid (LC) chromatographic step. For example, centrifugal partition chromatography (CPC) provides an excellent means of fractionation because there is no loss of sample, no solid support and a good separation of polar and apolar components.

A saponin with an unusual secoursene skeleton was isolated from the aerial parts of *Sesamum angolense* (Pedaliaceae) by the liquid-liquid/liquid-solid approach. After initial open-column chromatography of the methanol extract, the fraction of interest was subjected to CPC on a multilayer countercurrent chromatography apparatus (P.C. Inc.) with the lower layer of the solvent system chloroform-methanol-isopropanol-water (5:6:1:4). Thus 1.25 g of extract was injected into a 350 ml (2.6 mm i.d.) column rotating at 700 rpm. The saponin, alatoside A (**1**), was obtained after subsequent gel filtration (LH-20, MeOH) and LPLC on a Lobar RP-8 (size B) column with the solvent methanol-water 6:4 (Potterat et al. 1992).

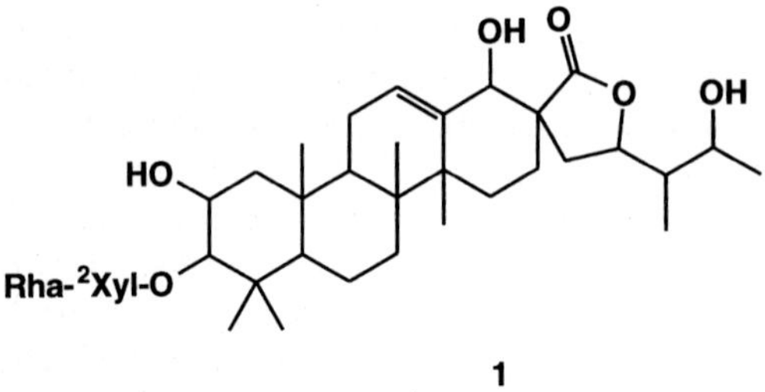

1

The traditional method of isolating saponins, by open-column chromatography on silica gel, with chloroform-methanol-water mixtures (for example, Hiller et al. 1987) gives satisfactory results but is time-consuming and involves loss of sample by irreversible adsorption.

10.1.2
Combinations Involving Liquid-Liquid Partition and Size Exclusion Chromatography

Size exclusion chromatography/gel filtration on supports such as Sephadex LH-20 and TSK Toyopearl HW (see Chap. 8) plays an important rôle in the isolation of natural products. The gel enables the separation of compounds according to their molecular weights and is, therefore, very useful for the separation of high molecular weight and polymeric material from a sample. Since these are often the components which cause the most trouble during chromatography on solid supports, a size exclusion chromatographic step is useful as a preliminary step to both countercurrent chromatography and pressure LC.

In the separation of saponins from starfish by Minale and co-workers, a strategy involving DCCC and gel filtration is systematically used. To illustrate this, the isolation of an asterosaponin from the Caribbean starfish *Echinaster brasiliensis* (Echinasteridae) will be described. The aqueous extract of the starfish was subjected to an initial purification and solvent partition procedure before chromatography on a column of Sephadex LH-60 with methanol-water (2:1) as eluent. This latter step separated the crude asterosaponins from polyhydroxysteroid mono- and diglycosides and anthraquinone pigments. The asterosaponin mixture (327 mg) was fractionated by DCCC using n-butanol-acetone-water (3:1:5) (descending mode; flow rate 12 ml/h) to give four main fractions. One of these (98 mg) contained brasiliensoside (**2**) and required a final HPLC step to provide the pure product (Iorizzi et al. 1993).

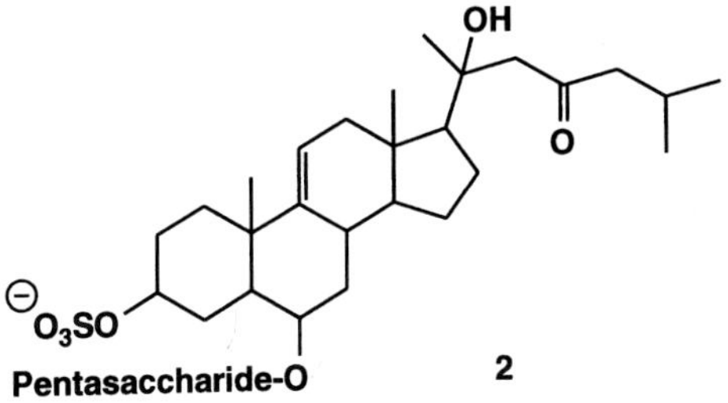

2

Xanthone glycosides and secoiridoid glucosides have been isolated from the South American plant *Halenia campanulata* (Gentianaceae) by the same strategy. Powdered whole plant was extracted first with dichloromethane and then with methanol. Two xanthones were obtained from the apolar (dichloromethane) extract. The methanol extract was subjected to Sephadex LH-20 gel filtration (elution with methanol). Subsequent CPC of one fraction (Fig. 10.1) gave two secoiridoid glucosides, vogeloside (**3**) and *epi*-vogeloside (**4**). This liquid-liquid fractionation step was remarkable in that it separated the two isomers. Another fraction from the gel filtration run was purified first by CPC and then by semi-preparative HPLC (LiChroprep RP-18; 7 µm; 250×16 mm), yielding the three xanthone glycosides **5**–**7** (Recio-Iglesias et al. 1992).

Sephadex LH-20 gel filtration is not only an efficient preliminary fractionation step but it can also be employed as the very *last step* in isolation work, to remove last traces of solid material, salts or other extraneous matter (e.g. Domon and Hostettmann 1984). One of the reasons for its popularity at this stage of an isolation procedure (when amounts of pure product are often rather small), is probably that it causes minimal material losses.

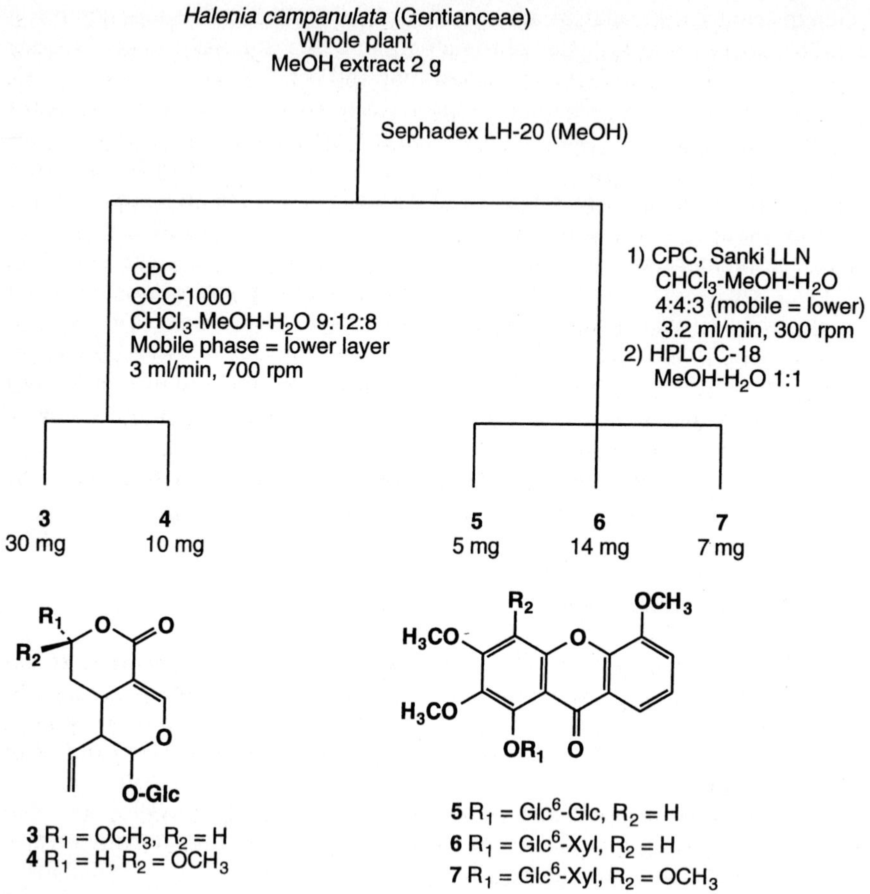

Fig. 10.1. Isolation of secoiridoid glucosides and xanthone glycosides from *Halenia campanulata* (Gentianaceae)

10.1.3
Combinations Involving Size Exclusion Chromatography and Liquid Chromatography

Highly polar glycosides have been purified by size exclusion chromatography followed by LC. In one application, flavone glucuronides from the aerial parts of *Lycopus virginicus* (Lamiaceae) were isolated by this strategy. A 50% ethanol extract of the plant material was subjected to *ultrafiltration* on a 1000 Dalton Amicon diaflo YM1 membrane and then chromatographed on Sephadex LH-20 (acetone-water 4:6). After further separation on Fractogel TSK HW40F (Merck) with methanol-water 8:2, the 7-*O*-glucuronides of apigenin, acacetin and luteolin were obtained by semi-preparative HPLC on a C-18 column with a THF-methanol-acetonitrile-water gradient (Bucar and Kartnig 1995).

Tannins are conveniently purified by a combination of size exclusion chromatography and liquid chromatography. In the case of 5α-reductase and aromatase inhibitors from *Epilobium* (Onagraceae) species, two ellagitannins were isolated from *E. capense* by this approach. The aerial parts were extracted with dichloromethane, methanol and finally methanol-water 1:1. The aqueous methanol extract (9.5 g) was chromatographed over a column of Sephadex LH-20 with a solvent gradient (water, water-methanol 9:1 to methanol in 10% increments and then acetone). The active fraction was purified by semi-preparative HPLC on a 5-μm LiChrospher 100 endcapped C-18 column, eluting first with acetonitrile-water 16:84 and then in a second run with acetonitrile-water 14:86. Two tannins, oenothein A (34 mg) and oenothein B (284 mg), with activity against the enzymes, were isolated (Ducrey et al. 1997).

Similarly, dimeric, trimeric and tetrameric ellagitannins have been isolated from *Reaumuria hirtella* (Tamaricaceae). The procedure involved extraction of the leaves with 70% acetone and subsequent chromatography over Toyopearl HW-40 (70–80% aqueous methanol and methanol-water-acetone 7:2:1 → 6:2:2 → 5.2.3). The trimers and tetramers were obtained after further purification with preparative HPLC on a C-18 column, using 0.01 mol/l KH_2PO_4-0.01 mol/l H_3PO_4-EtOH-EtOAc (41.5:41.5:12:5) (Ahmed et al. 1994).

10.1.4 Combinations Involving Polymeric Supports

A preliminary passage of a polar sample over a polymeric support is an excellent means of removing unwanted hydrophilic contaminants (amino-acids, carbohydrates etc.). This is the approach employed by many Japanese groups for the pre-purification of saponins (Hostettmann and Marston 1995) and other plant glycosides.

The procedure typically involves chromatography on highly porous polymer (Diaion HP-20 or similar), followed by open column chromatography on silica gel and/or C-18 semi-preparative HPLC (Mizui et al. 1990).

An alternative technique for the preliminary purification of polar compounds is the use of Amberlite XAD-2. This resin has been routinely employed for the reversible adsorption of drugs from body fluids and it is known to adsorb polyphenolic compounds. As a result, it is helpful for the purification of flavonoid fractions (Rosler and Goodwin 1984). After a flavonoid fraction has been purified, other adsorbents are required for the separation of individual compounds.

Amberlite XAD-2 has also been used as a first step in the isolation of saponins from marine organisms. For instance, an aqueous extract of a starfish can be loaded onto a column of the resin and eluted with water to remove unwanted water-soluble material. The desired components are then eluted by the passage of methanol before further chromatography (Iorizzi et al. 1993).

The isolation of antitumour cyclic hexapeptides from *Rubia cordifolia* (Rubiaceae) has been achieved by a complex strategy, including two XAD-2 steps. These were combined with silica gel open-column chromatography, DCCC, Sephadex LH-20 and reversed-phase low-pressure LC, in order to obtain the pure active constituents (Itokawa et al. 1984).

10.1.5
Combinations Involving Different Liquid Chromatographic Steps

Flash chromatography in conjunction with *MPLC* and semi-preparative *HPLC* has been employed for the isolation of a new secoiridoid glycoside, 2′-(*o*,*m*-dihydroxybenzoyl)sweroside, from *Gentiana algida* (Gentianaceae). An acetone extract of the whole herb was subjected to flash chromatography on silica gel (63–200 μm; chloroform-methanol gradient) before MPLC (C-18; methanol-water 1:3 → 7:13) and final HPLC (LiChrosorb RP-18; 7 μm; 25 × 1.6 cm; methanol-water 63:37) (Tan et al. 1996).

10.2
Lipophilic Compounds

10.2.1
Combinations Involving Liquid Chromatography and Planar Chromatography

For reasons of economy and ease of operation, an extremely popular method for the separation of lipophilic compounds has been the combination of silica gel open column chromatography and preparative TLC.

A newer and more rapid procedure is the use of vacuum liquid chromatography *(VLC)*, followed by centrifugal TLC *(CTLC)*. The VLC step gives a crude fractionation and then the resolution of the planar chromatographic method is sufficient to provide the pure compounds. This is illustrated by the isolation of clerodane diterpenes from the bark of *Casearia tremula* (Flacourtiaceae). A petrol ether extract (10 g) was fractionated over silica gel by VLC using petrol ether containing increasing amounts of ethyl acetate for elution. By CTLC (silica gel; toluene-ethyl acetate 9:1) of the 30% ethyl acetate in petrol ether, pure **8** (100 mg) was obtained. A further five diterpenes were isolated by a similar procedure (Gibbons et al. 1996).

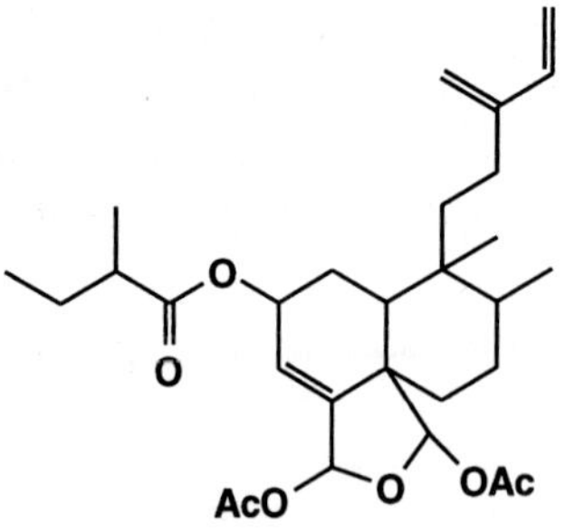

8

The combination VLC/CTLC has also found application in the purification of diterpene alkaloids. In some instances, silica gel CTLC was employed and in other cases, aluminium oxide was the support of choice. For the isolation of norditerpene alkaloids from *Dephinium ajacis* (Ranunculaceae), several CTLC steps were reported. The alkaloid mixture (1.7 g) was fractionated by VLC (aluminium oxide, 66.5 g, Merck 1085), with a gradient of hexane, diethyl ether and methanol. Delectine (8 mg) (**9**), was obtained by CTLC purification of the diethyl ether VLC fraction (108 mg): a rotor of Al_2O_3 (1 mm thick) was eluted with 10% ethanol in diethylether (Liang et al. 1991). Numerous other examples have been reported by Pelletier and co-workers. In one of these, Venkateswarlu et al. (1995) report initial purification by VLC and final fractionation by either CTLC or CPC on a Sanki 1000 LLB-M (solvent system: benzene-chloroform-methanol-water 5:5:7:2; lower layer as mobile phase).

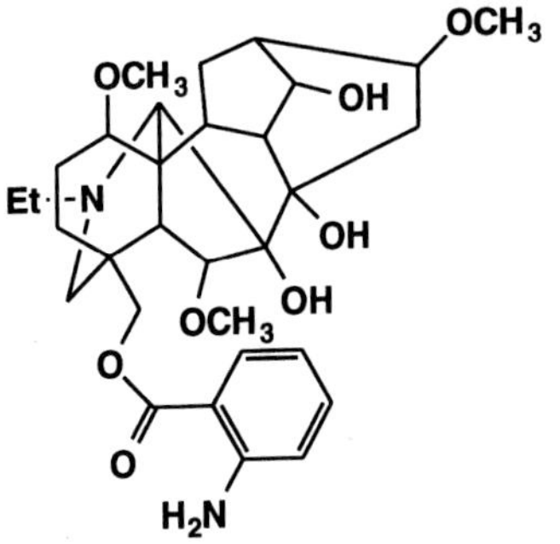

9

10.2.2 Combinations Involving Different Liquid Chromatographic Techniques

Flash chromatography is ideal for the rapid preliminary purification of crude samples but lacks the resolution of other modern chromatographic techniques. Consequently, it is often necessary to combine flash chromatography with a higher resolution separation technique in order to obtain pure compounds. In the isolation of prenylated xanthones from *Garcinia livingstonei* (Guttiferae), flash chromatography was used as the first fractionation operation before a gel filtration step. For obtaining the pure products, *LPLC* or semi-preparative HPLC was sufficient as the final operation. By this means, the new antifungal xanthone **10** was purified by LPLC on a Lobar RP-8 column with methanol-water 8:2 as eluent (Sordat-Diserens et al. 1992).

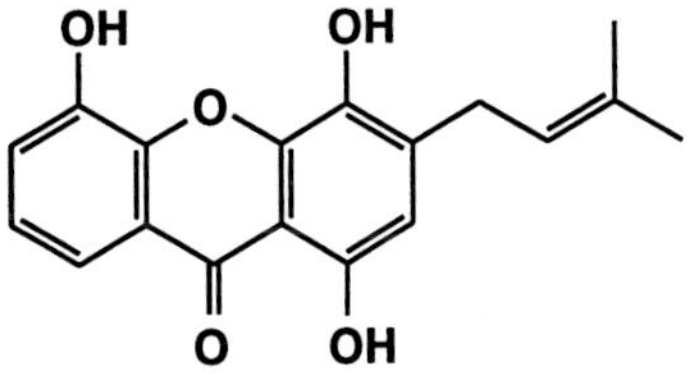

10

In the investigation of limonoid antifeedants from the root bark of Chinese *Melia azedarach* (Meliaceae), it was found that these compounds were very sensitive to traces of acid and decomposed during silica gel open-column chromatography (Nakatani et al. 1994). As a more rapid separation method was required, a combination of *flash* chromatography and semi-preparative *HPLC* was used instead (followed by a bioassay of antifeedant activity with *Spodoptera exigua*). A diethyl ether extract of the root bark was flash chromatographed twice on silica gel: the first column was eluted with 1–3% methanol in dichloromethane and the second column with 20% hexane in diethyl ether. Each limonoid fraction was then separated on a μBondapak C-18 HPLC column (eluent 25–40% water in methanol) to give the three new 12α-acetoxy limonoids **11**–**13**, together with three known limonoids.

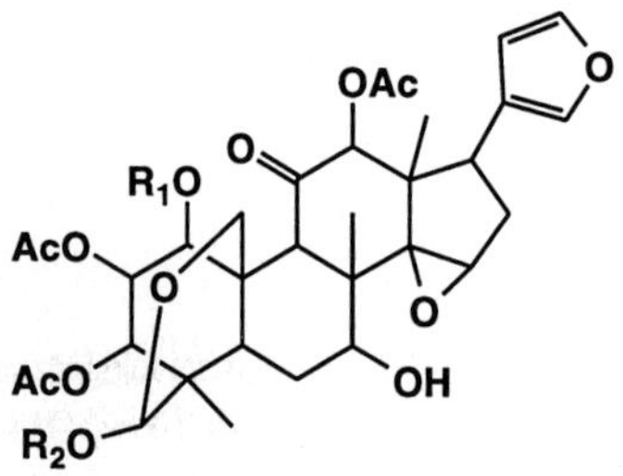

11 $R_1 = H$, $R_2 = COCH(CH_3)CH_2CH_3$
12 $R_1 = Ac$, $R_2 = COCH(CH_3)CH_2CH_3$
13 $R_1 = H$, $R_2 = COCH(CH_3)_2$

A very similar procedure was adopted for the isolation of limonoids from *Trichilia rubra* (Meliaceae). Silica gel flash chromatography was performed with a hexane-isopropanol gradient; a combination of silica gel HPLC (hexane-isopropanol gradient) and C-18 HPLC (methanol-water gradient) was used as the final purification step (Musza et al. 1995).

In another approach, a combination of *VLC*, *MPLC* and *HPLC* was used for the separation of certain apolar constituents of *Piper aduncum* (Piperaceae). Leaves were extracted with petrol ether and the oily residue obtained was treated by the following procedure:

1) VLC (RP-18 40–63 μm; methanol-water step gradient)
2) MPLC (silica gel HF_{254} 15 μm; ethanol-dichloromethane-TBME-hexane 0.3:0.3:0.4:99)
3) HPLC (Spherisorb S5 ODS II 5 μm, 250×15 mm; methanol-water 4:1)

Finally, two monoterpene-substituted dihydrochalcones were obtained: adunctin A (**14**) and adunctin B (**15**) (Orjala et al. 1993).

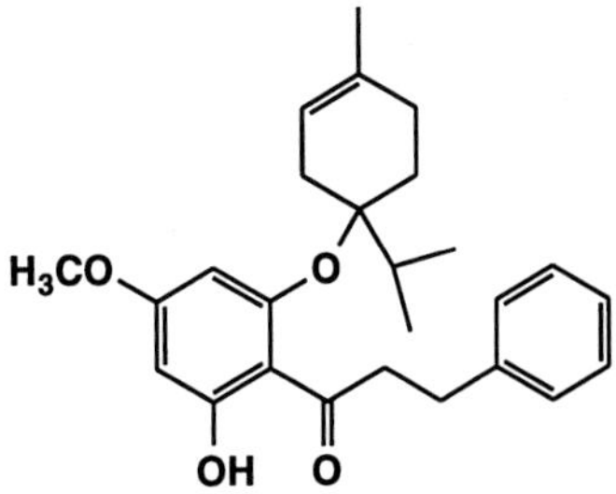

14

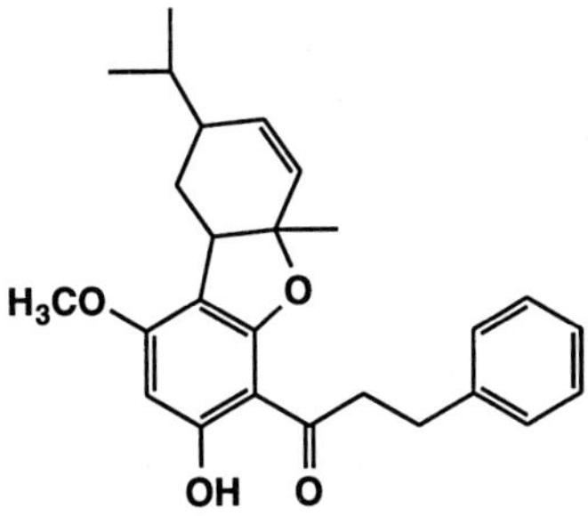

15

10.3 Conclusion

The introduction of modern separation techniques has revolutionized the science of separation of natural products. These new methods allow faster separations, facilitate the resolution of complex mixtures and often avoid the problems involved with the isolation of unstable substances. The actual separation method or methods depend(s) on a number of factors relevant to the separation problem but a judicial choice of strategy enables most targets to be reached. The complementarity of combinations such as liquid-liquid partition/liquid chromatography or gel filtration/liquid chromatography provides a very powerful tool for the separation scientist. Even when considering HPLC alone, the differences in selectivities between normal silica gel and reversed-phase columns enable mixtures of closely-related compounds to be successfully resolved (provided the sample is soluble in the eluents concerned). New methods and improvements are continually being introduced, with the result that the number of combinations available is steadily expanding – hopefully leading to a progressive simplification of the ever more complex separation problems that are being undertaken.

10.4 References

Ahmed AF, Yoshida T, Okuda T (1994) Chem Pharm Bull 42:246
Blunt JW, Calder VL, Fenwick GD, Lake RJ, McCombs JD, Munro MHG, Perry NB (1987) J Nat Prod 50:290
Bucar F, Kartnig T (1995) Planta Med 61:378
Domon B, Hostettmann K (1984) Helv Chim Acta 67:1310
Ducrey B, Marston A, Göhring S, Hartmann RW, Hostettmann K (1997) Planta Med 63:111
Gibbons S, Gray AI, Waterman PG (1996) Phytochemistry 41:565
Hiller K, Bardella H, Schulten HR (1987) Pharmazie 42:622
Hostettmann K, Marston A (1995) Saponins. Cambridge University Press, Cambridge
Iorizzi M, De Riccardis F, Minale L, Riccio R (1993) J Nat Prod 56:2149
Itokawa H, Takeya K, Mori N, Hamanaka T, Sonobe T, Mihara K (1984) Chem Pharm Bull 32:284

Liang X, Ross SA, Sohni YR, Sayed HM, Desai HK, Joshi BS, Pelletier SW (1991) J Nat Prod 54:1283
Mizui F, Kasai R, Ohtani K, Tanaka O (1990) Chem Pharm Bull 38:375
Musza LL, Killar LM, Speight P, Barrow CJ, Gillum AM, Cooper R (1995) Phytochemistry 39:621
Nakatani M, Huang RC, Okamura H, Naoki H, Iwagawa T (1994) Phytochemistry 36:39
Orjala J, Wright AD, Erdelmeier CAJ, Sticher O (1993) Helv Chim Acta 76:1481
Potterat O, Hostettmann K, Stoeckli-Evans H, Saadou M (1992) Helv Chim Acta 75:833
Recio-Iglesias MC, Marston A, Hostettmann K (1992) Phytochemistry 31:1387
Rosler KH, Goodwin RS (1984) J Nat Prod 47:188
Schwind P, Wray V, Nahrstedt A (1990) Phytochemistry 29:1903
Shimizu Y (1985) J Nat Prod 48:223
Sordat-Diserens I, Rogers C, Sordat B, Hostettmann K (1992) Phytochemistry 31:313
Tan RX, Wolfender JL, Ma WG, Zhang LX, Hostettmann K (1996) Phytochemistry 41:111
Venkateswarlu V, Srivastava SK, Joshi BS, Desai HK, Pelletier SW (1995) J Nat Prod 58:1527

Subject Index

Printing: Saladruck, Berlin
Binding: Buchbinderei Lüderitz & Bauer, Berlin